EMERGENCY RESPONSE PLANNING

EMERGENCY RESPONSE PLANNING

For Corporate and Municipal Managers

Second Edition

Paul A. Erickson, Ph.D.

Anna Maria College
Paxton, Massachusetts

AMSTERDAM • BOSTON • HEIDELBERG • LONDON
NEW YORK • OXFORD • PARIS • SAN DIEGO
SAN FRANCISCO • SINGAPORE • SYDNEY • TOKYO

Butterworth-Heinemann is an imprint of Elsevier

Senior Acquisitions Editor	Mark Listewnik
Project Manager	Philip Korn
Acquisitions Editor	Jennifer Soucy
Acquisitions Editor	Pamela Chester
Editorial Assistant	Kelly Weaver
Marketing Manager	Christian Nolin
Cover Design	Eric DeCiacco
Composition	SPI Technologies Company
Cover Printer	Phoenix Color
Printer	the Maple-Vail Book Manufacturing Group

Elsevier Butterworth–Heinemann
30 Corporate Drive, Suite 400, Burlington, MA 01803, USA
Linacre House, Jordan Hill, Oxford OX2 8DP, UK

Recognizing the importance of preserving what has been written, Elsevier prints its books on acid-free paper whenever possible.

Library of Congress Cataloging-in-Publication Data
Erickson, Paul A., 1941–
    Emergency response planning / Paul A. Erickson.— 2nd ed.
        p. cm.
    Rev. ed. of: Emergency response planning for corporate and municipal managers. c1999.
    Includes index.
    ISBN-13: 978-0-12-370503-7
    ISBN-10: 0-12-370503-7 (hardcover : alk. paper)   1. Emergency management— United States— Handbooks, manuals, etc.   I. Erickson, Paul A., 1941–Emergency response planning for corporate and municipal managers.   II. Title.
        HV551 3.E75 2006
        658.4'77—dc22
                                                                                            2005028628

British Library Cataloguing-in-Publication Data
A catalogue record for this book is available from the British Library.

ISBN 13: 978–0–12–370503–7   ISBN 10: 0–12–370503–7

For information on all Elsevier Butterworth–Heinemann publications visit our Web site at www.books.elsevier.com

Transferred to Digital Printing 2010

For

Dato' Soh Chai Hock, Kuala Lumpur, Malaysia

Augustine Koh, Tokyo, Japan

Rowan Amalia Luff (no other address than in grampa's heart)

CONTENTS

PREFACE

Effective emergency response planning, whether for incidents that derive from natural causes or for those that derive from human actions, demands both persistent and consistent liaison and coordination among a large diversity of governmental agencies, response services, and community support resources.

Although it remains true that professional fire services continue to bear the day-to-day responsibility of responding to local disasters, it is also true that several factors in modern society converge toward a broader expansion of this responsibility into both public and private domains. These factors include: (a) the complexity of modern industries and technologies that are dependent on the continual development of new materials that, beyond their functional roles, also become new sources of physical, chemical, and biological hazards and risks; (b) the on-going merging of dense population areas with diverse technology and production centers; and (c) the increasing availability of hazardous technologies and agents to those sociopaths otherwise known as terrorists.

The increasingly global recognition that any incident—whether a hurricane or a bombing, an accidental release of a toxic industrial gas or a purposeful contamination of items essential for daily commerce—can easily be magnified in its toll of human life by the very way we conduct our lives and structure our societies has profound implications for the emergency planning process. This book addresses some of the more important of these implications, especially with regard to industrial and municipal planning and response.

As a second edition to my previous *Emergency Response Planning for Corporate and Municipal Managers* (1999), I have made changes in the original sequence of chapters and in the content of appendices. I have also corrected (all too many) previous grammatical and syntactical errors. Of course, I also have updated some materials, especially with regard to recent developments in both law and public events. However, the basic scope and much of the detail of the first edition remain intact. Like the previous edition, this edition focuses on proactive and reactive aspects of emergency planning, and on the need for partnerships among federal, state, and local governmental agencies as well as among public and private community sectors.

For the development of this edition, I am indebted to Dato' Soh Chai Hock, for sharing his professional insight and for affording me his always enthusiastic support; to Augustine Koh, who continues to sustain (by means of his patient counsel and personal example) my best professional effort; and to Dr. Christine Holmes for her encouragement as well as for her active participation in the effort.

I also remain indebted to a large number of governmental agencies and divisions for the information they provide in both electronic and printed formats, including the U.S. Environmental Protection Agency; Centers for Disease Control; U.S. Public Health Services; Federal Emergency Management Agency: National Response Team; National Institute of Occupational Safety and Health; Occupational Safety and Health Administration; and (with special respect and admiration) U.S. Fire Administration.

SCOPE OF EMERGENCY RESPONSE

INTRODUCTION

Emergency response is an integral component of routine corporate and municipal management that, while directly influenced by diverse and long standing regulations at all levels of government, is also influenced by nonregulatory considerations that predate the infamous terrorism attacks of more recent history, including (a) obligations imposed by corporate insurance policies, (b) corporate and municipal stakeholder concerns over tort liability, and (c) the demands of both *ad hoc* and formal in-plant safety committees engendered by both regulatory and societal concerns over workplace health and safety. Overt and spectacular terrorists acts throughout the world over the past several decades, of course, have given particular public impetus to the importance of both emergency planning and emergency management practices.

At the municipal level, emergency response planning and management, accordingly, have become increasingly complex tasks that, despite a long and distinguished historical development, are continually compounded by social, technical, and political developments including (a) jurisdictional confusion among federal, regional, state, county, and municipal authorities; (b) the economic burden of maintaining adequately staffed, trained, and provisioned emergency response teams; (c) the sheer structural and operational complexity of modern municipalities; (d) the proliferation of sources and agents of potential public hazard; and most recently, (e) the widespread anxiety regarding the terrorist acts of politically (or otherwise) motivated groups and individuals (see Image 1.1).

In the United States, the primary federal influence on corporate emergency response planning is through legislation governing the workplace generation of hazardous waste (Resource Conservation and Recovery Act; RCRA) and activities associated with uncontrolled hazardous waste sites (Comprehensive Emergency Response, Compensation and Liability Act © CERCLA; also known

IMAGE 1-1

NEW YORK, NY, OCTOBER 20, 2001: URBAN SEARCH AND RESCUE TEAM AT THE SITE OF THE WORLD TRADE CENTER

Source: Andrea Booher/FEMA News Photo

as *Superfund*) and the Superfund Amendments and Reauthorization Act (SARA), although other legislation and regulations also establish emergency response requirements, including the Clean Water Act (CWA), the Hazardous Materials Transportation Act (HMTA), and the Chemical Process Safety Regulations (29 CFR 1910.119).

With respect to the health and safety of American workers involved in emergency response (see Tables 1.1–1.3), key baseline regulations include 29 CFR 1910.120 (Hazardous Waste Operations and Emergency Response) and 29 CFR 1910.38 (Employee Emergency Plans), which contain appropriate cross-references to additional regulatory requirements (e.g., respiratory protection, alarm systems, eye and foot protection). Under 29 CFR 1910.120, a written

TABLE 1-1

KEY OSHA STANDARDS RELATED TO EMERGENCY RESPONSE (CFR: U.S. CODE OF FEDERAL REGULATIONS)

Reference	Topic
29 CFR 1910	Table of contents
29 CFR 1910.119	Process safety management of highly hazardous chemicals
29 CFR 1910.119 App C	Compliance guidelines and recommendations
29 CFR 1910.119 App D	Sources of further information; non-mandatory
29 CFR 1910.120	Hazardous waste operations and emergency response
29 CFR 1910.120 App A	Personal protective equipment test methods
29 CFR 1910.120 App C	Compliance guidelines
29 CFR 1910.120 App D	References
29 CFR 1910.120 App E	Training curriculum guidelines; non-mandatory
29 CFR 1910.1027 App B	Substances technical guidelines for cadmium
29 CFR 1910.1051	1,3-Butadiene
29 CFR 1910.1052	Methylene chloride
29 CFR 1926	Table of contents
29 CFR 1926.64	Process safety management of highly hazardous chemicals
29 CFR 1926.64 App C	Compliance guidelines and recommendations
29 CFR 1926.64 App D	Sources of further information; non-mandatory
29 CFR 1926.65	Hazardous waste operations and emergency response
29 CFR 1926.65 App A	Personal protective equipment test methods
29 CFR 1926.65 App C	Compliance guidelines
29 CFR 1926.65 App D	References
29 CFR 1926.65 App E	Training curriculum guidelines; non-mandatory

emergency response plan must describe how an actual emergency will be handled to minimize risks to three groups of personnel:

1. Employees engaged in cleanups at uncontrolled hazardous waste sites.
2. Employees engaged in routine operations and corrective actions at RCRA facilities.
3. Employees engaged in emergency response without regard to location.

If an employer does not allow employees to respond to an emergency in any manner except by evacuating the premises, that employer must develop a written *emergency action plan* which, in compliance with 29 CFR 1910.39, includes the following minimum information:

- Emergency escape procedures and routes
- Procedures to be followed by employees who remain to operate critical plant operations before they evacuate
- Procedures to account for all employees after emergency evacuation has been completed
- Rescue and medical duties for those employees who are to perform them
- The preferred means of reporting fires and other emergencies

TABLE 1-2

KEY OSHA STANDARDS RELATED TO PROTECTION OF PERSONNEL (CFR: U.S. CODE OF FEDERAL REGULATIONS)

Reference	Topic
29 CFR 1910 Subpart I App B	Non-mandatory compliance guidelines
29 CFR 1910.120	Hazardous waste operations and emergency response
29 CFR 1910.120 App A	Personal protective equipment test methods
29 CFR 1910.120 App B	General description and discussion
29 CFR 1910.120 App C	Compliance guidelines
29 CFR 1910.120 App E	Training curriculum guidelines; non-mandatory
29 CFR 1910.132	General requirements
29 CFR 1910.183	Helicopters
29 CFR 1910.261	Pulp, paper, and paperboard mills
29 CFR 1910.266	Logging operations
29 CFR 1910.268	Telecommunications
29 CFR 1910.269	Electric power generation, transmission, and distribution
29 CFR 1910.335	Safeguards for personnel protection
29 CFR 1910.1001 App H	Medical surveillance guidelines for asbestos
29 CFR 1910.1027	Cadmium
29 CFR 1910.1030	Bloodborne pathogens
29 CFR 1910.1047	Ethylene oxide
29 CFR 1910.1048	Formaldehyde
29 CFR 1910.1050	Methylenedianiline
29 CFR 1910.1052	Methylene chloride
29 CFR 1915	Table of contents/authority for 1915
29 CFR 1915 Subpart I App A	Non-mandatory guidelines
29 CFR 1915.12	Precautions and the order of testing
29 CFR 1915.1001 App I	Medical surveillance guidelines for asbestos
29 CFR 1926	Table of contents
29 CFR 1926.28	Personal protective equipment
29 CFR 1926.60	Methylenedianiline
29 CFR 1926.65	Hazardous waste operations and emergency response
29 CFR 1926.65 App A	Personal protective equipment test methods
29 CFR 1926.65 App B	General description and discussion
29 CFR 1926.65 App C	Compliance guidelines
29 CFR 1926.65 App E	Training curriculum guidelines; non-mandatory
29 CFR 1926.95	Criteria for personal protective equipment
29 CFR 1926.300	General requirements
29 CFR 1926.302	Power-operated hand tools
29 CFR 1926.551	Helicopters
29 CFR 1926.1101	Asbestos
29 CFR 1926.1101 App I	Medical surveillance guidelines for asbestos
29 CFR 1926.1127	Cadmium

- Names or job titles of persons or departments who can be contacted for further information or explanation of duties associated with emergency response

Depending on relevant regulatory requirements, the overall in-plant responsibility for emergency response planning and implementation may be assigned to the *primary emergency response coordinator* (i.e., under RECRA regulations), the *site safety and health supervisor* (i.e., under 29 CFR 1910.120), or to any number of variously titled personnel having specialized knowledge and

TABLE 1-3

KEY OSHA STANDARDS RELATED TO MEDICAL SURVEILLANCE OF PERSONNEL (CFR: U.S. CODE OF FEDERAL REGULATIONS)

Reference	Topic
29 CFR 1910.120	Hazardous waste operations and emergency response
29 CFR 1910.120 App E	Training curriculum guidelines; non-mandatory
29 CFR 1910.1001	Asbestos
29 CFR 1910.1001 App H	Medical surveillance guidelines for asbestos
29 CFR 1910.1001 App I	Medical surveillance guidelines for asbestos
29 CFR 1910.1003	13 Carcinogens (4-nitrobiphenyl, etc.)
29 CFR 1910.1018 App C	Medical surveillance guidelines
29 CFR 1910.1025	Lead
29 CFR 1910.1025 App B	Employee standard summary
29 CFR 1910.1025 App C	Medical surveillance guidelines
29 CFR 1910.1027	Cadmium
29 CFR 1910.1027 App A	Substance safety data sheet for cadmium
29 CFR 1910.1028 App C	Medical surveillance guidelines for benzene
29 CFR 1910.1043	Cotton dust
29 CFR 1910.1044	1,2-dibromo-3-chloropropane (DBCP)
29 CFR 1910.1044 App C	Medical surveillance guidelines for DBCP
29 CFR 1910.1045	Acrylonitrile
29 CFR 1910.1045 App C	Medical surveillance guidelines for acrylonitrile
29 CFR 1910.1045 App C	Medical surveillance guidelines for ethylene oxide
29 CFR 1910.1048	Formaldehyde
29 CFR 1910.1048 App A	Substance technical guidelines for formaldehyde
29 CFR 1910.1048 App C	Medical surveillance for formaldehyde
29 CFR 1910.1050	Methylenedianiline (MDA)
29 CFR 1910.1050 App C	Medical surveillance guidelines for MDA
29 CFR 1910.1052	Methylene Chloride
29 CFR 1910.1052 App B	Medical surveillance for methylene chloride
29 CFR 1926.60	Methylenedianiline (MDA)
29 CFR 1926.60 App C	Medical surveillance guidelines for MDA
29 CFR 1926.62	Lead
29 CFR 1926.62 App B	Employee standard summary
29 CFR 1926.62 App C	Medical surveillance guidelines
29 CFR 1926.65	Hazardous waste operations and emergency response
29 CFR 1926.65 App C	Compliance guidelines
29 CFR 1926.65 App E	Training curriculum guidelines; non-mandatory
29 CFR 1926.1101	Asbestos
29 CFR 1926.1101 App I	Medical surveillance guidelines for asbestos
29 CFR 1926.1127	Cadmium
29 CFR 1990.151	Model standard
29 CFR 1990.152	Model emergency temporary standard

experience (see Tables 1.4–1.7). In many facilities, the facility manager or operations manager assumes all responsibility for emergency response activities. The key regulatory objective in assigning overall responsibility is to ensure that corporate authority is in fact commensurate with that responsibility—a requirement that is increasingly reflected in the consolidation of emergency response management duties within a corporate executive level function.

At the national level and reflecting the consistent and widespread concern of the American public regarding chemical hazards, the Emergency Planning and Community Right-to-Know Act of 1986 (EPCRA: SARA Title III) requires municipal authorities to:

TABLE 1-4

ESSENTIAL ON-SITE EMERGENCY RESPONSE PERSONNEL (ADAPTED FROM NIOSH, USCG, AND EPA, 1985: OCCUPATIONAL SAFETY AND HEALTH GUIDANCE MANUAL FOR HAZARDOUS WASTE SITE ACTIVITIES)

Title	General Description	Specific Responsibilities
Project Team Leader	Reports to upper-level management; has authority to direct response operations; assumes total control over site activities.	■ Prepares and organizes the background review of the situation, the Work Plan, the Site Safety Plan, and the field team. ■ Obtains permission for site access and coordinates activities with appropriate officials. ■ Ensures that the Work Plan is completed and on schedule. ■ Briefs the field teams on their specific assignments. ■ Uses the Site Safety and Health Officer to ensure that safety and health requirements are met. ■ Prepares the final report and support files on the response activities. ■ Serves as the liaison with public officials.
Site Safety and Health Officer	Advises the Project Team Leader on all aspects of health and safety on site; recommends stopping work if any operation threatens worker or public health or safety.	■ Selects protective clothing and equipment. ■ Periodically inspects protective clothing and equipment. ■ Ensures that protective clothing and equipment are properly stored and maintained. ■ Controls entry and exit at the Access Control Points. ■ Coordinates safety and health program activities with the Scientific Advisor. ■ Confirms each team member's suitability for work based on a physician's recommendation. ■ Monitors the work parties for signs of stress, such as cold exposure, heat stress, and fatigue. ■ Monitors on-site hazards and conditions. ■ Participates in the preparation of and implements the Site Safety Plan. ■ Conducts periodic inspections to determine if the Site Safety Plan is being followed. ■ Enforces the "buddy" system. ■ Knows emergency procedures, evacuation routes, and the telephone numbers of the ambulance, local hospital, poison control center, fire department, and police department. ■ Notifies, when necessary, local public emergency officials. ■ Coordinates emergency medical care.

Continued

TABLE 1-4—*Continued*

ESSENTIAL ON-SITE EMERGENCY RESPONSE PERSONNEL (ADAPTED FROM NIOSH, USCG, AND EPA, 1985: OCCUPATIONAL SAFETY AND HEALTH GUIDANCE MANUAL FOR HAZARDOUS WASTE SITE ACTIVITIES)

Title	General Description	Specific Responsibilities
Field Team Leader	May be the same person as the Project Team Leader and may be a member of the work party; responsible for field team operations and safety.	■ Manages field operations. ■ Executes the Work Plan and schedule. ■ Enforces safety procedures. ■ Coordinates with the Site Safety Officer in determining protection level. ■ Enforces site control. ■ Documents field activities and sample collection. ■ Serves as a liaison with public officials.
Command Post Supervisor	May be the same person as the Field Team Leader; responsible for communications and emergency assistance.	■ Notifies emergency response personnel by telephone or radio in the event of an emergency. ■ Assists the Site Safety officer in a rescue, if necessary. ■ Maintains a log of communication and site activities. ■ Assists other field team members in the clean areas, as needed. ■ Maintains line-of-sight and communication contact with the work parties via walkie-talkies, signal horns, or other means.
Decontamination Station Officer(s)	Responsible for decontamination procedures, equipment, and supplies.	■ Sets up decontamination lines and the decontamination solutions appropriate for the type of chemical contamination on site. ■ Controls the decontamination of all equipment, personnel, and samples from the contaminated areas. ■ Assists in the disposal of contaminated clothing and materials. ■ Ensures that all required equipment is available. ■ Advises medical personnel of potential exposures and consequences.
Rescue Team	Used primarily on large sites with multiple work parties in the contaminated area.	■ Stands by, partially dressed in protective gear, near hazardous work areas. ■ Rescues any worker whose health or safety is endangered.
Work Party	Depending on the size of the field team, any or all of the field team may be in the Work Party, but the Work Party should consist of at least two people.	■ Safely completes the onsite tasks required to fulfill the Work Plan. ■ Complies with Site Safety Plan. ■ Notifies Site Safety Officer or supervisor of unsafe conditions.

TABLE 1-5

OPTIONAL ON-SITE EMERGENCY RESPONSE PERSONNEL (ADAPTED FROM NIOSH, USCG, AND EPA, 1985: OCCUPATIONAL SAFETY AND HEALTH GUIDANCE MANUAL FOR HAZARDOUS WASTE SITE ACTIVITIES)

Title	Specific Responsibilities
Scientific Advisor	▪ Provides advice for field monitoring, sample collection, sample analysis, scientific studies, data interpretation, and remedial plans.
Logistics Officer	▪ Plans and mobilizes the facilities, materials, and personnel required for the response.
Photographer	▪ Photographs site conditions. ▪ Archives photographs.
Financial/Contracting Officer	▪ Provides financial and contractual support.
Public Information Officer	▪ Releases information to the news media and the public concerning site activities.
Security Officer	▪ Manages site security.
Recordkeeper	▪ Maintains the official records of site activities.

TABLE 1-6

OFF-SITE EMERGENCY RESPONSE PERSONNEL (ADAPTED FROM NIOSH, USCG, AND EPA, 1985: OCCUPATIONAL SAFETY AND HEALTH GUIDANCE MANUAL FOR HAZARDOUS WASTE SITE ACTIVITIES)

Title	General Description	Specific Responsibilities
Senior Level Management	Responsible for defining project objectives, allocating resources, determining the chain-of-command, and evaluating program outcome.	▪ Provide the necessary facilities, equipment, and money. ▪ Provide adequate personnel and time resources to conduct activities safely. ▪ Support the efforts of on-site management. ▪ Provide appropriate disciplinary action when unsafe acts or practices occur.
Multi-Disciplinary Advisors	Includes representatives from upper-level management and onsite management, a field team member, and experts in such fields as: ▪ Chemistry ▪ Engineering ▪ Industrial hygiene ▪ Information/public relations ▪ Law ▪ Medicine ▪ Pharmacology ▪ Physiology ▪ Radiation health physics ▪ Toxicology	▪ Provide advice on the design of the Work Plan and the Site Safety Plan.

Continued

TABLE 1-6—*Continued*

OFF-SITE EMERGENCY RESPONSE PERSONNEL (ADAPTED FROM NIOSH, USCG, AND EPA, 1985: OCCUPATIONAL SAFETY AND HEALTH GUIDANCE MANUAL FOR HAZARDOUS WASTE SITE ACTIVITIES)

Title	General Description	Specific Responsibilities
Medical Support	Consulting physicians	■ Become familiar with the types of materials on site, the potential for worker exposures, and recommend the medical program for the site.
	Medical personnel at local hospitals and clinics	■ Provide emergency treatment and decontamination procedures for the specific type of exposures that may occur at the site; obtain special drugs, equipment, or supplies necessary to treat such exposures.
	Ambulance personnel	■ Provide emergency treatment procedures appropriate to the hazards on site.

TABLE 1-7

ADDITIONAL PERSONNEL THAT MAY BE NEEDED FOR HAZARDOUS WASTE OPERATIONS (ADAPTED FROM NIOSH, USCG, AND EPA, 1985: OCCUPATIONAL SAFETY AND HEALTH GUIDANCE MANUAL FOR HAZARDOUS WASTE SITE ACTIVITIES)

Title	General Description	Specific Responsibilities
Bomb Squad Explosion Experts		■ Advise on methods of handling explosive materials ■ Assist in safely detonating or disposing of explosive materials
Communication Personnel	Civil Defense organizations; local radio and television stations; local emergency service networks	■ Provide communication to the public in the event of an emergency ■ Provide communication links for mutual aid
Environmental Scientists	Consultants from industry, government, universities, or other groups	■ Predict the movement of released hazardous materials through the atmosphere, soil, and water resources ■ Assess the effect of this movement on air, groundwater and surface water quality ■ Predict the exposure of people and the ecosystem to the materials
Evacuation Personnel	Federal, state, and local public safety organizations	■ Help plan for public evacuation ■ Mobilize transit equipment ■ Assist in public evacuation
Firefighters		■ Respond to fires that occur on site ■ Stand by for response to potential fires ■ Perform rescue

Continued

TABLE 1-7—*Continued*

ADDITIONAL PERSONNEL THAT MAY BE NEEDED FOR HAZARDOUS
WASTE OPERATIONS (ADAPTED FROM NIOSH, USCG, AND
EPA, 1985: OCCUPATIONAL SAFETY AND HEALTH GUIDANCE
MANUAL FOR HAZARDOUS WASTE SITE ACTIVITIES)

Title	General Description	Specific Responsibilities
Hazardous Chemical Experts	Consultants from industry, government, universities, or other groups	■ Advise on the properties of the materials on site ■ Advise on contaminant control methods ■ Advise on the dangers of chemical mixtures that may result from site activities ■ Provide immediate advice to those at the scene of a chemical-related emergency
Health Physicists	Experts in radiation health from industry, government, universities, or other groups	■ Evaluate radiation health hazards and recommend appropriate action
Industrial Hygienists	Consultants from industry, government, universities, or other groups	■ Conduct health hazard assessments ■ Advise on adequate health protection ■ Conduct monitoring tests to determine worker exposures to hazardous substances
Meteorologists	Consultants from government or other local organizations	■ Provide meteorological information
Public Safety Personnel	County Sheriff, industrial security forces, National Guard, police, etc.	■ Control access to the site
Toxicologists	Consultants from industry, government, universities, or other groups	■ Advise on toxicological properties and health effects of substances on site ■ Provide recommendations on protection of worker health

1. Prepare for emergency releases of hazardous substances by appointing a Local Emergency Planning Committee (LEPC).
2. Immediately notify the LEPC of any release of hazardous substances in quantities greater than prescribed levels.
3. Prepare an inventory of hazardous substances to be submitted to the LEPC.
4. Prepare an annual report detailing the amounts of hazardous substances released to the environment or transported as waste.

Under EPCRA, the LEPC must include, at the minimum, elected state and local officials; police, fire, civil defense, and public health professionals; environmental, hospital, and transportation officials; as well as representatives of facilities subject to emergency planning requirements, community groups, and the media. A primary responsibility of the LEPC is to develop an *emergency response plan* that:

1. Identifies facilities and transportation routes involved in the storage, use, or transport of specified hazardous substances.
2. Describes comprehensive emergency response procedures to be implemented both on- and off-site of any emergency incident.
3. Designates a community coordinator and facility coordinator to implement the plan.
4. Outlines emergency notification procedures.
5. Describes methods for determining the occurrence of a release and the probable affected area an population.
6. Describes community- and industry-owned emergency equipment and facilities and identifies personnel responsible for these resources.
7. Outlines evacuation plans.
8. Describes a training program for emergency response personnel.
9. Presents methods and schedules for exercising emergency response plans.

The promulgation of federal requirements under EPCRA, which effectively extends a national concern and responsibility down to local communities and, at that level, promotes an integration of regional and local governmental as well as private resources toward the objective of emergency planning and response, clearly reflects an ongoing change in paradigm regarding historical distinctions between federal and local interests and also between natural disasters and human-made emergencies (see Figure 1.1).

Whereas the Federal Emergency Management Agency (FEMA) is most commonly known for its responsibility as the lead federal agency within a consortium (national Emergency Management System) of 27 federal agencies (and the American Red Cross) devoted to providing aid and assistance after major natural disasters (e.g., floods, storms, earthquakes), FEMA is today probably best understood as a key partner in the National Mitigation Strategy—a federal programmatic initiative devoted to the development of additional partnerships among federal, state, and local governments and private sector constituents, including the general public, for the express purpose of promoting local community safety. Although the focus is still directed at so-called natural hazards (see Image 1.2), FEMA is also the lead agency of the National Arson Prevention Initiative (NAPI), a partnership that also includes the U.S. Department of Housing and Urban Development, the U.S. Department of Justice, and the U.S. Department of the Treasury. The objectives of NAPI are to increase public awareness regarding practical means for preventing arson and to provide appropriate resources to individuals and communities throughout the nation.

Although there can be doubt that emergency-related partnerships between diverse governmental agencies are sometimes the result of the need for coordinated intelligence gathering—as, for example, in a case of international terrorism, which may require the coordination of efforts of personnel from numerous agencies e.g., Office of Homeland Defense, Federal Bureau of

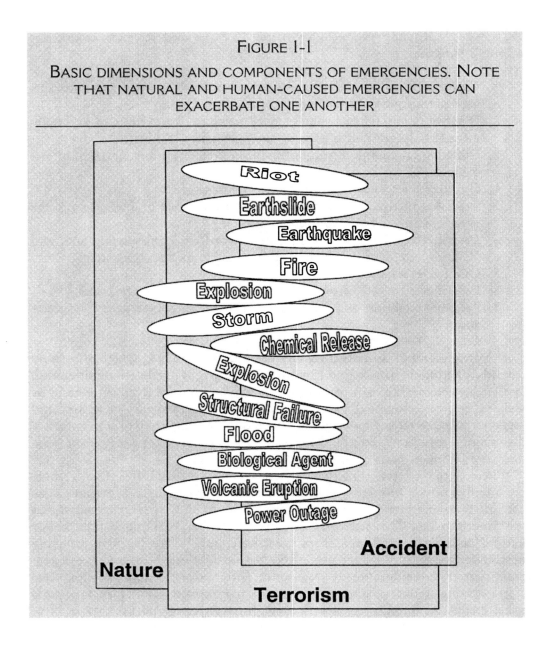

FIGURE 1-1

BASIC DIMENSIONS AND COMPONENTS OF EMERGENCIES. NOTE THAT NATURAL AND HUMAN-CAUSED EMERGENCIES CAN EXACERBATE ONE ANOTHER

Investigation (FBI), National Security Agency (NSA), Central Intelligence Agency (CIA), and others—other factors also promote intergovernmental coordination as well as governmental and private sector partnerships. For example, Civil Emergency Planning (CEP), though long an integral (albeit little noticed) part of the capability of the North Atlantic Treating Organization (NATO), has emerged as an increasingly important resource for NATO partners in their effort to prevent human-made disasters, mitigate the consequences of natural calamities, and protect the population, national wealth, and the environment.

The rapidly expansive trend in the United States to conceptualize emergency response in terms of the requirements for practical prevention, efficient

IMAGE 1-2

MT. ST. HELENS, WA, MAY 18, 1980: DISASTERS ARE DEVASTATING TO THE NATURAL AND HUMAN-MADE ENVIRONMENT. FEMA PROVIDES FEDERAL AID AND ASSISTANCE TO THOSE WHO HAVE BEEN AFFECTED BY ALL TYPES OF DISASTER

Source: NOAA News Photo

response, and effective mitigation, rather than in terms of type or source of threat (e.g., natural vs. human-made hazards) or jurisdictional mandate (e.g., national, regional, local) is clearly paralleled in European, Pacific rim, and other nations as well as international organizations. Of course, the seemingly ever-escalating scourge of both home-grown and international terrorism only underscores the necessity of a continual integration of international, national, local, and private sector emergency planning and response efforts. Today, such an integration of historically distinct jurisdictional authorities and responsibilities continues to drive the invention of new federal agencies as well as the reorganization of existing agencies (see "National Response Plan" in Chapter 12).

KEY ELEMENTS OF EMERGENCY RESPONSE PROGRAMS

Essentially synonymous with *crisis management, disaster planning and management, civil emergency response,* and *contingency planning,* emergency response (which is inclusive of planning, management, and response functions) is subject to a wide range of social, economic, and technical factors. In most recent years, perhaps the more significant of these factors include (a) public concern over hazardous chemicals, (b) international and domestic terrorism, (c) ongoing development of a global economy, and (d) rapid developments in electronic communications.

Hazardous Chemicals

Of roughly 16 million known chemical substances (including naturally occurring and humanly made chemicals), about 60,000 are in daily commercial use in any technologically developed country. Until the mid 1970s, primary concern was focused on only a small number of these chemicals—specifically, petrochemicals—even though the Federal Waste Pollution Control Act (1974) did establish clear federal concern regarding the discharge of environmental hazardous substances into the nation's waterways. In 1975, Congress enacted the Federal Hazardous Materials Transportation Act (HMTA), which was the first comprehensive attempt to regulate the transport of hazardous chemicals, following in 1976 with the Resource Conservation and Recovery Act (RCRA), which established a strong federal initiative to exert "cradle-to-grave" management of hazardous wastes. Within two years of the enactment of RCRA, the public became fully aware of the potential risks of hazardous wastes through the incident at Love Canal (Niagara Falls, New York), where residents finally had to be evacuated from houses built over an abandoned dumping ground used from 1947 to 1953 to bury industrial chemical wastes. In response to this incident, in 1980 Congress enacted CERCLA (Superfund). Unlike RCRA, which focuses on waste management by existent facilities, CERCLA deals with chemically contaminated sites that are abandoned.

Although the governmental and public consciousness of the potential risks of chemicals, whether those chemicals are of commercial value and legally defined as "materials" or have no commercial value and are defined as "wastes," expanded greatly between 1975 and 1980, that consciousness was

> Confined primarily to the cleanup of hazardous waste sites and to determining the long-term effects of exposure to hazardous wastes. That changed on December 3, 1984, when a cloud of methyl isocyanate from a Union Carbide manufacturing plant in Bhopal, India, killed more that 2,500 and injured an estimated 200,000 people. When the Bhopal tragedy was followed by an accidental chemical release on August 11, 1985, at another Union Carbide plant at Institute, West Virginia, public concern turned to alarm. Although the West Virginia release was not serious, it underscored for many Americans the lack of information about hazardous substances in their communities and about the health hazards associated with exposure. It also focused attention of the inadequacies of emergency response capabilities.
> —Firefighter Safety Study Act Working Group, 1992

In 1986, as a direct result of the Bhopal incident, Congress passed the Emergency Planning and Community Right-to-Know Act of 1986 (EPCRA: SARA Title III), which, for the first time, raised emergency response to chemical release to National as well as local preeminence. It should also be noted, moreover, that the Occupational Safety and Health Administration (OSHA; established in 1972) was at the same time beginning to raise the consciousness of industry regarding the vital importance of integrating in-plant emergency response programs with concerns about chemical health and safety through specific regulations, especially regulations related to Hazardous Waste Operations and Emergency Response (29 CFR 1910.120) and Employee Emergency Plans (29 CFR 1910.38)—a trend that is manifest in OSHA regulations throughout the past 20 years, including OSHA's Laboratory Standard (29 CFR 1910.1450), Respiratory Protection regulations (29 CFR 1910.134), and Chemical Process Safety regulations (29 CFR 1010.119).

Terrorism

Until recently, the word terrorism in the United States was largely restricted to acts of violence committed by politically motivated groups or foreign national agents, such as the destruction of the U.S. Marine barracks in Beirut (1983) or the bombing of Pan Am Flight 103 over Lockerbie, Scotland (1988). Until 1993, most Americans perceived terrorism as something to worry about only when traveling in other countries. However, with the 1993 bombing of the World Trade Center in New York City, it was clear that terrorism had arrived in America—a fact grimly underscored by the devastation of the WTC twin towers in 2001. With the 1995 bombing of the Federal building in Oklahoma City, it became equally evident that not only could terrorism occur in America, but that it could be carried out by Americans on each other—again, a fact underscored in 1996 by the apprehension of the Unabomber, who, over a period of 17 years, had carried out his own brand of domestic terrorism.

Today, terrorism is recognized more by its immediate consequences than by any specific intent of its perpetrators (see Image 1.3). After all, the motivation of the perpetrator of any terrorist act is absolutely irrelevant to the dead to or their survivors. Whether committed with political intent, out of personal rage, revenge, psychopathic pleasure, or any other dimension of depravity, terrorism is premeditated, covert violence against an unknowing, unprepared, and unnumbered public.

While home-grown extremists have been categorized in terms made all too familiar by the mass media (e.g., white supremacist skinheads, neo Nazis, tax protester, crazed constitutionalist, Ku Klux Klanner, militias, environmental anarchist), it must be understood that such categories by no means exhaust the possibilities. The simple fact is that the physical wherewithal for achieving widespread public harm and injury is increasingly available to more and more people. If, historically, it has been the bomb in the hands of a political extremist that served as the primordial image of the terrorist, that image is already supplanted by that of a canister of nerve gas in the hands of a religious zealot—which, in turn, will probably soon be supplanted by that of a piece of software in the hands of a disgruntled but highly knowledgeable employee or even of a student simply "playing" with a computer. The National Response Plan (see Chapter 12) and the reorganization (ongoing) of federal agencies attendant to the creation of the Department of Homeland Security clearly underscores the extensive range of concerns regarding both foreign and domestic terrorism.

IMAGE 1-3

NEW YORK, NY., OCTOBER 16, 2001: VEHICLES/DEBRIS REMOVAL
FROM GROUND ZERO TO THE STATEN ISLAND LANDFILL... A
24-HOURS-A-DAY OPERATION

Source: Andrea Booher/FEMA News Photo

Global Economy

Traditionally, and despite the variability inherent in the political pluralism of individual societies, health and safety standards have been essentially the province of the nation. However, with the advent of a global economy and its consequent emphasis on an integrated paradigm of environmental quality and human health, national standards can be expected to become increasingly influenced by the realities of international business. Perhaps of particular relevance is the growing body of international manufacturing standards that encompass concern not only for quality assurance of products and services, but also for the impact of industrial processes and products on environmental quality and human health, and thus for appropriate planning and response regarding potential emergencies associated with those processes and products.

The broad goal of the International Standards Organization (ISO) is to promote the development of standardization and related activities in the world with a view to facilitating international exchange of goods and services, and to develop cooperation in the sphere of intellectual, scientific, technological, and economic activity.

In essence, the clear intent of ISO is to provision a company's entry into international trade on the basis of a facility audit, external confirmation of broad compliance with environmental quality and human health standards, and the public disclosure of managerial failings.

Typically known within the international community as the *harmonizing of international environmental quality criteria*, this objective can be expected to provide a major impetus to the examination and reevaluation of traditional paradigms that underlie conventional business, legal institutions, and contemporary national approaches to the management of health, safety, and the environment. Already there is significant internal movement to (a) consider health, safety, and the environment (HSE) as a holistic and integral component of *Total Quality Management*; (b) reexamine the constraints imposed upon the English Common Law (and its diverse, global legal progeny) by the now historically dated agricultural and early industrial preoccupation with questions of property, possession, and fault; (c) recast the goal of short-term profit to one of long-term sustainability; and (d) accomplish the wholesale expansion of the public's right of access to all information that impacts human health and safety. It cannot be expected that, in light of such considerations, either the substance or the philosophy of the health and safety standards of any individual nation will long remain unaffected.

Electronic Communication

The constantly expanding availability and affordability of sophisticated electronic communication and analytical devices present new opportunities as well as challenges to emergency management. At the level of in-plant prevention, computerized unit-process control and alarm systems are effective means for keeping dangerous production processes within safe operational limits and for providing both in-plant and community-wide emergency services with early warning of potential danger. During an actual emergency incident, these devices can play a key role in all phases of incident response, facilitating effective evacuation, on-site management of response personnel as well as off-site backup services, and essential data retrieval and processing, including the use of expert computer programs for forecasting the air/ground/water transport of released chemicals and for the deployment of search and rescue personnel.

Although the potential value of sophisticated electronic devices for both emergency planning and response cannot be overemphasized, it is all to no avail if industry does not employ it or train plant personnel in its effective use, of if community emergency services are not provided the appropriate hardware and software or sufficient funds for personnel training and equipment maintenance and upgrade.

Although there are notable exceptions, it is fair to say that the members of a local fire department, for example, typically have greater access to the most advanced electronic data processing devices and techniques at home than they do within their department. It is also obvious that the greatest know-how and capability are in industry which, though being the very source of the vast majority of community-wide emergency incidents, continues to focus that know-how and capability on plant productivity, with too little attention given the role of electronic information processing to on-site emergency prevention and containment.

A corollary to the value of modern electronic information processing to emergency planning and response, is of course, the misuse of that capability, whether by accident (as when a computerized process control and alarm system is mistakenly deactivated in the process of routing electrical cables by a contractor, or on purpose, in the case of a terrorist act.

In 1992, OSHA implemented its final rule that is most often referred to as the *Chemical Process Safety Regulation*, which is a much abridged name for the more formal appellation, Process Safety Management of Highly Hazardous Chemicals, Explosives, and Blasting Agents. The final rule actually consists of two major sets of regulations: one dealing with the management of explosive and blasting agents (29 CFR 1910.109), the other with the management of highly hazardous chemicals (29 CFR 1910.119). Section (1) of 29 CFR 1910.119 defines the objective of *Management of Change* (MOC).

Those companies that handle any of more than 135 listed chemicals at or above so-called *Threshold Quantities* (pounds of chemical) must comply with the provisions of 29 CFR 1910.119. However, even a company that does not fall within the regulatory purview of the process safety regulations is well advised to consider the development of a management of change program—especially if it employs electronic or computerized means for controlling or alarming dangerous production processes.

The basic objective of any management of change program is to ensure that good engineering principles and practices are always used in designing, constructing, operating, and maintaining facilities. This objective requires the recognition that even relatively small and seemingly innocuous changes associated with facility design, construction, operations, and maintenance may actually result in unacceptable health and safety hazards. The various techniques employed in a management of change program are basically those used in any hazard assessment and are crucial to ensuring the proper use and maintenance of electronically controlled and alarmed production processes. The practical implementation of a MOC program requires clear criteria for distinguishing between those changes in plant operations, design, and features that have no reasonable likelihood of resulting in a threat to health and safety and those that do.

Under 29 CFR 1910.119, specific exemption from MOC requirements is granted any change that is a *replacement in kind*, which is any replacement of a part (i.e., equipment, machinery, or material) or procedures that satisfies ongoing design specifications that pertain to the performance or role of that part or procedure in plant processes involving regulated chemicals. Within the limited context of this regulatory authority, MOC procedures must address the following issues with regard to any change that is not a replacement in kind.

- The technical basis for the proposed change
- The impact of the change on safety and health
- Modification of operating procedures appropriate to the change and its related risks
- The time period required for preparing and implementing the change
- Authorization requirement attendant to the change

The usual means for addressing these issues is the integration of MOC procedures with existing in-plant approval and authorization procedures, especially standard internal work request and work order procedures. This approach is eminently practical whether a company falls within the jurisdictional purview of 29 CFR 1910.119 or, if not subject to these regulations, simply chooses to implement a MOC policy as one component of a comprehensive health and safety program, an option that is increasingly exercised by companies in the United States and elsewhere. In fact, MOC is widely recognized as a state-of-the-art business management practice regardless of legal authority.

Where MOC is practiced routinely and regardless of regulatory jurisdiction, corporate decision-making procedures involving work requests and work orders provide specific lines of authority and responsibility for all potential changes, including those involving replacement in kind (most often called *change in kind*) as well as *changes not in kind*. The typology of changes, inclusive of some range of changes from negligible to severe risk to health and safety, is precisely reflected by the increased level of authority required to implement the change.

For example, Figure 1.2 is an overview of a MOC procedure that provides for three basic types of changes, each type being defined essentially by the level of authority required to implement it:

- Level 1: A change that may be authorized solely by the department supervisor who, on the basis of written criteria provided by the company, determines that the change is change in kind, and thereby presents negligible hazard or risk.
- Level 2: A change that does not meet the criteria for a Level 1 change and which, with the concurrence of the corporate safety officer, may be implemented by the department manager only after completion of a management *record of change*, which is essentially a checklist that directs the manager's assessment of the change and its implementation; in this case, a Level 2 change is known to present more than a negligible health and safety risk, but one that is relatively uncomplicated and easily controlled.
- Level 3: A change that does not meet the criteria for a Level 1 change and that, by its nature or complexity, is judged to require the attention of the highest corporate authority, including the safety officer, the safety committee, the facility manager, and other selected corporate personnel, and, possibly, external consultants and experts.

The criteria for identifying levels of change in Figure 1.2 are included in Figures 1.3 and 1.4.

EXTENDED PARTNERSHIPS

Historically, the first line of defense against community disaster was the local fire brigade. Most often, it still is, and those paramount virtues of the firefighter—bravery, self-sacrifice, and brotherhood—are among the most

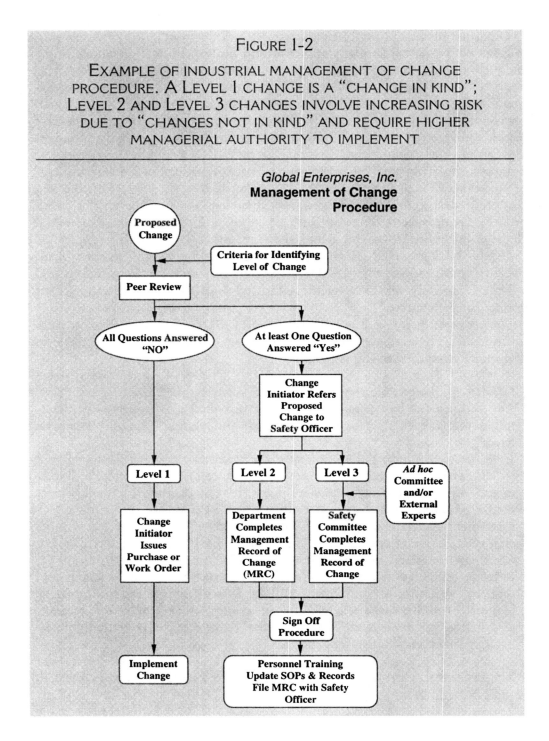

FIGURE 1-2

EXAMPLE OF INDUSTRIAL MANAGEMENT OF CHANGE
PROCEDURE. A LEVEL 1 CHANGE IS A "CHANGE IN KIND";
LEVEL 2 AND LEVEL 3 CHANGES INVOLVE INCREASING RISK
DUE TO "CHANGES NOT IN KIND" AND REQUIRE HIGHER
MANAGERIAL AUTHORITY TO IMPLEMENT

treasured values in any society. But, of course, modern societies are evermore
becoming increasingly complex and the range of potential emergencies
correspondingly expands, forcing greater specialization and bureaucratic
compartmentalization within the traditional armoratoreum of emergency
response. Today, no one group or organization is equipped by training, experience, knowledge, equipment, or legal mandate to deal with every type of

FIGURE 1-3

EXAMPLE OF A MANAGEMENT OF CHANGE FORM (REQUEST
AND DETERMINATION OF LEVEL)

Global Enterprises, Inc.

For Safety Officer's Use Only
Log Number
Process Code

**Management of Change
Request & Determination of Level**

Name of Person Initiating Change

Description of
Proposed Change

Individuals Consulted

Name | Department

I certify that I have consulted with the above individuals for the purpose of completing the reverse side of this form. On the basis of the information included on this form, I determine that the proposed change described above ...

☐ Is a Level 1 Change

☐ Must be referred to the Global Safety
Officer for his determination

Signature of Change Initiator   Date

I certify that I have reviewed this determination and hereby agree with the above determination.

Signature of Authorized Person   Date

Page 1 of 2

---

emergency. Even though still the pivotal organization in most all emergencies, the fire department is but one member of an extended partnership (see Figure 1.5) that, despite historic distinctions between natural and human-made disaster, between local, state, federal and even national jurisdictional authority, between public and private sectors, and between governmental and personal responsibility, must function as a smoothly integrated, albeit multifaceted, enterprise. The objectives of this extended partnership are, simply, to prevent, to prepare for, and to respond to situations that present serious risk to human health and safety.

FIGURE 1-4

EXAMPLE OF A MANAGEMENT OF CHANGE FORM (CRITERIA FOR
DETERMINATION OF LEVEL). THIS IS A CONTINUATION OF THE
FORM DEPICTED IN FIG. 1.3

Global Enterprises, Inc.   **Management of Change Request
& Determination of Level**

> **The supervisor or manager of the department responsible for carrying out
> the proposed change must complete this form and submit the completed form
> to the Global Safety Officer**

If the answer to each of the following questions is "no," the proposed change is a
Level 1 change. If the answer to *any* question is "yes," the proposed change must
be referred to the Global Safety Officer (or his designate).

Will the Proposed Change ...

	YES	NO
1. Require any modification of any of the following Global Programs:		
• Lockout/Tagout	☐	☐
• Confined Space Entry	☐	☐
• Hot Work Permit	☐	☐
• Respiratory Protection	☐	☐
• Bloodborne Pathogens	☐	☐
• Hazard Communication	☐	☐
• Laboratory Standard	☐	☐
• Electrical Safety	☐	☐
• Hazardous Waste Contingency Plan	☐	☐
• Hearing Conservation Program	☐	☐
• Process Safety Management	☐	☐
• Ergonomic Safety	☐	☐
• Good Manufacturing Processes	☐	☐
• Personal Protective Clothing & Equipment	☐	☐
• Stormwater Pollution Prevention	☐	☐
2. Require more than routine coordination with other departments	☐	☐
3. Result in any change in equipment or piping	☐	☐
4. Result in any change in structural design or physical layout	☐	☐
5. Result in any change in raw materials or by-products	☐	☐
6. Result in a significant change in energy consumption	☐	☐
7. Result in any interruption of automatic or manual signaling devices or alarms, automatic process controls, alarms or instrumentation	☐	☐
8. Interfere with the normal functioning of any safety or emergency equipment (e.g., sprinklers, ventilation, emergency lighting)	☐	☐
9. Significantly affect the routine on-site work of external contractors or consultants	☐	☐
10. Result in a significant change in operating procedures or process directions	☐	☐
11. Result in a change in process parameters (e.g., temperature, pressure) beyond documented operational limits	☐	☐

Page 2 of 2

Although federal, state, and local authorities (as well as certain international
authorities) have made significant advances in promoting cross-jurisdictional
partnerships, it is regrettably true that, by and large, industry has maintained
a "not my job" mentality regarding emergency response—despite the fact that,
under regulations pursuant to the Resource Conservation and Recovery Act of
1976, hazardous waste generators must prepare *contingency plans*. In the over

FIGURE 1-5

EMERGENCY RESPONSE CAPABILITY RESULTS FROM EFFECTIVE PARTNERSHIPS AMONG PUBLIC AND PRIVATE SECTOR ORGANIZATIONS

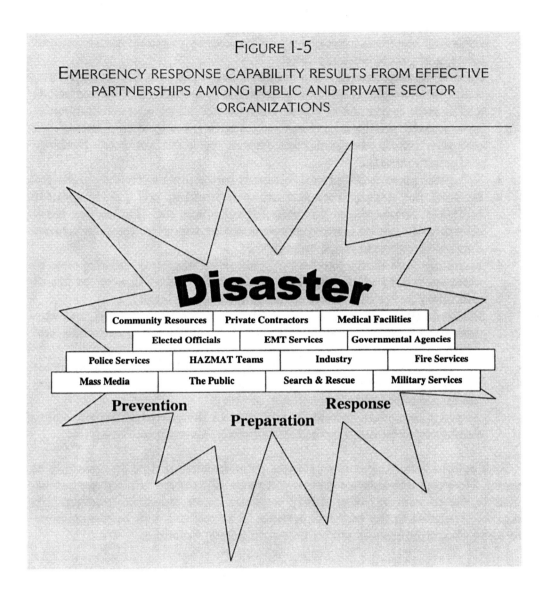

quarter-century since then, contingency planning, of course, has been continually expanded under various OSHA regulations, but it would appear that industry has typically substituted a rather complacent "paper-compliance" with these regulations in place of substantial commitment to the social objectives they represent.

The rather dismal performance of industry at large is best highlighted by those companies that stand out as state-of-the-art companies regarding commitment to employee and community health and safety. Acting well beyond the literal requirement of specific regulations, such companies:

best practice

1. Coordinate directly with local fire departments, ambulance and EMT services, and hazardous chemical specialists in the design and implementation of their in-plant emergency response plans. This coordination includes on-site inspections, the sharing of pertinent information

regarding chemical inventories and hazardous in-plant operations and areas, and practice-evacuations.

2. Review equipment and other material needs of local fire departments, with special emphasis on inadequacies due to specialized plant operations and hazardous materials. Where inadequacies are noted, such companies have actually provided local authorities with specialized equipment (e.g., computers; radio communication devices; specialized protective clothing) and other materials.

3. Volunteer plant facilities and resources for training exercises conducted by local fire departments and other authorities, and jointly sponsor table-top exercises (see Chapter 13) in which such authorities work directly with in-plant emergency response personnel in designing a coordinated response to mock emergencies.

4. Invite all community partners in emergency planning and response to plant-sponsored workshops and seminars that relate directly to health and safety issues of interest to the community at large.

5. Maintain contracts with private contractors, service providers, and vendors for 24-hour availability of specialized equipment, materials, and services that may be needed in case of a plant-related emergency.

6. Ensure by means of appropriate engineering controls and security measures that all sensitive hazardous operations and materials are fully safeguarded against unauthorized use or entry.

7. Actively encourage, promote, and reward the participation of facility employees in community-based emergency response activities.

Such actions reflect, of course, serious commitments of time and possibly of money. However, any state-of-the-art company fully recognizes that any investment in the prevention of or timely response to an industrial emergency is minuscule relative to the very real benefits to be realized both by the corporation and the community in which that corporation functions.

PROACTIVE AND REACTIVE DIMENSIONS

The on-going switch in paradigm of emergency response being solely the jurisdictional province of a particular agency to being the responsibility of an amalgam of governmental and private sector partnerships clearly reflects a growing realization that any emergency, whether natural or human-made, is not simply an event, but a societal phenomenon—not simply an isolated incident, but a product of social circumstance. From this perspective, a flood (for example) is not an emergency. A flood becomes an emergency only in terms of the devastation it causes in terms of human life and property. Moreover, key factors that influence the degree of human risk are not simply hydrological and meteorological variables, which are beyond human control, but also those political, economic, and social variables that influence the location of dwellings within flood plains, and land use policies that can dramatically affect runoff storage capacity—variables that are subject to human control.

As long as the focus of emergency response is only on the "incident" as opposed to the "circumstance" in which the emergency occurs, emergency response must be viewed as an essentially reactive exercise—an unacceptable approach that relegates a certain number of human deaths per year and certain degrees of human misery per community essentially to an act of God. However, if the focus is in fact more on the circumstances of an emergency than on the incident itself, emergency response is seen to encompass both proactive and reactive strategies that, in combination, can effectively reduce the needless loss of human life and all its attendant trauma.

Over the last three decades of the twentieth century, there was significant and steady progress in the long historical effort to transform emergency response into a combined proactive and reactive endeavor and, in combination with the new emphasis on intergovernmental and governmental and private sector partnerships, to imbue that endeavor with practical planning, managerial, legal, and enforcement tools. There can be no question that this progress has occurred as a result of what some see as an erosion, and others see as ongoing maturation of traditional western values regarding property and personal rights. Whether at the national or international level, it is not likely that such contrary perceptions will soon (or ever can) be resolved. Meanwhile, there is everywhere a growing understanding that the continually expanding complexity and interconnectedness of the social milieu in which we all live requires a significantly more holistic approach to emergency response than previously has been the case. Simply put, even the most rigid attitudes and philosophies tend to change in the midst of disaster.

STUDY GUIDE

True or False

1. Emergency response planning has become a basic component of modern business practice in the United States simply because of federal regulatory mandate.
2. The Emergency Planning and Community Right-to-Know Act was the first federal legislation dealing with chemical emergencies.
3. Terrorism is the primary reason that emergency response planning is today such a commonplace business activity in the United States.
4. Management-of-Change regulations are intended primarily for the safe management of selected hazardous chemicals.
5. In technologically developed countries, emergencies can be adequately handled by a single response authority.
6. RCRA, HMTA, and CERCLA are major federal acts that focus on the safe management of hazardous materials and chemicals.

7. The Resource Conservation and Recovery Act (RCRA) focuses on the cleanup of contaminated sites that are abandoned.
8. So-called "accidents" should not be considered to be "emergencies," which are restricted simply to natural calamities and acts of terrorism.
9. Emergency response planning requirements are essentially defined in the United States under federal RCRA and CERCLA regulations.
10. In Emergency response planning, it is essential that individual responsibility is commensurate with authority.
11. In the United States, regional and local emergency planning and response activities are solely the responsibility of governmental agencies; they do not involve the private sector.
12. The International Standards Organization (ISO) has no influence on American corporate activities related to emergency planning and response.
13. Health, safety, and environmental considerations are increasingly considered to be integral to corporate Total Quality Management practices.
14. Modern emergency response practices are inclusive of both planning and management functions as well as response functions.
15. At the heart of a modern, state-of-the-art approach to emergency planning is the integration of both proactive and reactive approaches to risk management.

Multiple Choice

1. Emergency response planning is
   A. increasingly a routine activity in American corporate life
   B. essentially the responsibility of governmental agencies
   C. not influenced by corporate stakeholders
2. The safety of personnel involved in emergency response activities in the United States is regulated by
   A. the Occupational Safety and Health Administration
   B. the U.S. Environmental Protection Agency
   C. the International Safety Organization
   D. none of the above
3. The written emergency response plan required by OSHA for emergency response workers describes how an emergency will be handled to minimize risks to
   A. workers engaged in cleanups at uncontrolled hazardous waste sites
   B. workers engaged in routine operations and corrective actions at RCRA facilities
   C. workers engaged in emergency response without regard to location
   D. all of the above
   E. none of the above
4. The Federal Emergency Planning and Community Right-to-Know Act of 1986 (EPCRA) requires municipal authorities to
   A. apply to appropriate state and federal authorities for guidance in the preparation of plans for responding to the release of hazardous substances

    B. contract with federally certified expert consultants for the preparation of emergency response plans

    C. appoint a Local Emergency Planning Committee (LEPC)

    D. none of the above

5. Under EPCRA, an emergency response plan must

    A. describe comprehensive emergency response procedures to be implemented on-site of any emergency incident

    B. describe comprehensive emergency response procedures to be implemented off-site of any emergency incident

    C. both of the above

6. The National Mitigation Strategy is a federal initiative to develop partnerships among

    A. only federal and state response resources

    B. only state and local response resources

    C. federal, state, and local governmental response resources as well as private sector resources

7. The basic strategy of the U.S. government is to specify response efforts most appropriate to

    A. the type of hazard (e.g., natural vs human-made hazards)

    B. different jurisdictional authorities (e.g., national, regional, state, and local)

    C. ensure practical prevention, efficient response, and effective mitigation of all emergencies

8. The ongoing development of modern emergency planning in the United States is strongly influenced by

    A. public concern over hazardous chemicals

    B. international and domestic terrorism

    C. global economic development

    D. high-tech electronic communications

    E. all of the above

    F. none of the above

9. An important factor influencing the Congressional enactment of the Emergency Planning and Community Right-To-Know Act (EPCRA) was

    A. the Love Canal incident

    B. the Bhopal, India incident

    C. the Lockerbie, Scotland incident

    D. the Oklahoma City incident

10. Management of Change regulations are intended to ensure that good engineering principles and practices are always used when

    A. designing facilities

    B. constructing facilities

    C. operating facilities

    D. maintaining facilities

    E. all of the above

    F. only C and D

    G. only A and B

Essays

1. Discuss the importance of establishing cross-jurisdictional partnerships for emergency planning and response activities.
2. Describe the various ways in which a highly localized incident (e.g., a chemical spill in a local manufacturing plant) in your community may become a significant community-wide emergency. Identify several reactive and proactive approaches to emergency management that might be relevant to keeping the local incident "contained."
3. Given the diversity of types of expertise required to adequately plan for and respond to any emergency, identify and discuss several major problems that must be addressed by any emergency response planner.

Case Study

Assume that you are a teacher of Emergency Response Management at a local college and that you are composing a curriculum for second year undergraduate students. On the basis of Chapter 1 materials and discussions:

1. Identify which topical areas you would give priority to.
2. What type of backgrounds would you look for in prospective faculty?
3. How would you monitor/measure the success of the curriculum?

ESSENTIALS OF HOLISTIC PLANNING AND MANAGEMENT

INTRODUCTION

Whether at the level of the community or individual corporation, emergency response planning and management require the coordination of a wide range of information, services, and materials. Clearly, much of the information required and the services and materials needed in the response phase are highly specific with regard to location and attendant circumstances. However, there are certain aspects of emergency response planning and management that are categorically appropriate to all emergency response efforts and that may therefore be called universal elements of holistic planning and management, including:

- *Scope of Emergency Planning*: It is essential that planning proceed from the premise that the objective of planning is inclusive of (a) the prevention of an emergency, (b) preparation for the occurrence of an emergency, and (c) actual response to an emergency.
- *Assessment of Hazard and Risk*: Hazard assessment is the identification of potential harm and injury; risk assessment involves the determination of the probability of harm and injury to specific persons and groups. Hazard assessment is a necessary first step toward realistic risk assessment, but the estimation of actual risks also depends upon the analysis of potential exposures of defined persons and groups to individual hazards.
- *On- and Off-Site Management*: The site of an actual emergency is typically taken to be the location of actual response efforts, but those efforts trigger many activities that affect off-site locations. In some instances, the off-site sequellae of a distantly located emergency can be more disastrous than on-site consequences. This is particularly the case when an act of terrorism is involved and the objective of the terrorist is to implement a series of interrelated but individually staged incidents.

- *Authority and Responsibility*: Whether in prevention, preparation, or response phases, emergency response planning must be built on absolutely clear lines of authority and responsibility. During the response phase, this usually is referred to as *incident command*.
- *Communications and Information Handling*: Communications and information handling are key activities throughout prevention, preparation, and response phases of emergency planning management.
- *Provisions and Support*: Provisions are the equipment, supplies, and materials that are immediately on-hand to prevent or contain an emergency incident. Support from off-site sources, including governmental agencies and private contractors, must also be provided for in any comprehensive emergency management program.
- *Medical Treatment and Surveillance*: Medical treatment and surveillance are inclusive of all first-aid and subsequent medical treatment of acutely affected victims as well as long-term follow-up examinations to monitor for emergent chronic health effects.
- *Remediation and Review*: Both during and after an emergency incident, appropriate remedial actions must be taken to protect against further harm or injury (see Image 2.1) due to secondary effects of the emergency (e.g., structural damage, runoff of hazardous materials, hazardous debris); remedial actions also include those taken to recharge or replace supplies and materials depleted in the initial response effort. All remediation efforts should be given first priority in the comprehensive review and evaluation of the emergency incident.

Each of these elements requires comprehensive examination of diverse types of information, including information and data related to procedures, operations, equipment, and materials. Within individual corporations, they may well require reconsideration of normal chains of command and procures that, after all, typically reflect the needs of productivity and quality control as opposed to the real needs of effective emergency management. At the level of the community, the complexity of these issues and concerns has (in the past) typically constrained by legal jurisdictional authority among governmental authorities and agencies, although the National Response Plan (see Chapter 12) resolves such jurisdictional constraints by ensuring the coordination and cooperation of federal, regional, local, and tribal authorities as well as private sector organizations.

SCOPE OF PRAGMATIC EMERGENCY PLANNING

The acronym HSE (Health, Safety, and Environment; see Chapter 1) is frequently used in the international literature to denote the paradigm of *Integrated Environmental Planning and Management*. This paradigm, essentially formulated by the United Nations Conference on Environment and Development (UNCED), commonly known as the Earth Summit (Rio, 1992), is based on a broad consensus that the management of human health and safety and the management of environmental quality can be carried out more efficiently when both efforts are integrated. This consensus is expressed in the

IMAGE 2-1

LA CONCHITA, CA, JANUARY 15, 2005: THE VENTURA COUNTY
FIRE DEPARTMENT URBAN SEARCH AND RESCUE TEAMS STABILIZE
STRUCTURES SO THAT THEY CAN SEARCH THEM IN LA CONCHITA,
CALIFORNIA, WHERE WINTER STORMS CAUSED FATAL LANDSLIDES
THAT DAMAGED HOMES AND ROADS

Source: John Shea/FEMA News Photo

1992 U.N. Earth Summit agenda and has important social ramifications for both public and private sector decision making:

> The primary need is to integrate environmental and developmental decision-making processes. To do this, governments should conduct a national review and, where appropriate, improve the processes of decision-making so as to achieve the progressive integration of economic, social, and environmental issues in the pursuit of development that is economically efficient, socially equitable and responsible, and environmentally sound.
>
> —1992 Earth Summit Agenda 21, p. 66

Although the terms *occupational safety and health* (OSH), *environmental safety and health* (ESH), and *environmental quality* (EQ) are still commonly used to focus, respectively, on workplace, non-workplace, and environmental contributions to human health and safety, it is widely understood that such distinctions may often obscure rather than clarify dynamic linkages (see Figure 2.1) between

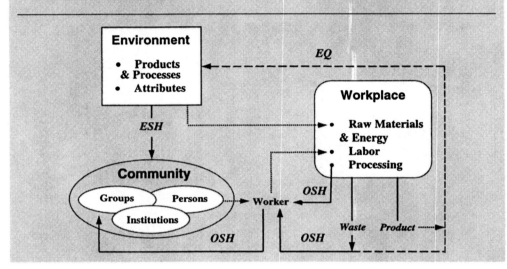

FIGURE 2-1

HOLISTIC VIEW OF DYNAMIC LINKAGES BETWEEN THE ENVIRONMENT, COMMUNITY, AND WORKPLACE (EQ, ENVIRONMENTAL QUALITY; ESH, ENVIRONMENTAL SAFETY AND HEALTH; OSH, OCCUPATIONAL SAFETY AND HEALTH). HEAVY SOLID LINES REPRESENT ISSUES HISTORICALLY ADDRESSED BY U.S. REGULATIONS HEAVY DOTTED LINES REPRESENT ISSUES LIKELY TO BE GIVEN INCREASED REGULATORY ATTENTION; LIGHT DOTTEDNES INDICATE BASIC OPERATIONAL OR FUNCTIONAL RELATIONSHIPS

environmental processes and attributes, human activities (e.g., work), and human health and safety. It is precisely these dynamic linkages that must inform any comprehensive approach to emergency planning and response.

For example, if the focus of planning efforts becomes simply the knockdown of a structural fire in a manufacturing plant, it is likely that relatively little attention will be given to the management of the water runoff that results from firefighting activities. However, such runoff may contain hazardous chemicals that, if allowed to infiltrate into ground- or surface-water supplies, may result in a subsequent emergency involving contaminated drinking water.

Another example might be an emergency incident involving the release into the community of toxic vapors or particulates as a result of a highway accident involving a chemical tank truck. The obvious emergency incident is the accident scene, but less obvious is the possibility that some of those vapors or particulates may become entrained into the ventilation systems of even distantly located downstream plants and other facilities, which may result in subsequent acute and/or chronic effects on workers.

By focusing our attention upon interconnected environmental, community, and workplace dynamics, the HSE paradigm requires emergency planning and response to give specific and detailed attention to the fact that *any emergency incident* is also a *systemic threat*, with possible consequences that can extend well beyond the particular spatial or temporal coordinates of the incident itself.

HAZARD AND RISK ASSESSMENT

The term hazard (see Figure 2.2) is sometimes used to define a source of potential harm or injury and, sometimes, the potential harm or injury itself. Thus, a silo containing plastic chips or grain or any other raw materials may be said to be a hazard because, having entered the silo, a worker might become engulfed and subsequently asphyxiated; the hazard may also be defined as the actual engulfment or asphyxiation. This double meaning of the word hazard (i.e., the silo itself or the dangers that exist within a silo) often results in a confusion of cause and effect. However used, the world hazard always denotes a possibility or potential. This is what differentiates a hazard from a risk. Whereas a hazard is a *possible* (or *potential*) harm or injury (or an immediate precursor to harm or injury), risk is the *probability* that a person or group will actually experience a

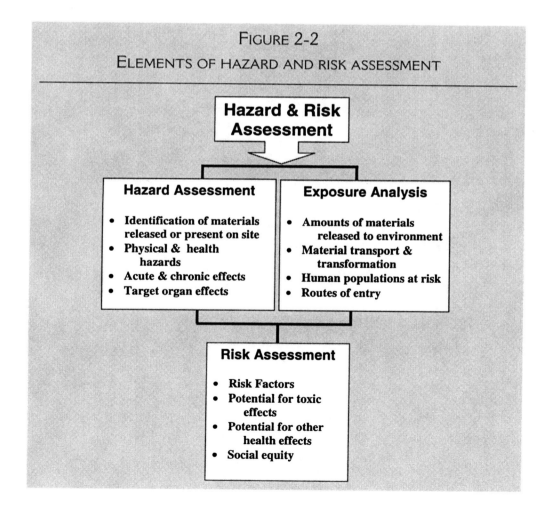

FIGURE 2-2

ELEMENTS OF HAZARD AND RISK ASSESSMENT

Hazard & Risk Assessment

Hazard Assessment
- Identification of materials released or present on site
- Physical & health hazards
- Acute & chronic effects
- Target organ effects

Exposure Analysis
- Amounts of materials released to environment
- Material transport & transformation
- Human populations at risk
- Routes of entry

Risk Assessment
- Risk Factors
- Potential for toxic effects
- Potential for other health effects
- Social equity

specific hazard. As the probability that individual persons or groups will experience the harm or injury of a specific hazard, risk depends upon *exposure* of those persons or groups to the actual hazard. Emergency response planning begins with a comprehensive inventory of hazards and proceeds in stepwise fashion to the analysis of exposure and, finally, the estimation of risk.

As shown in Figure 2.3, the assessment of hazards, exposure, and risk may be considered the first of three phases of decision making required for devising effective emergency response policies and procedures that form the basis of the emergency response or action plan. The approach depicted in Figure 2.3 is one

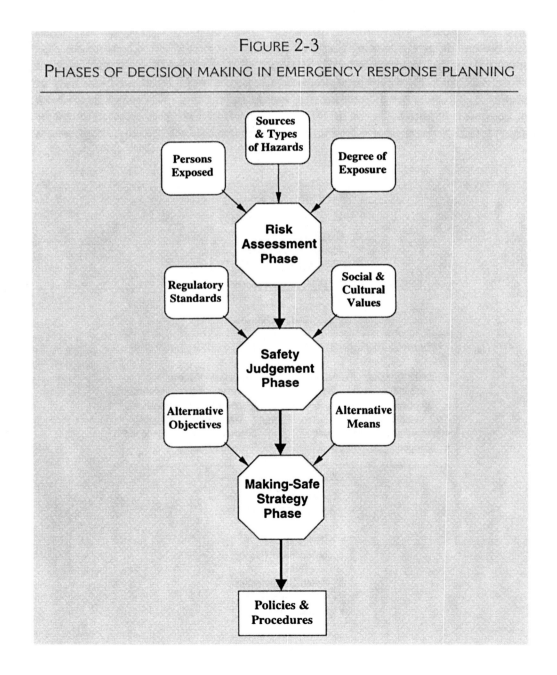

FIGURE 2-3

PHASES OF DECISION MAKING IN EMERGENCY RESPONSE PLANNING

historically developed by public health personnel concerned with the development of safe water supplies, but its depiction of the three phases of decision-making are at the heart of any emergency planning:

- *Risk Assessment Phase*: Identification of the potential sources or cause of emergencies and the types of degrees of risk to be experienced by the work force, the public at large, and emergency response personnel
- *Safety Judgment Phase*: Establishment of levels of protection to be provided to persons at risk during an emergency
- *Make-Safe Strategy Phase*: Formulation of specific procedures for achieving decided levels of protection

Risk Assessment Phase

The risk assessment phase is highly influenced by the concerns and considerations broadly attendant to the Bhopal (India) tragedy in which several thousand died outright and tens of thousands more were seriously injured due to a leak of toxic gas at a Union Carbide pesticide plant—an event that, in the United States, became a prime motivation for the development of the Chemical Process Safety Regulations (29 CFR 1910.119). Even where these regulations do not specifically apply, they provide an excellent overview of the broad scope of modern emergency planning and are thereby highly instructive for any emergency response manager.

Various analytical techniques are germane to this phase, each of them providing different means for identifying potential sources of workplace emergencies and persons potentially at risk. Standard techniques (see Figure 2.4) include (a) preliminary hazard analysis, (b) what-if analysis, (c) hazard and operability analysis, (d) failure modes and effects analysis, (e) fault and event tree analysis, and (f) human reliability analysis.

Preliminary hazard analysis focuses on the hazardous materials and major processing areas of a facility in order to identify hazards and potential accident situations. It requires consideration of facility equipment, the interface among facility components, the operational environment, specific facility operations, and the physical layout of the facility. The objective of this technique is to assign a criticality ranking to each hazardous situation that may be envisioned, even in the absence of specific information about facility design features or operational procedures. It is particularly useful for identifying broadly defined causal chains (e.g., fire in materials processing can lead to explosion and release of toxic vapors; release of toxic vapors to the ambient atmosphere may threaten homes abutting facility property) that can then be subjected to more detailed analysis.

What-if analysis requires experienced personnel to formulate a series of questions that must be evaluated with respect to potential hazards identified in the preliminary hazard analysis. Typical questions might be of the type, "What if pump 23b shuts off?" and "What if the operator forgets to empty the overflow tank at the end of the week?" The basic strength of this approach is to define more precisely specific causal chains that can lead to an emergency.

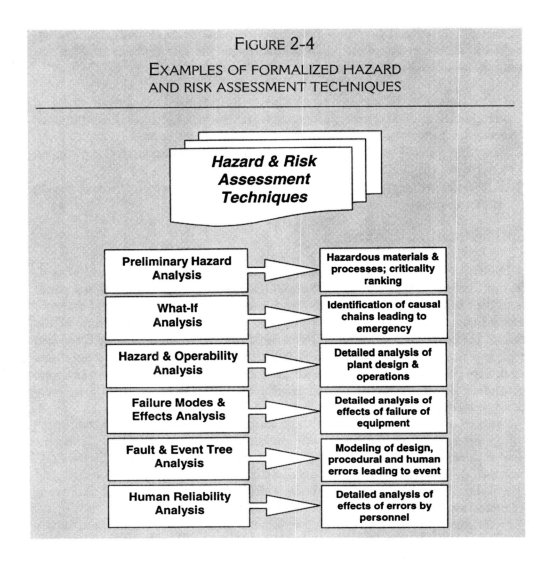

FIGURE 2-4

EXAMPLES OF FORMALIZED HAZARD
AND RISK ASSESSMENT TECHNIQUES

Hazard and operability analysis depends upon detailed information on the design and operation of the facility. In using this technique, the assessment team uses a standard set of guide words that, when combined with specific process parameters, lead to resultant deviations that may result in an emergency health and safety situation. For example, the guide word "less" might be combined with the process parameter "pressure" to produce the resultant deviation "low pressure." The assessment team may then focus on the possible causes of low pressure (e.g., in a reactor) and the possible consequence of that low pressure (e.g., change in the rate of chemical reaction in the reactor).

Failure modes and effects analysis is closely related to what-if analysis. This technique focuses on the various failure modes of specific equipment and the effects of such failures on plant operations and human health and safety. Examples of questions that reflect this type of analysis when applied, say, to a control valve in a reactor vessel might include: What are the possible consequences of the control valve failing in the open position? In the closed position?

What are the possible consequences if the control valve leaks while in the open or closed position?

Fault tree analysis and *event tree analysis* involve graphically modeling accidents and failures in equipment and personnel. In fault tree analysis, a specific accident or plant failure (e.g., release of a toxic gas) is defined and all design, procedural, and human errors leading to that event (called the top event) are graphically modeled in a fault tree. The fault tree allows the analysis to define and rank particular groupings of external factors, equipment failures, and human errors (called minimal cut sets) that are sufficient to lead to the top event. Whereas fault tree analysis focuses on failures in equipment or personnel that lead to the top event, event tree analysis focuses on how successes or failures of specific in-place safety equipment, devices, and procedures may contribute to a developing emergency. This type of analysis typically is used to analyze very complex processes that incorporate several layers of safety systems or emergency procedures.

Human reliability analysis generally is conducted in parallel with other techniques that tend to be equipment-oriented. This type of assessment focuses on factors that influence the actual job performance of personnel. In such an assessment, detailed descriptions of task requirements, the skills, knowledge, and capabilities necessary for meeting each requirement, and error-prone situations that may develop during task performance are combined to isolate specific factors that, if ignored, might result in an emergency. It is important that considered factors not be limited to those that are directly related to workplace conditions (e.g., ambient noise levels, which might affect a worker's concentration; work schedules, which can result in inattention due to fatigue), but are inclusive of the universe of factors that may influence workplace performance (e.g., personal financial difficulties; marital problems; substance abuse).

Regardless of the individual technique (or combination thereof) employed, the risk assessment process must consider potential sources of emergency that derive from other than plant operations, including storms and floods, area-wide fires and chemical releases, and terrorist acts. With regard to the latter, it is advisable that particular attention be given to the fact that a perceived emergency may well be "blind" to another.

For example, a telephoned bomb threat is likely to result in a facility evacuation within a matter of minutes, followed closely by the arrival of fire, police, and/or specialized investigatory and emergency response units. However, it may be the evacuating personnel or the emergency response personnel, not the physical facility, who are the real targets of the threat. Given this possibility, the prudent planner would ensure the implementation of appropriate procedures for detecting explosive or toxic charges that may be planted in evacuation assembly areas or precisely where emergency vehicles are likely to enter the premise.

Safety Judgment Phase

Having identified potential sources of emergencies as well as contributing factors and populations at risk, emergency planners must establish criteria regarding appropriate levels of protection for each at-risk population. This is a very difficult task precisely because it requires that judgments be made directly affecting the safety of human beings. The simple fact is that there is no such thing as 100% guaranteed protection for all. The mere act of evacuating a group of people from a building puts some of those people at real risk of suffering a heart attack or a fall-related injury. Panic can kill as effectively as fire. Of course, individual physical and psychological conditions that ensure some differential distribution of risks regardless of any effort to the contrary are not excuses for inaction. In fact, it is precisely the recognition of a differential distribution of risks that becomes the basis of an effective emergency response plan.

In the United States, regulatory guidance (OSHA and EPA) regarding the level of protection for personnel having specific responsibility in an emergency involving hazardous chemicals is based on the following typology of emergency responders, which includes member of so-called HAZMAT (for "hazardous materials") teams. The designation HAZMAT always denotes personnel who are expected to perform work in close proximity to a hazardous substance while handling or controlling actual or potential leaks or spills, and should not be confused with other emergency personnel, such as members of a fire brigade.

- Level 1: Responders who are not likely to witness or discover a hazardous substance release or to initiate an emergency response sequence by notifying the proper authorities
- Level 2: Police, firefighters, and rescue personnel who are part of the initial response to a real or potential release of hazardous substances
- Level 3: HAZMAT technicians, who are the first level specifically charged with trying to contain a release of hazardous substances
- Level 4: HAZMAT specialists, who respond with and provide support to HAZMAT technicians and have specific knowledge of hazardous substances
- Level 5: On-scene incident commanders or senior officials in charge, who assume control of the emergency response incident scene and coordinate all activities and communications

The various levels of responders indicated can be cross-referenced with various levels of protective ensembles (see Table 2.1) to meet regulatory requirements regarding personal protective clothing and equipment, such as the requirements of 29 CFR 1910.120. Although the preceding typology of responders gives heavy emphasis to protection from hazardous chemicals, other types of emergency situations and job tasks require other regulatory inputs to the planning process, such as 29 CFR 1910.156 standards that specifically apply to members of a fire brigade.

In the process of coordinating with community-based emergency responders, including local fire departments, particular attention should be given to the adequacy of protective clothing and equipment available to external responders with respect to the specific hazards associated with facility operations. This is often a critical concern because local firefighters or other local responders, who are typically the first on-scene responders, often do not have direct access to the kind of personal protection devices that are standard equipment

TABLE 2-1

PROTECTIVE CLOTHING THAT MAY BE INCLUDED IN ENSEMBLES FOR HAZARDOUS WASTE OPERATIONS (ADAPTED FROM NIOSH, USCG, AND EPA, 1985: OCCUPATIONAL SAFETY AND HEALTH GUIDANCE MANUAL FOR HAZARDOUS WASTE SITE ACTIVITIES)

Level of Protection and Equipment	Overview of Protection	Conditions for Use and Limitations
A	The highest available level of respiratory, skin, and eye protection	■ The chemical substance has been identified and requires the highest level of protection for skin, eyes, and the respiratory system based on either:
Recommended: ■ Pressure-demand, full-facepiece SCBA or pressure-demand supplied air respirator with escape SCBA ■ Fully encapsulating, chemical resistant suit		1. measured (or potential for) high concentration of atmospheric vapors, gases, or particulates, or 2. site operations and work functions involving a high potential for splash, immersion, or exposure to unexpected vapors, gases, or particulates of materials that are harmful to skin or capable of being absorbed through the intact skin.
■ Inner chemical-resistant gloves ■ Chemical resistant safety boots/shoes ■ Two-way radio		■ Substances with a high degree of hazard to the skin are known or suspected to be present, and skin contact is possible.
Optional: ■ Cooling unit		■ Operations must be conducted in confined, poorly ventilated areas until the absence of conditions requiring Level A protection is determined.
■ Coveralls ■ Long cotton underwear ■ Hard hat ■ Disposable gloves and boot covers		■ Fully encapsulating suit materials must be compatible with the substances involved.

Continued

TABLE 2-1—*Continued*

PROTECTIVE CLOTHING THAT MAY BE INCLUDED IN ENSEMBLES FOR HAZARDOUS WASTE OPERATIONS (ADAPTED FROM NIOSH, USCG, AND EPA, 1985: OCCUPATIONAL SAFETY AND HEALTH GUIDANCE MANUAL FOR HAZARDOUS WASTE SITE ACTIVITIES)

Level of Protection and Equipment	Overview of Protection	Conditions for Use and Limitations
B	The same level of respiratory protection but less skin protection than Level A.	■ The type and atmospheric concentration of substances have been identified and require a high level of respiratory protection, but less skin protection. This involves atmospheres
Recommended: ■ Pressure-demand, full facepiece SCBA or pressure-demand supplied air respirator with escape SCBA ■ Chemical-resistant clothing	This the minimum level recommended for initial site entries until the hazards have been further identified.	– with IDLH concentrations of specific substances that do not represent a severe skin hazard, or – that do not meet the criteria for use of air-purifying respirators.
■ Inner and outer chemical-resistant gloves ■ Chemical resistant safety boots/shoes ■ Hard Hat ■ Two-way Radio		■ Atmosphere contains less than 19.5 % oxygen. ■ Presence of incompletely identified vapors or gases is indicated by direct-reading organic vapor detection instrument, but vapors and gases are not suspected of containing high levels of chemicals harmful to skin or capable of being absorbed through intact skin.
Optional ■ Coveralls ■ Disposable boot covers ■ Face shield ■ Long cotton underwear		■ Use only when highly unlikely that the work will generate either high concentrations of vapors, gases, or particulates or splashes of material will affect exposed skin.
C	The same level of skin protection as Level B, but a lower level of respiratory protection	■ Atmospheric contaminants, liquid splashes, or other direct contact will not adversely affect any exposed skin.

Continued

TABLE 2-1—*Continued*

PROTECTIVE CLOTHING THAT MAY BE INCLUDED IN ENSEMBLES FOR HAZARDOUS WASTE OPERATIONS (ADAPTED FROM NIOSH, USCG, AND EPA, 1985: OCCUPATIONAL SAFETY AND HEALTH GUIDANCE MANUAL FOR HAZARDOUS WASTE SITE ACTIVITIES)

Level of Protection and Equipment	Overview of Protection	Conditions for Use and Limitations
Recommended: ■ Full-facepiece, air purifying, canister equipped respirator ■ Chemical resistant clothing ■ Inner and outer chemical resistant gloves ■ Chemical resistant safety boots/shoes ■ Hard hat ■ Two-way radio **Optional:** ■ Coveralls ■ Disposable boot covers ■ Face shield ■ Escape mask ■ Long cotton underwear		■ The types of air contaminants have been identified, concentrations measured, and a canister is available that can remove the contaminant. ■ All criteria for the use of air-purifying respirators are met. ■ Atmospheric concentration of chemicals must not exceed IDLH levels. ■ The atmosphere must contain at least 19.5% oxygen.
D **Recommended:** ■ Coveralls ■ Safety boots/shoes ■ Safety glasses or chemical splash goggles ■ Hard hat **Optional:** ■ Gloves ■ Escape mask ■ Face shield	No respiratory protection; minimal skin protection	■ The atmosphere contains no known hazard. ■ Work functions preclude splashes, immersion, or the potential for unexpected inhalation of or contact with hazardous levels of any chemicals. ■ This level should not be worn in the Exclusion Zone. ■ The atmosphere must contain at least 19.5% oxygen.

for HAZMAT team members who typically arrive on-site only well after an emergency has progressed.

In many cases, for example, local firefighters will not be equipped with chemically impervious protective clothing that would be required to retrieve personnel trapped within a facility where highly toxic chemicals are used or stored. In some situations, facility managers have purchased such clothing and maintain it for use by local firefighters. Sometimes a company may also supply the local fire department with additional materials and specialized equipment, including antidotes to toxic chemicals used on-site, specialized monitoring devices, and materials that firefighters can use to disinfect clothing and equipment contaminated by especially dangerous chemicals.

Making-Safe Strategy Phase

In this phase, the objective is to assess and select from alternative means for achieving the standards and objectives previously identified and, finally, to develop specific policies and procedures that govern all aspects of emergency response. As shown in Figure 2.5, policies and procedures should address three basic types of emergency response activities: (a) preparation activities, which are undertaken immediately upon discovery of a potential or actual emergency and prior to the initiation of any response; (b) response activities, which include all efforts to control the emergency and provide assistance to affected personnel; and (c) follow-up activities, which focus on post-emergency actions to bring the facility or emergency site back to a state of emergency readiness, including revisions to emergency plans necessitated by the experience of the now-past emergency. It must be emphasized here that the too-frequent tendency is for facilities to concentrate on the response to an emergency at the expense of attention given to both preparatory and follow-up actions, which is an extremely dangerous approach to emergency planning. Effective emergency planning always requires equally serious attention to all three types of actions. A checklist of basic issues that must be addressed in any comprehensive emergency response plan is included in Table 2.2.

ON- AND OFF-SITE MANAGEMENT

Proactive and reactive emergency response activities are essentially exercises in *risk management*. The basic generic tasks of risk management, shown in Figure 2.6, include:

1. Identification of hazard exposure
2. Evaluation of risk potential
3. Ranking and prioritization of risks
4. Determination and implementation of control actions
5. Evaluation and revision of actions and techniques

Each of these tasks is equally applicable to the specific site of an emergency incident (see Image 2.2) and to any off-site areas that may be impacted (directly or indirectly) by the known emergency.

Just as a comprehensive holistic approach to risk management requires consideration of both on- and off-site sources of hazard to the community and environment (see Figure 2.7), so it requires consideration of the potential impli-

FIGURE 2-5

BASIC TYPES OF EMERGENCY RESPONSE OPERATIONS

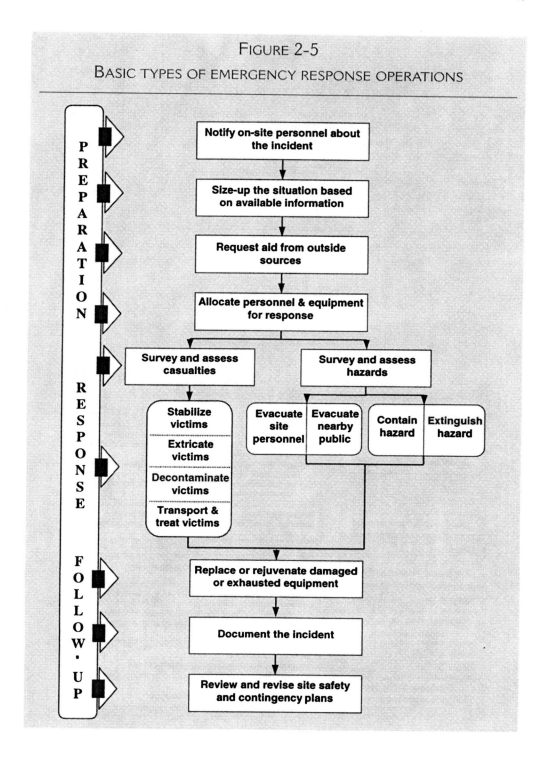

cations of a particular incident on distantly located community resources and dynamics. This approach is particularly important, of course, in cases of terrorism where one or more incidents may be planned primarily to draw attention away from a primary target or, otherwise, to lure a target (e.g., firefighters, police) into attack position. However, even where terrorism is not involved,

TABLE 2-2

CHECKLIST OF BASIC ISSUES TO BE ADDRESSED IN DETAIL
IN ANY EMERGENCY RESPONSE PLAN

1. Description of type of emergency and minimum information required
2. On- and off-site notification requirements
3. On- and off-site responsible personnel
4. Criteria for evaluating levels of emergency
5. Evacuation requirements
6. PPC/PPE by Task
7. Personnel Monitoring Requirements
8. Communication and Information Processing
9. On- and off-site control procedures
10. Emergency medical care and surveillance
11. Post-emergency actions and documentation

FIGURE 2-6

FIVE KEY STEPS IN DEVELOPING A RISK MANAGEMENT PROGRAM
(ADAPTED FROM U.S. FIRE 1966: ADMINISTRATION, 1966: RISK
MANAGEMENT PRACTICES IN THE FIRE SERVICE [FA-166])

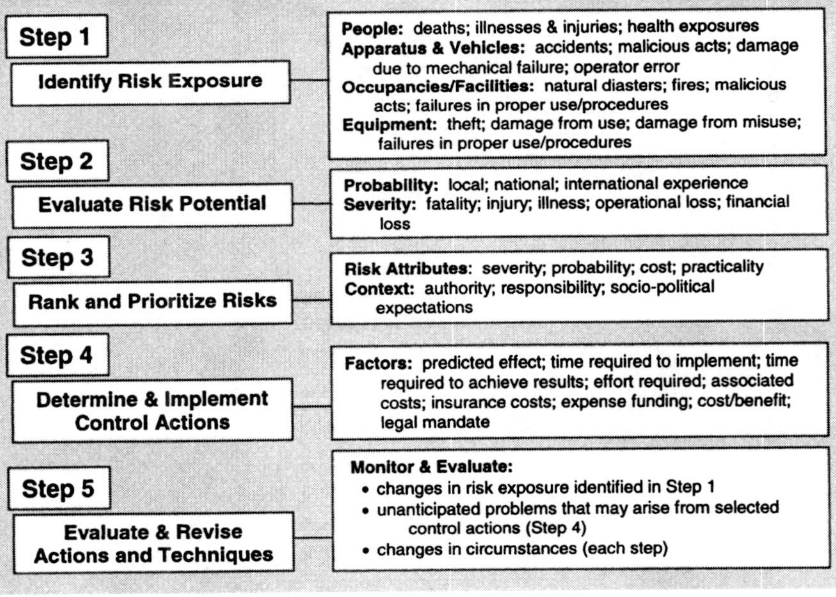

Step 1 Identify Risk Exposure	**People:** deaths; illnesses & injuries; health exposures **Apparatus & Vehicles:** accidents; malicious acts; damage due to mechanical failure; operator error **Occupancies/Facilities:** natural diasters; fires; malicious acts; failures in proper use/procedures **Equipment:** theft; damage from use; damage from misuse; failures in proper use/procedures
Step 2 Evaluate Risk Potential	**Probability:** local; national; international experience **Severity:** fatality; injury; illness; operational loss; financial loss
Step 3 Rank and Prioritize Risks	**Risk Attributes:** severity; probability; cost; practicality **Context:** authority; responsibility; socio-political expectations
Step 4 Determine & Implement Control Actions	**Factors:** predicted effect; time required to implement; time required to achieve results; effort required; associated costs; insurance costs; expense funding; cost/benefit; legal mandate
Step 5 Evaluate & Revise Actions and Techniques	**Monitor & Evaluate:** • changes in risk exposure identified in Step 1 • unanticipated problems that may arise from selected control actions (Step 4) • changes in circumstances (each step)

IMAGE 2-2

TWISTED STEEL AND OTHER DEBRIS AT THE WORLD TRADE
CENTER SITE POSE MANY POTENTIAL SAFETY HAZARDS.
OSHA EMPLOYEES PROVIDED AROUND-THE-CLOCK MONITORING
OF THE SITE TO IDENTIFY AND ALERT WORKERS TO SAFETY
AND HEALTH HAZARDS

Source: Shawn Moore/OSHA News Photo

FIGURE 2-7

DIRECT AND INDIRECT RISKS TO THE COMMUNITY DUE TO
INDUSTRIAL ON- AND OFF-SITE OPERATIONS. INDIRECT RISKS ARE
DUE TO IMPACTS OF INCIDENT ON ENVIRONMENTAL RESOURCES
(E.G., AIR, WATER) THAT CAN LEAD TO SUBSEQUENT EXPOSURES OF
THE PUBLIC TO HAZARDOUS CONTAMINANTS

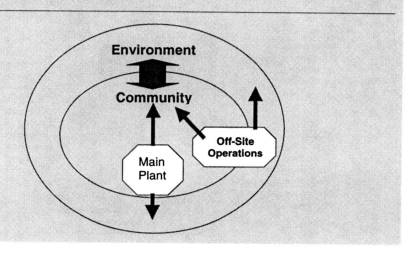

any incident in one location can increase risks at other locations—as, for example, when an extensive fire in one part of the community leaves the rest of the community with reduced fire response capacity.

Whereas the term *site control* typically refers to response activities undertaken on the site of (or in the immediate vicinity of) a specific incident, off-site management effort must be simultaneously directed not only to ensure the proper back-up of that on-going emergency response effort, but also to manage additional potential risks for the remainder of the community. Of the seven common sources of emergency response failure shown in Figure 2.8, most (if not all) become evident in a particular site-specific response. As serious as these failures may prove to be in terms of any particular incident, their importance is greatly magnified in terms of the potential needs of effective off-site management.

For example, while the lack of an established chain of command may impede if not prevent effective and efficient response to a particular, site-specific incident, it almost guarantees the vulnerability of the rest of the community to even greater devastation through a well-planned and coordinated sequence of terrorist acts.

Another example is when, due to the extent of a particular emergency that essentially depletes the response resources of an urban area (e.g., New York,

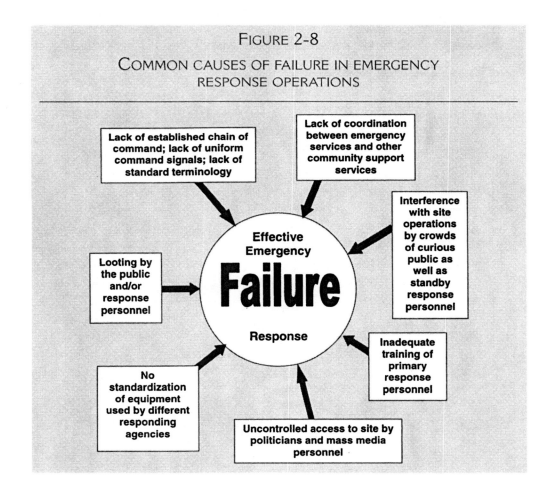

FIGURE 2-8

COMMON CAUSES OF FAILURE IN EMERGENCY RESPONSE OPERATIONS

2001; London, 2005), additional reserve resources must be drawn temporarily from surrounding communities to deal with any additional potential emergency. Depending upon the training, experience, and technology of these reserve resources, they may not be able to deal effectively with additional emergencies that may occur during the interval of the ongoing emergency.

AUTHORITY AND RESPONSIBILITY

At the heart of any bureaucracy, whether explicitly stated or, as most often the case, only to be inferred from organizational structure, is the distinction between authority and responsibility—the former being, in essence, the right to exact the obedience of others while exercising the prerogatives of independent determination and judgment, whereas the latter is the duty or obligation to be fulfilled through the exercise of that authority. One implies the other and, in consequence, the concepts of authority and responsibility become intimately interconnected in both the encultureted expectations of everyday life and the more formal principles and doctrines that guide institutional behavior.

Of course, the marvel of all cultural traditions is that often they are easily "short-circuited"—modified to meet the demands of new experience or, as may often be the case, simply ignored. With regard to corporate attitudes toward health and safety risks, it would appear that the traditional sense of the need for a commensurate balance between authority and responsibility has much more frequently been purposely ignored than usefully modified.

Despite a growing number of exceptions, the corporate employee who is assigned programmatic environmental or health and safety responsibility (and, therefore, responsibility for in-house emergency response) is typically a low-level manager, supervisor, or technician who has little if any discernible authority over—or measurable influence on—key corporate decision-making or over any substantive planning or production-related process. In such a situation, it is not surprising that the so called "safety officer" usually becomes preoccupied with actual health and safety incidents and regulatory compliance failures rather than effectively managing a comprehensive health, safety, and emergency response program—or that the workplace continues to be the focus of governmental and social concern about community health and safety as well as environmental quality.

The only practical way by which to ensure that the authority of corporate safety officials is in fact commensurate with their responsibility is *to extend that authority to whatever extent required* for the effective managerial control of the sources of health and safety hazards and of all circumstance, including emergency planning and response, that may contribute or be affected by potential human exposure to those hazards.

State-of-the-art companies today understand that this approach requires that significant health and safety responsibility, especially responsibility for emergency planning and response, be matched with high-level executive authority.

At the operational level of governmental agencies and community services, the concentrated effort to manage the diverse difficulties inherent in any bureaucratic structure of authoritative responsibility has been toward the implementation of the *Incident Command System* (ICS).

ICS is essentially a management system that can be used in any incident regardless of kind or size, including:

- Fires, HAZMAT incidents, and multicausal incidents
- Single and multiagency law enforcement efforts
- Multijurisdictional and multiagency disaster responses
- Search and rescue missions
- Oil spill response and recovery incidents
- Air, rail, water, and ground transportation accidents
- Planned events (e.g., celebrations, parades, concerts)
- Private sector emergency response

Capable of expansion or contraction, the ICS management system consists of five basic functions (see Figure 2.9) that are equally pertinent to emergency planning and response undertaken either by governmental agencies or by private sector facilities.

Perhaps the most important of the key features of ICS is that the individual designated as the Incident Commander (IC) has absolute responsibility for all functions, even if the IC chooses to delegate authority to perform selected functions to other persons. It is of vital importance, therefore, that the adoption of

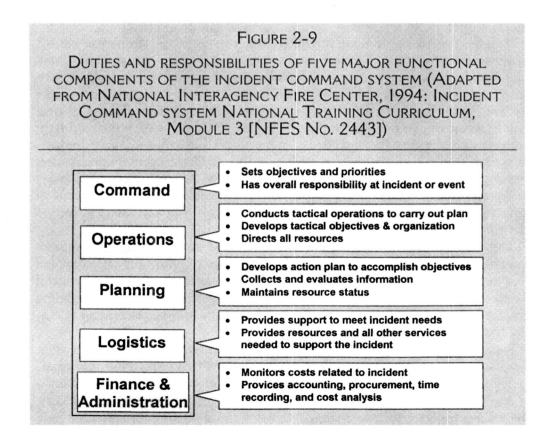

FIGURE 2-9

DUTIES AND RESPONSIBILITIES OF FIVE MAJOR FUNCTIONAL COMPONENTS OF THE INCIDENT COMMAND SYSTEM (ADAPTED FROM NATIONAL INTERAGENCY FIRE CENTER, 1994: INCIDENT COMMAND SYSTEM NATIONAL TRAINING CURRICULUM, MODULE 3 [NFES NO. 2443])

Command
- Sets objectives and priorities
- Has overall responsibility at incident or event

Operations
- Conducts tactical operations to carry out plan
- Develops tactical objectives & organization
- Directs all resources

Planning
- Develops action plan to accomplish objectives
- Collects and evaluates information
- Maintains resource status

Logistics
- Provides support to meet incident needs
- Provides resources and all other services needed to support the incident

Finance & Administration
- Monitors costs related to incident
- Provices accounting, procurement, time recording, and cost analysis

ICS for emergency planning and response be undertaken only when the IC is provided absolutely clear and comprehensive authority and support necessary to achieve the objective of emergency planning and response.

Much of the flexibility of ICS derives from the various options available to the IC (and, subsequently, to appropriate incident response managers) regarding the activation of subsidiary components of the ICS management organization. These options are based on the assessment of the ongoing developing

FIGURE 2-10

COMPARISON OF ALTERNATIVE EMERGENCY MANAGEMENT SYSTEMS UNDER ICS (ADAPTED FROM NATIONAL INTERAGENCY FIRE CENTER, 1994: INCIDENT COMMAND SYSTEM NATIONAL TRAINING CURRICULUM, MODULE 1 [NFES NO. 2468]; MODULE 16 [NFES NO. 2470])

Incident Command System (ICS)	The management system used **to direct all operations at the incident scene.** The Incident Commander (IC) is located at an Incident Command Post (ICP) at the incident scene.
Unified Command (UC)	An application of ICS used when there is more than one agency with incident jurisdiction. **Agencies work together through their designated Incident Commanders at a single ICP to establish a common set of objectives and strategies and a single Incident Action Plan.**
Area Command/ Unified Area Command AC/UAC	Established as necessary **to provide command authority and coordination for two or more incidents in close proximity.** Area Command works directly with Incident Commanders. Area Command becomes Unified Area Command when incidents are multi-jurisdictional. **Area Command may be established at an EOC facility or at some location other than an ICP.**
Multi-Agency Coordination Systems (MACS)	An activity or formal system used **to coordinate resources and support between agencies or jurisdictions.** A MAC Group functions with the MACS. MACS interact with agencies or jurisdictions, not with incidents. MACS are useful for regional situation. **A MACS can be established at a jurisdiction EOC or at a separate facility.**
Emergency Operations Center (EOC)	Also called Expanded Dispatch or Emergency Command and Control Centers. EOCs are used in varying ways at all levels of government and within private industry **to provide coordination, direction, and control during emergencies.** EOC facilities can be used to house Area Command and MACS activities as determined by agency or jurisdictional policy.

nature and extent of the incident. Another aspect of the flexibility of ICS derives from the different modes of coordinating and directing incident response on the basis of (a) multijurisdictional responsibilities, (b) the occurrence of two or more incidents in close proximity, and (c) multiagency and multijurisdictional responsibilities within an extended geographic region.

Because of the importance that must be given in the United States to jurisdictional responsibilities of different agencies at federal, regional, state, and local levels, the ICS must be flexible enough to accommodate these differences without, at the same time, sacrificing efficiency and effectiveness. This is accomplished by extending and adapting ICS to meet the needs of (a) Unified Command, (b) Area Command, (c) Multiagency Coordination Systems, and (d) Emergency Operations Centers (see Figure 2.10). The full range of these adaptations of ICS not only ensures proper involvement of diverse responsible authorities in incident response, but also ensures that the response to a particular incident will not unduly detract from local and regional resources that may be needed in response to other incidents.

It must be understood that there is no (nor should there necessarily be any one) compelling consensus regarding all possible relationships (or even terminology) regarding multiagency coordination and management of incident response. The appropriate relationships among response components (as well as precise definitions) ultimately depend upon the specific procedures in place in particular agencies and organizations at the time and place of the incident. The ICS therefore provides a flexible framework of management rather than a definitive algorithm.

For example, the U.S. National Response Team has promulgated guidelines for adapting the ICS to meet the needs of a *Unified Command* (see Figure 2.11). As noted in these guidelines, no attempt is made "to prescribe specifically how a particular organization or individual fits within a given response structure." Despite the lack of prescriptions regarding specific assignments, however, these guidelines do give explicit directions for ensuring that the accommodation of diverse jurisdictional interests of various public agencies and private organizations do not detract from those clear lines of authority, responsibility, and accountability (see Figure 2.12) that are firmly established by the ICS.

It is very likely that, upon a first introduction, the apparent intricacies of ICS management seem overly complex. That is one way to look at it. Another way is to step back and consider, for a moment, just what is involved in mounting any effort—any effort composed of variously trained and available personnel, a huge potential arsenal of different equipment and services that may or may not be available, and a collage of diverse public and primate authorities— toward the time-constrained objective of minimizing the loss of life and property suddenly at risk in the midst of total confusion and manifest anguish. The fact of the matter is that it is *crisis* that is complex!

The ICS is a practical and comprehensive framework that provides for flexible, efficient, authoritative, and accountable response to the extraordinary challenge of life-threatening crisis. But even in the best of circumstances, it can accomplish absolutely nothing in the absence of serious, preincident planning and training.

Whether implemented by public or private organizations, ICS is based upon the premise that what is done during an emergency is no more important than

FIGURE 2-11

GENERAL GUIDELINES PERTAINING TO UNIFIED COMMAND AND THE ICS SYSTEM (ADAPTED FROM U.S. NATIONAL RESPONSE TEAM [NRT]. MANAGING RESPONSE TO OIL DISCHARGES AND HAZARDOUS SUBSTANCE RELEASES UNDER THE NCP: TECHNICAL ASSISTANCE DOCUMENT, NRT ELECTRONIC REFERENCE LIBRARY)

Guidelines: Unified Command and the Incident Command System

1. For the ICS/UC(Incident Command System/Unified Command) to be effective, the following elements should be in place well before an incident occurs:

   - The structure must be formalized in the planning stages and must be accepted by all parties concerned,
   - Specific functions and responsibilities must be well defined,
   - Individuals must be designated for each function and the reporting mechanisms defined and accepted,
   - The participating organizations must make a committed effort to respond as a team,
   - Area Contingency Plans (including facility/vessel response plans) must address training and ensure familiarity with ICS utilizing a Unified Command, and
   - Relationships to entities outside the ICS but relevant to the response structure (e.g., Regional Response Team, Natural Resource Trustees) must be defined.

2. The NCP does not attempt to prescribe specifically how a particular organization or individual fits within a given response structure. The FOSC (Federal On-Scene Coordinator) and the Area Committee are responsible for developing, adopting, and implementing a response management system, through the ACP (Area Contingency Plan). A NIIMS-based (National Interagency Incident Management System), ICS/UC can be used as the model for response management in the ACP to ensure an effective response. Because key players differ from area to area, however, Area Committees must have flexibility to adapt the ICS/UC in order to be effective in each specific area.

3. In addition, when developing an ICS/UC, it is important to recognize that the key players in the response management system maintain a separate internal management infrastructure during a response; they do not relinquish authority, responsibility, or accountability.

4. The following items should be considered when developing the Area Contingency Plans:

   - Jurisdictional responsibilities,
   - Roles of all levels of government in the Unified command,
   - Relationship between the OSC (On-Scene Commander) and other officials who also have decision-making authority but are not part of the UC (Unified Command),
   - Financial agreements,
   - Information dissemination,
   - Communications,
   - Training and exercises,
   - Logistics, and
   - Lessons learned.

5. When plans and procedures are understood, agencies can support each other effectively. However, each response results in new lessons learned, which necessitates a continuing need to refine the procedures and processes, develop better methods, and mesh agency needs and actions.

6. Planners and responders at all levels need to understand the authorities and resources each response organization brings to a specific incident. ICS/UC is an important concept to practice as part of response exercises and to include in local and area contingency plans. Such exercising and planning will facilitate coordination and cooperation between federal, state, local and private party responders when the ICS/UC is implemented for a specific incident.

what is done before and after the emergency—before the crisis, because any adequate response must be conditioned by previous planning (see Figure 2.13), and after the crisis, because we must both learn from our mistakes and, even in the process of learning, prepare for the next crisis.

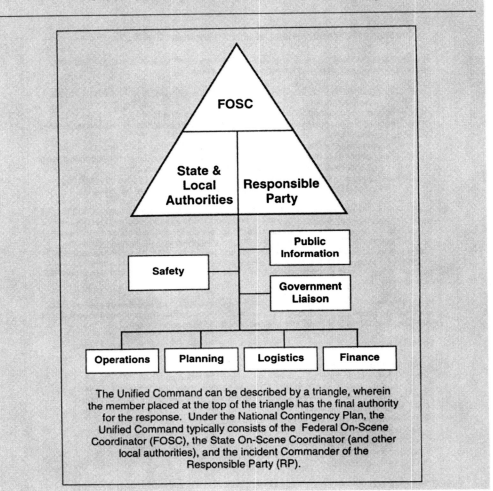

FIGURE 2-12

RELATIONSHIP BETWEEN UNIFIED COMMAND AND ICS (ADAPTED
FROM U.S. NATIONAL RESPONSE TEAM [NRT]. MANAGING
RESPONSE TO OIL DISCHARGES AND HAZARDOUS SUBSTANCE
RELEASES NDER THE NCP: TECHNICAL ASSISTANCE DOCUMENT,
NRT ELECTRONIC REFERENCE LIBRARY)

The Unified Command can be described by a triangle, wherein
the member placed at the top of the triangle has the final authority
for the response. Under the National Contingency Plan, the
Unified Command typically consists of the Federal On-Scene
Coordinator (FOSC), the State On-Scene Coordinator (and other
local authorities), and the incident Commander of the
Responsible Party (RP).

COMMUNICATION AND INFORMATION PROCESSING

In both planning and response phases of emergency management, there must
finally be reliance upon human judgment. Whereas the soundness of human
judgment can be assessed by various criteria that typically pertain to the specific

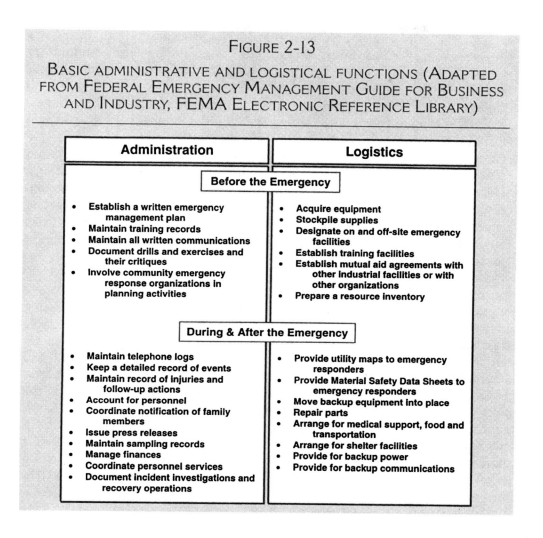

FIGURE 2-13

BASIC ADMINISTRATIVE AND LOGISTICAL FUNCTIONS (ADAPTED FROM FEDERAL EMERGENCY MANAGEMENT GUIDE FOR BUSINESS AND INDUSTRY, FEMA ELECTRONIC REFERENCE LIBRARY)

Administration	Logistics
Before the Emergency	
• Establish a written emergency management plan • Maintain training records • Maintain all written communications • Document drills and exercises and their critiques • Involve community emergency response organizations in planning activities	• Acquire equipment • Stockpile supplies • Designate on and off-site emergency facilities • Establish training facilities • Establish mutual aid agreements with other industrial facilities or with other organizations • Prepare a resource inventory
During & After the Emergency	
• Maintain telephone logs • Keep a detailed record of events • Maintain record of injuries and follow-up actions • Account for personnel • Coordinate notification of family members • Issue press releases • Maintain sampling records • Manage finances • Coordinate personnel services • Document incident investigations and recovery operations	• Provide utility maps to emergency responders • Provide Material Safety Data Sheets to emergency responders • Move backup equipment into place • Repair parts • Arrange for medical support, food and transportation • Arrange for shelter facilities • Provide for backup power • Provide for backup communications

person exercising judgment (e.g., range of practical experience, demonstrated theoretical expertise, flexibility of approach in different contexts), the essence of sound judgment becomes most often clearly evident in the manner in which information is specifically and efficiently marshaled toward the actual achievement of objectives. In this sense, there can be no effective emergency planning and response without fastidious attention to communication and to information processing, including the mechanical and electronic wherewithal as well as those components related to substantive content and format, software, and computerized databases.

Examples of key communication and informational needs for emergency planning and response include:

■ Alarm and alert systems and devices (i.e., facility evacuation alarms, process control alarms, safety alert devices for SCBA ensembles, automatic notification of community services, in-plant monitoring system)

- Line, radio, and oral communication devices/procedures (e.g., inter- and intra-response teams and units; incident command and community resources; public address and mass-media; communication reliability, security, redundancy, and backup devices)
- Chemical databases (e.g., on-site chemical inventories, chemical attributes, required personal protective clothing (PPC) and equipment (PPE), accepted disposal techniques)
- Computerized modeling (e.g., air dispersion models) and information retrieval systems (e.g., virtual reality system for directed entry into buildings)
- Hardcopy information files (e.g., Material Safety Data Sheets, plant and responder personnel rosters, inventory of at-risk persons and resources in general area, location of access roads and entrances, location of sensitive environmental resources, structural components and attributes of facilities, location of in-plant hazards)

In any particular incident, much of the information required for effective emergency response actually exists, but either is located or formatted in such a manner as to preclude its timely use. For example, Figure 2.14 is an example of a chemical inventory containing information about a company's stock chemicals as required under OSHA Hazard Communication (29 CFR 1910.1200) and Laboratory (29 CFR 1910.1450) standards. The format of information depicted in Figure 2.14 typically would not be useful to an Incident Commander in the midst of an actual emergency. However, such information is most often computerized, and therefore the database could easily be managed so as to produce, for example, a printed list of all flammable and corrosive liquids—a list that could be of immediate use to emergency response teams. The production and availability of such a printed list depends, of course, on appropriate liaison between corporate officials and emergency responders prior to an actual in-plant emergency.

PROVISIONS AND SUPPORT

In the vast majority of emergency incidents, it is highly unlikely that sufficient supplies of materials resources (provisions) or appropriate personnel (support) will be immediately available to meet the needs of emergency response—especially when that response requires highly specialized expertise, materials, or equipment, or when the incident involves potential risk to large numbers of people or extends over an extensive geographic area. Moreover, because even a seemingly small and well-contained incident may suddenly escalate to a major incident or develop in an unforeseen manner, it is necessary that emergency planning include detailed plans for obtaining, as necessary, provisions and support beyond that readily available and under direct control of the Incident Commander. In this regard, emergency planning (and training) must be guided by the assumption of a *worst-case scenario* for the incident.

Just what constitutes the worst-case scenario for a potential incident cannot, of course, be precisely defined except in the specific context of an actual inci-

FIGURE 2-14

SAMPLE PAGE FROM A CORPORATE CHEMICAL INVENTORY THAT INCLUDES CHEMICALLY SPECIFIC HEALTH AND SAFETY INFORMATION

Global Enterprises, Inc.
Chemical Inventory

Department	Date	Authorization
Quality Control Laboratory	July 19, 1996	Elizabeth Kohl

1, 3-Phenylguanidine
- *Route(s):* Inhalation; Absorption; Surface Contact
- *Hazard(s):* Irritant; Sensitizer; Toxic
- *Target Organ(s):* Skin; Eye; Mucous Membranes; Respiratory Tract

2-Butoxyethanol
- *Route(s):* Inhalation; Ingestion; Absorption; Surface Contact
- *Hazard(s):* Combustible; Irritant; Toxic; Teratogen
- *Target Organ(s):* Skin; Eye; Mucous Membranes; Kidney; Liver; Blood; Respiratory Tract; Reproductive System; Lymphatic System

2, 4, 6-Trichlorophenol
- *Route(s):* Inhalation; Ingestion; Absorption; Surface Contact
- *Hazard(s):* Irritant; Toxic; Carcinogen
- *Target Organ(s):* Skin; Eye; Mucous Membranes; Respiratory Tract

Acetophenetidin
- *Route(s):* Inhalation; Ingestion; Absorption; Surface Contact
- *Hazard(s):* Irritant; Toxic; Carcinogen; Teratogen; Mutagen
- *Target Organ(s):* Skin; Eye; Mucous Membranes; Lung; Kidney; Bladder; Respiratory Tract, GI Tract; Reproductive System; Nervous System

Ceric Ammonium Nitrate
- *Route(s):* Inhalation; Ingestion; Surface Contact
- *Hazard(s):* Oxidizer; Irritant; Toxic
- *Target Organ(s):* Skin; Eye; Mucous Membranes; Respiratory Tract

Page 3 of 65

---

dent. However, certain categorical circumstances must always be considered, including:

- In-plant work-shift schedules resulting in variable availability of personnel who can serve as initial responders (or, otherwise, as potential victims)
- Holidays, local events (e.g., parades, concerts), and highway traffic congestion
- Severe weather conditions
- Concurrent disasters in local area (or the region)

- Local or regional power failure
- Disruption of primary means of transportation
- Significant risk to large portions of the public
- Significant risk to environmental resources (e.g., public water supplies)
- Overwhelmed local medical treatment or temporary housing facilities
- Potential involvement of special facilities/populations (e.g., hospitals, schools, day-care centers, nursing homes)

Table 2.3 includes a variety of resources that may have to be obtained from external sources, including public and private sources. In addition to these sources, both municipalities and corporations should consider establishing formal *mutual assistance agreements* among local organizations to ensure the timely availability of necessary resources. This approach is particularly useful where there is a local concentration of similar or related industries, such as deep-water ports, technology parks, and industrial centers.

In addition to material resources, informational resources must also be identified and effectively integrated into the emergency planning process. Information resources include public and private services of local, regional,

TABLE 2-3

POTENTIAL RESOURCES FOR RESCUE TEAMS (ADAPTED FROM U.S. FIRE ADMINISTRATION, 1995: TECHNICAL RESCUE PROGRAM DEVELOPMENT MANUAL [FA-159])

Supplier	Resource
Construction/heavy equipment companies; state and local public works agencies	Backhoes; cranes; air compressors; dewatering pumps; dozers; loaders; welders; bobcats; generators; cherry pickers; tractor trailers; lighting; heavy tools; cutting and breaching equipment
Rental companies	Light tools; lighting; generators; air compressors
Lumber yards	Lumber; cutting equipment
Association of Engineers	Civil engineers; electrical engineers; fire protection engineers
Communications	Television stations; radio stations; ham radio groups
Emergency equipment suppliers	Sandbags; hazardous waste removal firms; vacuum trucks
Schools, churches, Red Cross, food suppliers	Disaster centers; food; shelter
Funeral homes and medical examiners	Morgue services
Helicopter terminals	Medevac; rescues; aerial photography; personnel and supply transport
Military/National Guard	Personnel; equipment
Transport companies	Equipment and supply transport; refrigerated trucks
Utility companies	Utility shut-off
Bottled water companies	Bottled water

national, and (increasingly) international scope (see Table 2.4). Many of these services are organized on the basis of specific types of hazards (e.g., poison, biological hazards, pesticides), but a growing number of professional organizations provide and share information on the basis of specific types of industry (e.g., pharmaceutical manufacturers, electroplaters).

At the local level, an important pool of valuable information and resources is available for purposes of emergency response planning through individual corporate plans (see Table 2.5) already developed to meet specific regulatory requirements (e.g., Chemical Hygiene Plan, Hazard Communication Plan). In some nations, some of these corporate plans are required to be filed with community authorities. For example, in the United States, the corporate contingency plan developed in compliance with U.S. EPA hazardous waste regulations must be made available to the local fire chief and other authorities (including corporate-designated medical facilities). However, regardless of specific regulatory requirements regarding coordination with local authorities, most corporations develop extensive information and detailed plans that, if made more generally available, would significantly expand local emergency response capacity. The sharing of such information and related material resources beyond the requirements of individual laws and regulations is probably most feasible through individual mutual assistance agreements among corporations as well as between corporations and local community authorities.

TABLE 2-4

INFORMATION SERVICES (ADAPTED FROM U.S. FIRE ADMINISTRATION, 1994: EMS SAFETY: TECHNIQUES AND APPLICATIONS [FA-144/APRIL 1994])

Information Services

CHEMTREC
  Tel: 800-424-9300
  A private service providing information about chemicals involved in
    transportation accidents
ATSDR (Agency for Toxic Substances and Disease Registry)
  Tel: 404-488-4100
  A 24-hour service that provides toxicological information and HAZMAT
    incident guidance
CDCP (Centers for Disease Control and Prevention)
  Tel: 404-633-5313
  Provides information about biological and disease-related hazards
Local Poison Center
  Regional centers that cover all regions in the United States
NPTN (National Pesticide Telecommunications Network)
  Tel: 800-858-7378
  A 24-hour service for information related to pesticide exposures and accidents
NRC (Nuclear Regulatory Commission)
  Tel: 301-951-0550
  A 24-hour service for information regarding radioactive materials

TABLE 2-5

EXAMPLE OF TABLE OF CONTENTS FOR CORPORATE CHEMICAL HYGIENE PLAN PURSUANT TO 29 CFR 1910.1450

Chemical Hygiene Plan

Table of Contents

1. Introduction
2. Responsibility
3. Standard Operating Procedures (general)
   - Personal Preparation & Behavior
   - Preparation of Work Area & Equipment
   - Maintenance of Work Area
   - Emergencies
   - Ordering Chemicals
   - Receiving Chemicals
   - Transporting Chemicals
   - Storing Chemicals
   - Using Chemicals
   - Using Extremely Hazardous Chemicals
4. Standard Operating Procedures (Chemically Specific)
   - Compressed Gases
   - Corrosive Chemicals
   - Flammable Chemicals
   - Extremely Hazardous Chemicals
   - OSHA Listed Chemicals
5. Methods for Limiting Exposure
   - Engineering Controls
   - Fume Hood Inspection
   - Personal Protective Clothing & Equipment
   - Emergency Equipment
6. Availability of Data & Information
7. Personnel Training
8. Medical Surveillance
   - Exposure Determination
   - Methods of Surveillance
   - Documentation
9. Determination of Health Hazards
   - Acute Hazards
   - Chronic Hazards
10. Special Issues
    - Prior Approval for New Chemicals
    - Chemicals Generated in Laboratory

Appendices

MEDICAL TREATMENT AND SURVEILLANCE

In any incident, there are several distinct groups of potential victims that must be considered for possible medical treatment and surveillance:

- On-site victims (i.e., persons who are present on-site at the time of the incident and who are immediately at risk)
- On-site emergency response personnel (i.e., initial or subsequent responders who, though prepared to respond to the emergency, are nevertheless subject

to risk (see Image 2.3); this group includes firefighters, HAZMAT, and other specialized (e.g., EMT) teams

- Off-site emergency response personnel (personnel who, though off-site, may be at risk due to chemical/biological contamination of evacuated victims (e.g., hospital personnel; ambulance personnel) or to fugitive toxic fumes/particles)
- Off-site general public (any other off-site person who may become at risk due to air or water contamination, or through contact with contaminated persons, or through contact with facilities used as temporary shelters/housing for contaminated victims)

IMAGE 2-3

ARLINGTON, VA, SEPTEMBER 14, 2001: A FEMA URBAN SEARCH AND RESCUE TEAM WORKS TO BUILD COLUMNS AND STRENGTHEN SUPPORTS IN THE PENTAGON FOLLOWING THE ATTACK

Source: Jocelyn Augustino/Fema News Photo

Historically, emergency response planning typically has focused on providing on-site victims the medical treatment and surveillance appropriate for acute physical injury and psychological trauma. However, it has become increasingly evident (especially with respect to terrorism) that treatment and surveillance related to chemical and biological contamination of both on- and off-site victims are critical, and demand equal regard for acute and chronic injury and disease.

Even in the absence of terrorism, the increasing global dependence on industrial chemicals and the rapid development of biotechnology increases the probability that industrial incidents will result in increasing numbers of the general public being at risk due to dangerous chemical and biological exposure. In this sense, our historic experience with providing medical treatment to large populations simultaneously subjected to the geographically extensive acute risks of storms, floods, and earthquakes provides limited instruction for providing medical treatment and surveillance to large populations simultaneously subjected to both the acute and the chronic hazards of chemical and biological agents.

REMEDIATION AND REVIEW

Whatever the operational or even regulatory definition of an emergency response incident, the actual incident is multidimensional, consisting of (a) preceding events, (b) the primary event that precipitates emergency response, (c) all actions taken during the response effort, and finally (d) all circumstances resultant from the response effort.

Remediation is inclusive of all actions undertaken during and after the response effort (c and d, earlier) to minimize harm and injury due to the secondary effects of the emergency. For example, the incident may be essentially defined as a structural fire; however, runoff waste water resultant from fighting that fire may contain toxic chemicals, and remedial actions must be taken to contain and properly dispose of that contaminated water. Also, it may be necessary to demolish remaining structures that may have become unsound.

With similar regard for managing risk both during and after the incident, it is incumbent upon the emergency planner to consider a wide range of additional issues, including:

- Off-site management of crowds that might not only interfere with the specific response effort, but also result in other public and personal risks
- Implementation of facility and local community evacuation
- Assessment of injuries and allocation of victims to first-aid and subsequent medical treatment
- Decontamination of personnel, equipment, materials, and facilities that may have become contaminated during the response effort
- Containment and ultimate disposal of contaminated soil, water, structural and inventory materials

- Preservation of scene for subsequent criminal investigation
- Recharge and/or replacement of exhausted, damaged, or proven-inadequate emergency response materials, supplies, and equipment
- Documentation of all response efforts
- Documentation of possible exposures of response personnel (for use in subsequent long-term medical surveillance of response personnel)
- Retraining of personnel in light of actual response performance
- Review and revision of existing emergency response plans and response procedures in light of the incident

Perhaps the major difficulty in ensuring a comprehensive assessment of remediation and review efforts is that, in an actual incident, various authorities play diverse roles and are subject to different jurisdictional constraints. It is therefore essential (a) to conduct a comprehensive and intensive post-incident debriefing of all responding agencies and authorities, and (b) to integrate resultant findings, information, and recommendations into subsequent emergency response planning and training activities.

STUDY GUIDE

True or False

1. Emergency planning must give the highest priority to response efforts.
2. Off-site management during an emergency effort is conducted only to ensure the proper transport of equipment, supplies, and personnel to the incident site.
3. A comprehensive plan for proactive and reactive emergency response is based on an integrated understanding of how environmental, workplace, and community dynamics actually interact.
4. Risk Assessment is inclusive of both hard assessment and exposure analysis.
5. In the Risk Assessment Phase of emergency planning, the focus is on the identification of the potential sources or causes of emergencies and the types and degrees of risk to be experienced by the workforce, the public at large, and emergency response personnel.
6. Hazard and Operability Analysis requires detailed information on the design and operation of a facility that produces or uses hazardous materials.
7. The failure of any emergency response effort typically is due to circumstances beyond the control of any response authority.
8. The Incident Command System (ICS) is essentially a response management system that can be used in any incident regardless of kind or size.
9. Chemical databases that might be useful in case of an emergency release of chemicals in a local plant typically are lacking.
10. Emergency planning should be guided by the assumption of a worst-case scenario for a potential incident.

11. Even in the absence of terrorism, the increasing global dependence on industrial chemicals and the rapid development of biotechnology increase the probability that industrial incidents will result in increasing numbers of the general public being at risk due to an industrial emergency incident.
12. Remediation is inclusive of all actions undertaken during and after the response effort to minimize harm and injury due to the secondary effects of the emergency.
13. Documentation of possible exposures of response personnel is important for use in subsequent, long-term medical surveillance of response personnel.
14. At the local level, an important pool of information and resources is available for purposes of emergency response planning through corporate plans developed to meet specific regulatory requirements.
15. In all instances, any bomb threat to a building should result in the immediate evacuation of that building.

Multiple Choice

1. The primary objective of emergency response planning is to:
   A. respond to the emergency
   B. prevent the emergency
   C. both A and B
2. Risk and Hazard Assessment
   A. mean essentially the same thing
   B. refer to different aspects of harm or injury
   C. are key mechanisms of any emergency response plan
   D. are not typically included as key components of emergency planning
3. Failure Modes & Effects Analysis is a hazard and risk assessment technique that requires detailed analysis of
   A. effects of errors by personnel
   B. effects of failure of equipment
   C. plant design and operations
4. In the United States, regulatory guidance regarding the level of protection for response personnel is given by
   A. OSHA and EPA
   B. NIOSH and USCG
   C. all of the above
5. Proactive and reactive emergency response activities are essentially exercises in
   A. risk management
   B. minimization of hazard
   C. toxic use reduction
6. Within the Incident Command System (ICS), a key function of the Planning component is to
   A. provide support to meet incident needs
   B. collect and evaluate information

  C. conduct tactical operations
  D. monitor costs related to the incident
7. The environmental context in which a hazardous plant is located is the link between a plant emergency and the surrounding community due to
  A. possible air and water contamination
  B. return of contaminated plant and respondent personnel to the larger community
  C. both A and B

Essays

1. Social equity is the disproportionate distribution of risks and benefits among the community. Considerations of social equity are an essential part of any emergency planning effort. How might you address this issue in your emergency response plans?
2. Exposure analysis requires consideration of different human populations at risk. Some of these differences include differences in general state of health (e.g., high blood pressure, use of medications). How might you approach this problem in terms of your emergency response plan?

Case Study

Assume that, as a safety officer in a local hazardous facility, you have been invited to speak at a local high school to explain the steps your company has taken to protect the community at large from any harm due to an emergency at the plant.

1. How do you establish a reasonable degree of credibility?
2. What linkages between the community and your plants might you give some emphasis?
3. Describe the types of day-to-day measures that are taken in-plant to ensure community safety.

THE EMERGENCY RESPONSE PLAN

INTRODUCTION

There are numerous formats for emergency response plans promulgated by various agencies, laws, and regulations (e.g., U.S. Fire Administration, U.S. Environmental Protection Agency, EPCRA, RCRA, 29 CFR 1910.1200), but it is clear that no single format can be sufficient for preparing for and responding to all potential emergency incidents. In this regard, legally mandated formats typically include minimum categorical requirements, with specific details left to be decided by the operational organization having primary responsibility for site-specific facilities and situations. This not to say that there are no objective criteria for evaluating an emergency response plan. Certainly, any reasonable plan must include certain provisions, including such vital elements as emergency evacuation procedures, methods for accounting for personnel, rescue and medical services, reporting requirements, and chain of command. However, of paramount importance in a plan is that specificity of circumstance that can be addressed adequately only by on-site personnel.

Unfortunately, the flexibility necessarily afforded by regulatory authority is all too often misinterpreted. The tendency, especially where industry perceives itself to be over-regulated, is to conform to the minimum regulatory requirements (if at all).

Where *de minimus* regulatory compliance becomes the measure of the attainment of regulatory objectives, the usual self-fulfilling prophecy unfolds—more and more detailed regulations are developed by frustrated governmental agencies, and industry, perceiving itself increasingly burdened by unreasonable regulators, focuses more and more on simple minimal compliance. During this ongoing and (most often) increasingly heated political confrontation, lives are lost that otherwise might have been saved.

In the United States as well as in some other nations having highly diverse, overlapping political jurisdictions and operational authorities, it is perhaps understandable that persons tend to see the prevention of and response to community emergencies as the primary responsibility of public agencies and public services. However, other nations take a different approach.

For example, in Malaysia, the prime responsibility for preventing and responding to a community at risk is specifically and legally assigned to the persons who cause that risk. Malaysian fire and rescue services, as well as all other public emergency response services, respond in the same professional manner as these services do in the United States—the difference is that industry knows that industry (and not a public agency or service) has primary responsibility to do whatever is necessary, first, to prevent an incident and second, to provide appropriate initial response; hence industry, being so clearly accountable, adopts a *de minimus* approach to human health and safety only at certain and severe financial and criminal risk.

Regardless of political system, cultural tradition, or stage of economic development, nations today are partners in a global village that is increasingly subject to the risks of sophisticated technologies employed by specialized industries typically located in major metropolitan areas. Whatever the first cause of a particular emergency, whether earthquake, typhoon, terrorist act, or simple accident, that emergency is highly likely to involve (directly or indirectly) hazardous industrial chemicals or materials, with potentially devastating effects on whole communities. But no emergency appears instantaneously. As shown in Figure 3.1, any potential emergency evolves over longer or shorter periods of time, most often presenting warnings that, if recognized and properly acted upon, can preclude the development of an actual crisis.

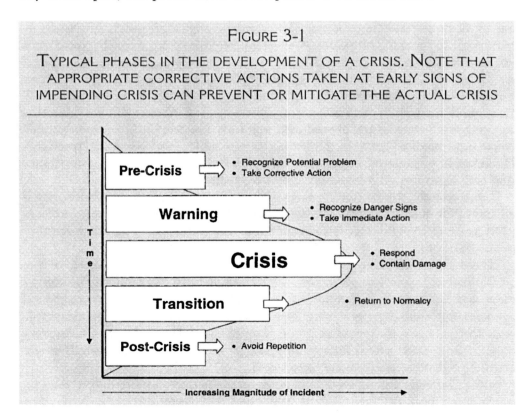

FIGURE 3-1

TYPICAL PHASES IN THE DEVELOPMENT OF A CRISIS. NOTE THAT APPROPRIATE CORRECTIVE ACTIONS TAKEN AT EARLY SIGNS OF IMPENDING CRISIS CAN PREVENT OR MITIGATE THE ACTUAL CRISIS

Those who are in the best position to recognize the early and progressive warning signals of an impending crisis are precisely those people on-site where the emergency begins. Typically, these are not the personnel of some governmental agency or public emergency response service but, rather, the workers and administrators in the facility wherein the crisis begins or in the facility that may be affected by an out-of-control emergency that develops elsewhere.

Even where industry realizes and accepts its primary responsibility for preventing, preparing for, and responding to emergencies, there is too often a misconception of the purpose of a fully developed, written emergency response plan. The purpose of that plan is not simply to document what should be done but, rather, to train personnel how to perform their specific emergency responsibilities. Prevention, preparation, and response requirements that are defined by the emergency response plan must be translated by training into on-the-job performance and, as necessitated by revisions to the plan, by retraining (see Figure 3.2). After all, a crisis is not the appropriate time to read a manual—a crisis demands immediate, premeditated, and pretested action.

Regarding the development of an emergency response plan that is truly proactive as well as reactive and also one that is effective and practical, it is useful to emphasize as well as expand upon five basic principles. These principles focus on the issues of responsibility and training discussed earlier, but also on important correlates to responsibility and training, including communication, practice, and command:

1. *Proper emergency planning begins with the owners, operators, and managers of the facility involved in or contributing to a potential or actual emergency.*

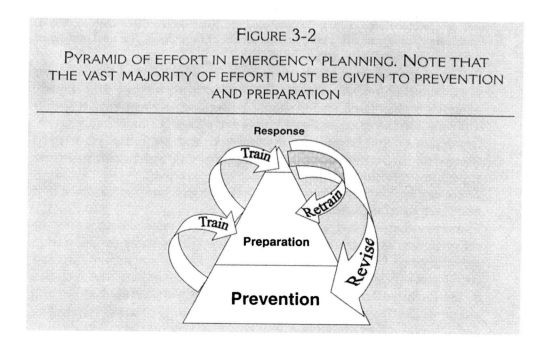

FIGURE 3-2

PYRAMID OF EFFORT IN EMERGENCY PLANNING. NOTE THAT THE VAST MAJORITY OF EFFORT MUST BE GIVEN TO PREVENTION AND PREPARATION

Employers must take primary responsibility for the workplace health and safety of their employees and for the potential risks that their operations present to the larger community. Meeting this responsibility requires the development, implementation, and periodic testing of an emergency response plan that details (a) the potential sources of hazards, (b) specific steps to be taken by employees to prevent or respond to an emergency, and (c) necessary coordination and liaison with local, national, and other competent authorities.

2. *An emergency response plan is only as good as the training given to the personnel who must implement the plan.*

   Periodic, in-depth training and practice drills must be conducted with the objective of ensuring that personnel respond immediately and appropriately to potential and actual emergencies. This training must include the evacuation of nonessential personnel as well as the proper use of equipment and procedures by first-responders. It is particularly important to coordinate selected in-plant training exercises with local competent authorities, such as the fire services and medical/ambulance services. Where time constraints on community services permit, it is highly desirable to involve such services in actual in-plant training sessions, especially sessions that focus on procedures for coordinating on- and off-site response.

3. *Communication plays an especially vital role in both the prevention of and the response to any emergency.*

   In-plant communication among personnel, managers, and first-responders as well as external communication with competent local authorities must be specifically channeled for the purpose of immediately providing key information to persons authorized to make decisions. During an actual emergency, such information includes (but is not limited to) the in-plant location of hazardous chemicals and other potential hazards (such as electrical or radioactive hazards), the number and location of trapped and evacuated personnel, and immediate medical needs.

4. *Ongoing facility audits and practice drills are essential for updating and refining an emergency response plan.*

   Audits of the physical plant, procedures, equipment, and the behavior of personnel should be continually conducted to identify and remedy potential emergencies. Audits should be conducted by personnel specifically authorized to ensure that corrective action is immediately taken. Corrective actions should be documented and periodically reviewed along with the results of evacuation and first-responder drills to ensure appropriate revisions of the emergency response plan.

5. *There can be no proper emergency response without the existence of a practiced, on-site chain of command.*

   Primary and alternate facility emergency response coordinators must be designated so as to ensure their earliest possible presence on-site during an emergency. It is the responsibility of the facility emergency response coordinator to (a) implement and direct all response activities included in the facility emergency response plan and (b) provide whatever aid, assistance, and information that may be required by external responding authorities, such as fire services and police. Whenever the external emergency response authority may take command of the emergency, the facility emergency response coordinator will ensure that he/she will act only upon the specific direction (or with the approval of) that authority.

The development of an emergency response plan begins with a comprehensive overview of the facility (see Figure 3.3) that includes (a) all on-site and off-site operations, (b) the industrial, commercial, community, and environmental surroundings of the facility and its operations, (c) the analysis of potential hazards associated with facility operations as well as of potential human environmental targets of those hazards, and (d) the identification of all community and in-plant emergency response resources available for managing both potential and actual emergencies. The corporate plan for conducting emergency response must integrate these considerations into clear, concise directions for implementing immediate and effective response.

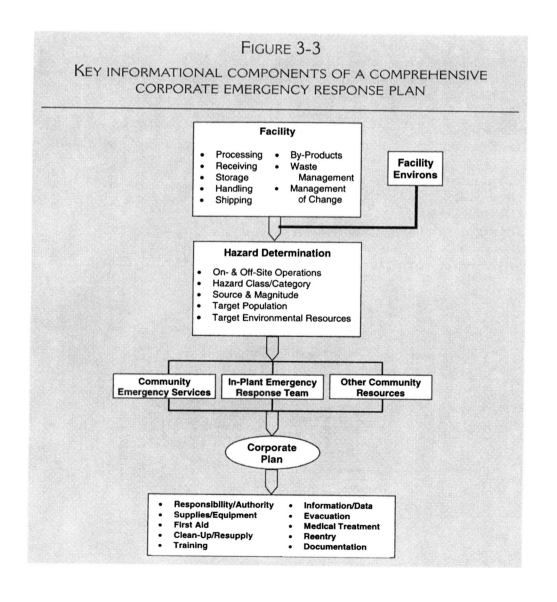

FIGURE 3-3

KEY INFORMATIONAL COMPONENTS OF A COMPREHENSIVE
CORPORATE EMERGENCY RESPONSE PLAN

CONTENTS OF PLAN

The organization and contents of emergency response plans can vary considerably, depending upon the responsibility and needs of the particular facility or organization that develops the plan. Basic emergency services (e.g., fire, search and rescue, medical) often adapt essentially generic plans (see Figure 3.4) to conform to individual circumstances. Different industrial facilities may also do the same, modifying generally available basic formats to meet specific requirements regarding type of industry, number of personnel, and nature of industrial hazard. Excellent guides for emergency planning and response for industry are provided by:

- Federal Emergency Management Agency (FEMA; *Emergency Management guide for Business and Industry*): http://www.fema.gov/
- Occupational Safety and Health Administration (OSHA; *Principal Emergency Response and Preparedness: Requirements and Guidance*): http://www.osha.gov/
- National Fire Prevention Association (NFPA; *NFPA 1600: Standard on Disaster/Emergency Management and Business Continuity Programs—2004 Edition*): http://www.nfpa.org/

Regardless of the type of facility, an emergency response plan for any facility should contain (at a minimum) the following basic categories of information:

- Objectives
- Responsibility and authority
- Distribution of plan
- Emergency equipment and supplies
- Location of data/information
- Assessment of hazards

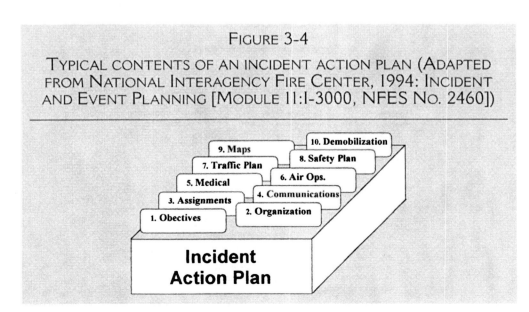

FIGURE 3-4

TYPICAL CONTENTS OF AN INCIDENT ACTION PLAN (ADAPTED FROM NATIONAL INTERAGENCY FIRE CENTER, 1994: INCIDENT AND EVENT PLANNING [MODULE 11:1-3000, NFES No. 2460])

- General procedures
- Notification procedures
- Evacuation procedures
- Containment procedures
- Special procedures (e.g., fire, explosion, flood, toxic gas release)
- Equipment shutdown
- Return to normal operations
- Training
- Documentation
- Informational appendices

An example of the more detailed information typically included under each of these broad headings is included in Table 3.1. Although the detailed format and informational contents of any emergency response plan must finally be based on site-specific details, it is worth examining each of the sections included in Table 3.1, as follows, in order to underscore certain key considerations.

Introduction

Whereas one of the objectives of the corporate emergency response plan may well be regulatory compliance, it is important that industry understand that

TABLE 3-1

EXAMPLE OF DETAILED TABLE OF CONTENTS FOR A COMPREHENSIVE EMERGENCY RESPONSE PLAN

Global Enterprises, Inc.

Emergency Response Plan

Table of Contents

1. **Introduction**
   A. Objectives: Regulatory Compliance
   B. Objectives: In-Plant Safety and Health Program
   C. Objectives: Personnel Training
   D. Objectives: Community Health and Safety
2. **Responsibility and Authority**
   A. Preparation, Review and Update of Plan
   B. Primary and Alternate Response Coordinators
   C. Liaison with Community Services
   D. Liaison with Local Industry
   E. In-Plan Emergency Response Team
   F. Communication with Media
   G. Personnel Training Program
   H. Facility Audits
   I. In-Plant Hazard and Risk Assessment
   J. Hazardous Waste Management
   K. Liaison with Contractors (for special emergency related services)
   L. Maintenance of Documentation
   M. Overview of Emergency Response Organization
3. **Distribution of Plan**
   A. Facility Personnel
   B. Community Services

Continued

TABLE 3-1—*Continued*

EXAMPLE OF DETAILED TABLE OF CONTENTS
FOR A COMPREHENSIVE EMERGENCY RESPONSE PLAN

4. **Emergency Equipment and Supplies**
   A. Emergency Containment Equipment and Supplies
   B. Personal Protective Clothing and Equipment
   C. Fire Fighting Equipment
   D. Medical Supplies
   E. Monitor and Alarm Systems
   F. Equipment and Supplies Available through Mutual Assistance Programs
   G. Equipment and Supplies Available through Contractors
5. **Location of Data/Information**
   A. Chemical Inventory
   B. Material Safety Data Sheets
   C. Layout of Facility (with access points)
   D. Floor Plans
   E. Location of Hazardous Areas/Materials
   F. Catchments and Drains
   G. Site and Area Topography
   H. Sensitive Natural Resources in Plant Vicinity
6. **Assessment of Hazards**
   A. Hazardous Stock Chemicals, Energy, Materials & Conditions
   B. Materials Safety Data Sheets & Other Specifications
   C. Hazards & Risks Associated with Facility Environs
   D. Hazards & Risks Associated with Off-Site Operations
   E. Process By-Products
   F. Summary of Hazards & Risks: Types, Potential Target Populations, Potential Target Environmental Resources, and Management Strategies
7. **Potential Emergency: General Procedures**
   A. Audits (type; frequency; responsibility; documentation)
   B. General Personnel (responsibility; chain-of-command; documentation)
8. **Actual Emergency: Notification Procedures**
   A. Notification of Emergency Response Coordinator and Team
   B. Activation of Evacuation Signal
   C. Notification of Community Services
   D. Notification of Other Potentially Affected Facilities/Persons
   E. Notification of Contractors (for emergency supplies/equipment)
   F. Notification of Mutual Assistance Partners
9. **Actual Emergency: Evacuation Procedures**
   A. Primary and Secondary Routes of Evacuation
   B. Location of Alternate Assembly Points
   C. Communication Requirements
   D. Monitoring Personnel During Evacuation & Assembly Points
   E. Decontamination and Medical Service Procedures
   F. Temporary Shelter/Housing
   G. Post Evacuation Procedures
10. **Actual Emergency: Containment Procedures**
    A. Communication Requirements
    B. Ventilation Systems
    C. Berms
    D. Absorbent Materials
    E. Fire Barricade
    F. Temporary Runoff Storage
    G. Temporary Storage of Other Hazardous Materials
    H. Follow-Up Procedures
11. **Actual Emergency: Special Procedures**
    A. Fire/Explosion
    B. Flood
    C. Storm

Continued

TABLE 3-1—*Continued*
EXAMPLE OF DETAILED TABLE OF CONTENTS
FOR A COMPREHENSIVE EMERGENCY RESPONSE PLAN

proper emergency response planning is an integral component of plant design and operations. In this regard, the emergency response plan must be viewed as a key means of meeting facility obligations regarding employee health and safety, personnel training, and community (including both human and environmental) health and safety—obligations that may be made subservient to other production and business objectives only at ever-increasing financial and criminal risk to corporate management. The emergency response plan therefore must begin with clear statements with regard to corporate recognition of and commitment to these objectives. In some jurisdictions, such statements are considered by legal authority to establish legally binding, contractual commitments between the facility and regulatory authority, employees, and the general public.

Responsibility and Authority

Over the last two decades, regulatory agencies (in the United States as well as in an increasing number of other countries) have given much emphasis to the importance of identifying specific individuals who both bear the responsibility and are the corporate authority for ensuring compliance with regulatory requirements.

There can be no question that proper emergency planning and response demand effective and efficient management of a myriad of detailed information, diverse personnel and skills, and precisely defined procedures in perilous and confusing circumstances.

Just who must do what, and when? Where and how must he or she do it? These are the quintessential questions to be immediately answered and acted upon in emergency response; it cannot be done where there is no clear designation of responsibility and authority, or where there is no commensurate authority for any given responsibility.

Distribution of Plan

In some nations, regulations require the distribution of emergency response plans to specific individuals and organizations (e.g., in the United States, regulations regarding contingency plans developed under RCRA). Regardless of jurisdictional authority, it is recommended that emergency response plans be distributed among all persons and organizations having primary and support responsibility in order to (a) ensure the proper sharing of important information and standard procedures, (b) provide a basis for continual feedback regarding proposed revisions and refinements, and (c) provide an essential tool for conducting coordinated training and practice among diverse facility and community responders. However, the distribution of emergency response plans should also be influenced by the following two considerations:

1. As plans are revised and refined, it is possible that various members of facility and community response services and teams will maintain different versions of the same plan. This could result in disastrous confusion. It is therefore necessary that the distribution process include

a means of recalling and destroying versions of the plan that have been superseded.

2. Various members of facility and community response services and teams do not need copies of the complete plan. For example, community hospitals certainly do not require information on in-plant evacuation routes, whereas information on potential in-plant chemical exposures is absolutely essential. Provisions should be made, therefore, to provide appropriate components of the plan to individual services. It is especially important to coordinate with individual services to ensure not only that they receive appropriate information, but also that they receive the information in a format that facilitates efficient use.

Emergency Equipment and Supplies

The number, type, description, and location of all emergency equipment and supplies must be clearly identified, including on- and off-site equipment and supplies. Where specific items are available through mutual assistance agreements or through prearranged contractor services, realistic estimates of availability (i.e., time-to-site) must also be included. Additional considerations include:

1. The availability of any item is a function not only of its location but also of its state of readiness. The actual availability of an item must therefore be estimated in light of documented (or otherwise assured) adherence to testing, maintenance, and replacement schedules.

2. Depending upon the development of an actual crisis, different types of equipment and supplies (see Image 3.1) may be needed, as well as different numbers or amounts. Appropriate emergency response planning therefore must take into account a range of potential emergency scenarios and corresponding demands on equipment and supplies. The typical approach is to designate equipment and supplies as being on-line, reserve, and backup. Such designators may also be used with regard to personnel as well as other emergency services.

3. Other factors governing the actual availability of off-site resources include, of course, inclement weather, traffic congestion, power and communication failures, and the simultaneous occurrence of local and regional multiemergencies. Emergency response planning for an industrial facility that does not take these possible factors into account cannot be considered realistic. To ensure realistic planning, any facility is well advised to include a worst-case scenario among the various response scenarios to be addressed by the plan.

Location of Data/Information

Data and information drive emergency response. Put another way, emergency response conducted in the absence of data and information is simply well intended guesswork that will most likely result in significant loss of human life.

During an emergency, the first duty of facility management is to provide appropriate and precise data and information to responders. Planning for an

IMAGE 3-1

NEW YORK, NY, OCTOBER 28, 2001: WORKERS SPRAY THE SMOLDERING RUBBLE WITH WATER ON THE DAY OF THE MEMORIAL SERVICE AT THE WORLD TRADE CENTER

Source: Andrea Booher/FEMA News Photo

emergency therefore requires concentrated effort to ensure that needed data and information will be immediately available to responders regardless of circumstance.

Section 5 of the Table of Contents depicted in Table 3.1 focuses on data and information most pertinent to plant layout, contents, and physical environs that must be used by incident commanders to choose among alternative strategies. Much of this information typically is included in various formats as appendices to the written emergency response plan—different formats being used to meet different needs.

During the planning phase, careful attention must be given to the following questions:

1. Is provision made for locating the information where it can be immediately and safely retrieved during even the worst-case scenario for an actual emergency? (Note: It may be necessary to provide for alternative locations, and for specially designed storage areas to protect contents from loss or damage due to the emergency.)
2. Are data and information up to date? (Note: Special effort must be given to ensuring that data and information regarding structural features of the facility and the location of specific types of hazards are accurate.)

3. Are data and information in a format that is immediately usable for responders? (Note: This requires previous coordination and liaison with incident commanders. Format, here, refers not only to the organization of the data and information, but also to the physical medium containing the information. For example, computer disks are not likely to be appropriate in the midst of an actual emergency; neither are hard copy materials that, under heavy rain, will quickly become tissue paper or, in the case of maps and diagrams, varicolored smears of water-soluble multicolor inks.)

Assessment of Hazards

Just as data and information drive the actual emergency response, so does the assessment of hazards (see Table 3.2) drive the entire process of planning for emergency response. If the assessment of hazards is inadequate, the emergency response plan is also inadequate, regardless of any seeming sophistication of the plan.

Minimal criteria for evaluating the assessment of hazards include:

1. Comprehensiveness of Assessment
   - Is there consideration of not only hazardous stock materials but also byproducts (i.e., materials produced as intermediaries of operational

---

TABLE 3-2

IMPORTANT PRINCIPLES FOR THE GUIDANCE OF HAZARD IDENTIFICATION AND ASSESSMENT (ADAPTED FROM ORGANIZATION FOR ECONOMIC COOPERATION AND DEVELOPMENT, 1992: GUIDING PRINCIPLES FOR CHEMICAL ACCIDENT PREVENTION, PREPAREDNESS AND RESPONSE, ENVIRONMENTAL MONOGRAPH NO. 51)

Hazard Identification And Assessment

1. When planning, designing and modifying installations and processes, management should ensure that critical examination techniques such as hazard analysis, hazard and operability studies, and fault tree and event tree analysis are utilized in order that hazards are identified and ranked as early as possible at the various stages of the project and the most suitable means of eliminating or reducing the hazards are instituted.
2. The nature and extent of the consequences which could result from each significant hazard and their likelihood should also be assessed, using techniques such as consequence analysis to ascertain the potential for harm. Reducing either the hazard or its probability of occurrence reduces the risk and increases the inherent safety of the design.
3. For existing installations which have not been subject to critical safety examinations, the appropriate hazard studies should be carried out in retrospect.
4. The management of hazardous installations should collate all safety-related information on the process and associated equipment concerning, for example, design, operation, maintenance and foreseeable emergencies.
5. Safety measures should be incorporated at the earliest conceptual and engineering design stages of an installation, to enhance the intrinsic safety of the installation wherever practicable.

Continued

TABLE 3-2—*Continued*

IMPORTANT PRINCIPLES FOR THE GUIDANCE OF HAZARD IDENTIFICATION AND ASSESSMENT (ADAPTED FROM ORGANIZATION FOR ECONOMIC COOPERATION AND DEVELOPMENT, 1992: GUIDING PRINCIPLES FOR CHEMICAL ACCIDENT PREVENTION, PREPAREDNESS AND RESPONSE, ENVIRONMENTAL MONOGRAPH NO. 51)

Hazard Identification And Assessment

6. In designing new installations and significant modifications to existing installations, industry should use the relevant, most up-to-date international standards, codes or practice and guidance established by public authorities, enterprises, industry and professional associations, and other bodies in order to achieve a high level of safety.
7. Existing installations should be assessed to determine whether they meet these standards, codes and guidance. Appropriate improvements should be carried out as soon as practical.
8. The design of a hazardous installation should integrate the appropriate equipment, facilities and engineering procedures that would reduce the risk from hazards as far as is reasonable the practicable (i.e., all measures to reduce the risk should be taken until the additional expense would be considered far to exceed the resulting increase in safety).
9. Processes should be designed to contain, control and minimize the quantity of hazardous intermediate substances to the extent that this would increase safety. Where this is not possible, the quantity of hazardous intermediates produced should be reduced to that required for use in the next stage of production so that quantities held in storage are kept to a minimum.
10. Systems should be designed so that individual component failures will not create unsafe process conditions and/or will be capable of accommodating possible human errors.
11. Although emphasis should be on inherent safety in design, consideration should be given to the need for "add-on" protective systems, thereby assuring safety through mitigation measures.
12. In the design phase, management should ensure there is adequate consideration of the site layout as guided by overall safety goals. Particular regard should be given to: the establishing of safe separation distances to minimize any "domino effects"; the location of hazardous processes and substances relative t the location of critical safety-related equipment and instruments; and the local community and environment.
13. Relevant personnel who will be involved in the operation of a hazardous installation should also be involved in the planning, design and construction phases of the installation. Employees, and their representatives, should participate in decisions concerning the design of their workplace, and should be given the opportunity to provide input in the design, application and improvement of equipment in order to utilize employee "know-how" and experience.
14. The management of a hazardous installation should pay particular attention to quality assurance during the construction phase of the project.
15. Safety checks should also be carried out at the commissioning and startup phases of a project to ensure that the design intent has been completely fulfilled. Functional tests should be carried out for all components, controls and safety devices critical to the safety of the installation.
16. An enterprise should purchase equipment only from reputable suppliers, and should formally inspect equipment to ensure that it conforms to design specifications and safety requirements before being put into use.
17. In the construction of a hazardous installation, an enterprise should do business with only those contractors who are able to satisfy the enterprise that their services will be carried out in compliance with all applicable laws and regulations, as well as in compliance with relevant safety standards and policies of the enterprise, so as not to increase the risk of an accident involving hazardous substances. Contractors should work to the standards set by the management of the installation and, to the extent appropriate, under the direct surveillance of management.

processes) and combustion products (i.e., vapors, fumes gases, mists, and particles) produced as a result of burning stock materials or byproducts?

- Is there consideration of not only the hazards associated with chemicals and other materials (biological) that may present risk to employees and/or the surrounding public and environment, but also those hazards (e.g., structural, mechanical, electrical, cryogenic) that may present risk to emergency responders?
- Is there consideration of hazards associated with off-site as well as on-site operations of the facility?
- Is there consideration of hazards and risks to employees and incident responders that may emanate from other neighboring facilities or sources?

2. Application of Assessment
   - How are hazards related to on- and off-site target human populations?
   - How are hazards related to on- and off-site target environmental resources?
   - What alternative risk management practices (e.g., administrative, engineering, personal protective clothing and/or equipment) are considered and evaluated with respect to controlling what sources of hazards?
   - Are hazards prioritized with respect to type of potential emergency incident (e.g., fire, explosion, flood, terrorism, earthquake)?

In any well-managed facility, hazard assessment is an integral component of a facility-wise vulnerability analysis, which is a methodical attempt to integrate information on (a) types of potential emergency, (b) likely impacts and probability of occurrence of each type, and (c) resources available for use in an actual emergency response (see Figure 3.5).

General Procedures

Because prevention must be the first objective of any emergency response planning, particular attention must be given to specific managerial methods and techniques as well as early warning procedures and devices that can decrease the probability of the occurrence or minimize the magnitude of an incident.

Both on- and off-site audits of facility operations can be an effective means of identifying and correcting situations that, if left untended, will result in or exacerbate an actual emergency. Consideration should be given to implementing a variety of audits, each having its own focus, including:

- Compliance with specific regulations (e.g., hazardous waste, laboratory chemical, hot work permits)
- General housekeeping
- Plant access and egress
- Receiving, handling, storage, and disposal of chemicals
- Tank and reactor maintenance
- Flood prevention procedures
- Plant and property security measures
- In-plant signing and labeling of hazardous areas and materials
- Employee on-the-job behavior

FIGURE 3-5

VULNERABILITY ANALYSIS CHART (ADAPTED FROM FEDERAL EMERGENCY MANAGEMENT AGENCY (FEMA). EMERGENCY MANAGEMENT GUIDE FOR BUSINESS & INDUSTRY [FEMA ELECTRONIC REFERENCE LIBRARY])

Type Of Emergency	Probability High Low 5 ←→ 1	Human Impact High 5 ←	Property Impact	Business Impact Low → 1	Internal Resources Weak 5 ←	External Resources Strong → 1	TOTAL*

* Plan to minimize Total

Audits should be carried out by persons having full corporate authority to implement appropriate corrections immediately. The results of all audits, including findings and corrections, should be fully documented and specifically used to review and, as necessary, amend company policy statements, written protocols and procedures, and employee job descriptions, as well implement personnel actions.

General facility procedures (e.g., chain of command, employee information and training program, purchase and inventory control, on-site monitoring of contractor services, response to system alarms, operations monitoring) should all be reviewed on a regular basis and, as appropriate, revised to ensure consistency with emergency response objectives and plans.

Notification Procedures

Notification procedures are inclusive of all procedures designed to inform all responsible persons of the event or progress of a potential or actual incident, and to provide those persons with information required for their proper performance of emergency-related functions. In the event of an actual emergency, notification also implies the activation of the facility evacuation signal.

Even where specific notification procedures are successfully implemented, all too often too little consideration is given to the precise information to be conveyed to the personnel or organization notified. After all, in an emergency, fear, panic, confusion and/or frustration may result in people giving garbled messages. It is therefore strongly recommended that the emergency planning team compose specific formats (and, as appropriate, prewritten texts) to be used by notifying parties.

Another major problem typically encountered in actual emergencies is that the devices (alarms, telephones, automatic electronic alarm and notification devices) relied upon to effect communication become inoperative (e.g., due to power outage, overload, and/or interference). Alternative backup means are therefore necessary.

Finally, because many communication devices and systems devoted entirely to notification of responsible persons and organizations during an emergency typically are used infrequently, regularly scheduled testing and maintenance of these devices and systems is an absolute requirement.

Evacuation Procedures

Although many people tend to view evacuation as a rather elementary procedure, it is one that, if not planned and accomplished correctly, can lead to even more deaths and injury than those resulting from the primary source of an emergency. This is because, during an evacuation, evacuating personnel are subject not only to risk due to the emergency itself (e.g., fire and smoke), but also to risks due to panic and hysteria (e.g., stampeding) and to personal stress (e.g., heart attack), and to risks that arise from the physical features and encumbrances associated with plant design and normal operations (e.g., narrow, steep stairways; barrier to (or temporary blocking of) egress by work—including the efforts of emergency responders—in progress).

It is generally not sufficient to rely upon a single mode (e.g., sound, flashing light) of evacuation alarm. For example, an audible alarm may not be heard in more isolated areas of the facility (e.g., toilets, storage areas); certainly, no audible alarm, regardless of volume or placement, can give warning to a deaf employee.

Primary and secondary routes of evacuation, which are clearly marked and identifiable at all times and in all potential circumstances (e.g., heavy smoke), are absolute minimum requirements; depending upon facility layout, as well as specific disabilities of personnel, additional routes (e.g., ramps, guiding handrails) may be necessary.

Although it may not be possible to monitor evacuating personnel for signs and symptoms of exposure (e.g., chemical exposure, burns) or of personal stress (hyperventilation, heart attack, broken limb, cuts), concentrated effort must be made to identify personnel who may require immediate medical attention.

Alternative assembly points must be managed by personnel specifically designated and trained to ensure (a) ongoing communication with incident command (e.g., regarding movement of personnel to avoid interference with emergency vehicles), (b) proper accounting of assembled personnel, (c) identifying need of evacuees for decontamination, first aid, and/other medical treatment, and (d) assignment of evacuees to other on- and off-site facilities for temporary shelter and housing, food and water, and sanitary facilities.

Finally, evacuated personnel must be managed effectively to ensure proper compliance with corporate post-evacuation procedures, including procedures regarding the control of personal vehicles, medical consultation, and follow-up notification of employee families.

Containment Procedures

Emergency containment is a key mitigative measure that may, in fact, prevent a minor spill (e.g., of a flammable material) from becoming a major incident (e.g., facility fire); containment may also be necessary during a full-fledged emergency response, as when contaminated firefighting runoff water must be contained on-site to protect downstream environmental resources (surface and groundwater resources, soil) from subsequent contamination.

All facility personnel should be trained in the proper use of containment materials. However, it is vital that facility personnel understand that they must use these materials only if they can do so without undue risk to themselves. Facility and community responders must be provided with appropriate training and protective clothing and equipment to implement containment measures with appropriate control of attendant risks.

In addition to the use of containment materials, emergency planning must take into account the proper disposal of contaminated containment materials, as well as of the hazardous chemicals contained by those materials.

Special Procedures

Special procedures are those procedures implemented in response to specifically designated hazards and risks (e.g., fire/explosion, flood, storm, toxic gas release) or to specific chemicals/materials (e.g., water-reactive chemicals, teratogenic chemicals). Such procedures may be developed for implementation by different levels of personnel (e.g., corporate safety officer, facility first-responders, facility and/or community-based specialized teams/consultants), with due regard for different levels of knowledge, personal protective clothing, and personal protective equipment associated with different personnel who implement those procedures.

The text of a written emergency response plan contains all information relevant to special procedures; however, it is recommended that basic procedures also be reduced to simple, stepwise directives that can easily be reduced to small placards, signs, or poster-boards, or even to wallet-size cards. This approach will help to ensure that these procedures are readily available to personnel as needed—being either in their personal possession (as with wallet-size cards) or easily observable as appropriately located signs or posters.

Equipment Shutdown

The designation of specific equipment and production processes to be shut down during emergency response and procedures and protocols for effecting shutdown should be done only in close liaison with potential responding authorities and facility engineers.

As a preventive action, shutdown should be implemented only by designated employees who are fully trained as to the precise circumstances that require shutdown. This will prevent actions that may inadvertently turn a minor, manageable accident (e.g., a minor spill of flammable liquid) into a major incident, as when the electric arc produced by a nonexplosion proof light switch serves as a source of ignition to an explosive room atmosphere.

In no circumstance should any facility officer or employee actually undertake to shut down equipment or processes during an actual emergency except by the express order to do so from the incident commander. This approach will ensure that shutdown does not interfere with the proper functioning of protective devices or systems, or of emergency response equipment.

Return to Normal Operations

Once the emergency is terminated, all responsibilities assumed by community emergency response services cease. It is therefore the responsibility of facility owners and managers to determine that the facility is fit to return to normal operations.

There is often such a single-minded determination of facility owners and managers to resume operations as quickly as possible that they can easily overlook potential consequences of both the emergency and the response effort, which could result in not only additional risk to employees, but also long-term financial and legal risk to themselves. Before returning the facility to normal operations, owners and managers, in coordination with facility engineers and external consultants, therefore should evaluate the following potential conditions:

1. The facility is structurally unsound as a direct consequence of the emergency.
2. The facility is contaminated with hazardous stock chemicals and/or byproducts, which were released and/or produced during the emergency.
3. Essential operational alarms and monitoring systems are nonfunctional or are in need of testing and adjustment.
4. Emergency containment and other incident response materials and supplies are depleted as a result of response activities.
5. On-site emergency equipment is in need of repair and/or decontamination as a result of previous response activities.
6. There are hazardous materials derived from the previous emergency that remain on-site and that must be properly disposed of.
7. Personnel are traumatized by the previous emergency.

The basic rule governing return to normalcy is that it can be attempted only after full assurance is gained that (a) the facility presents no unmanaged risks to employees, and (b) all alarm systems and emergency resources (including equipment and supplies) are fully replenished or replaced and functionally on-line.

Training

All provisions and procedures included in an emergency response plan must be considered as defining specific needs for personnel training (see Image 3.2). In no circumstance may the simple presentation of training programs and information to employees be construed as sufficient. The object of emergency response training is the actual behavior of personnel—which only comes with practice.

Because an emergency response plan assigns different responsibilities to different personnel and presumes the availability of diverse skills, emergency response training should be designed to meet the different skill performance levels and informational needs of employees—and training, retraining, and practice must be stubbornly pursued to achieve and maintain specified job-performance standards.

Joint training exercises involving both on-site personnel and community emergency services, including simulated emergencies (e.g., table-top exercise, field exercise) should be implemented following a regular schedule. Particular attention should be given to using external experts and professional response services to assess and evaluate the practical effectiveness of all on-site training.

IMAGE 3-2

EMMITSBURG, MD, MARCH 10, 2003: AN INCIDENT COMMAND SYSTEM COURSE IS HELD AT FEMA'S NATIONAL EMERGENCY TRAINING CENTER, ONE OF DOZENS OF COURSES OFFERED THERE EACH YEAR FOR FIRST RESPONDERS, EMERGENCY MANAGERS, AND EDUCATORS

Source: Jocelyn Augustino/FEMA News Photo

Documentation

The importance of documenting all aspects of emergency planning and response cannot be overemphasized because documentation is the primary (and often the only) means of adequately addressing the following issues and needs:

1. Potential regulatory and other legal proceedings related to regulatory compliance or other legal standards (e.g., Common Law doctrine of negligence).
2. On-going evaluation and revision of emergency response plans on the basis of actual experience with emergencies.
3. Medical assessment of long-term employee exposures to hazardous chemicals and materials.
4. Ongoing improvement of in-plant safety provisions and procedures on the basis of regularly conducted audits.
5. Ongoing assessment of the effectiveness of emergency-related personnel training.

All documentation (whether printed or electronic) should be maintained in an on-site "fail-safe" storage area, with duplicates regularly updated and safely warehoused in at least one off-site facility.

Appendices

Appendices typically contain information required for special purposes. As shown in Figure 3.6, much of this information pertains to the location of specific hazards and other items of particular importance to emergency response teams (e.g., chemical inventories, floor plans). However, appendices also often are used to compile specific procedures, determinations (e.g., plant vulnerability analysis), and lists of equipment and supplies (e.g., spill containment supplies, first aid supplies).

The information contained in appendices should be in a format that facilitates rapid access to the information as well as immediate use. In close liaison with community emergency response services, facility management should determine which of this information should be regularly disseminated among the various services and/or included in lockout boxes and other structures that guarantee responder access as needed.

IMPLEMENTATION OF PLAN AND OVERSIGHT

Given the great amount of careful effort necessary to develop and comprehensive emergency response plan, it is not surprising that, once compiled, the plan tends to become a "shrine" to that effort—a suitably jacketed, thick compendium of multicolored flowcharts, diagrams, and closely printed text. The problem, of course, is that such a tome is hardly a practical tool. It often

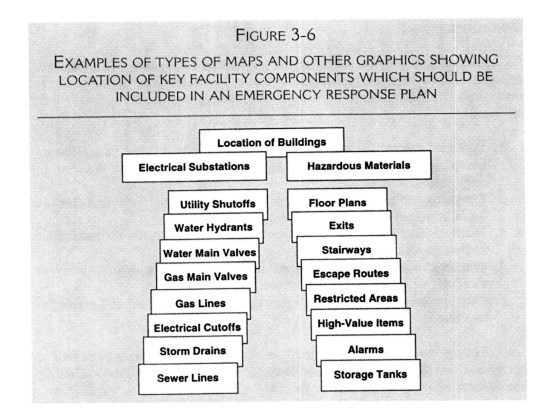

FIGURE 3-6

EXAMPLES OF TYPES OF MAPS AND OTHER GRAPHICS SHOWING LOCATION OF KEY FACILITY COMPONENTS WHICH SHOULD BE INCLUDED IN AN EMERGENCY RESPONSE PLAN

ends up, as with so many other bulky manuscripts, entombed on a dusty bookshelf, supposedly to be retrieved and consulted in the event of an actual emergency.

Certainly it is necessary to maintain a complete and detailed master copy of the emergency response plan, but the implementation of that plan requires that it serve primarily as the source of practical training materials, policies, and directives that, properly formatted, are better suited than a massive book to transforming a plan of action into real behavior (see Image 3.3). In an emergency, whether potential or real, one must act . . . not read!

Upon completion of the plan (which must always be subject to revision), the tasks that remain are therefore even more demanding than those performed during the development of the plan precisely because the goal of implementation is to direct (and/or change) human behavior—at best, a thoroughly difficult undertaking that demands persistent training and practice, followed by yet more training and more practice.

Whether by an industrial facility or a public agency, the implementation of any emergency plan ultimately depends upon the absolute commitment of the highest executive authorities to institute health and safety objectives and policies (see Tables 3.3 and 3.4). Without such executive and administrative commitment, there is little likelihood that adequate resources (including time, money, and personnel) will be made available to ensure effective implementation by such means as:

IMAGE 3-3

TALLAHASSEE, FL, MAY 4, 2005: FEMA, ALONG WITH STATE AND LOCAL EMERGENCY MANAGEMENT AGENCIES, JOIN WITH 64 HISTORICAL BLACK COLLEGES AND UNIVERSITIES TO EXPAND EMERGENCY PLANS AND PREPAREDNESS, AND TO KEEP CAMPUSES SAFER

Source: Butch Kinerney/FEMA photo

1. Using the overall emergency response plan.
   ▪ Define specific responder tasks (taking into account primary, secondary, and backup responsibilities of all members of the response team).
   ▪ Identify specific needs for information, types of skills, and necessary skill levels for individuals having responsibility for each responder task.
2. Develop appropriate training materials, informational packages, and summary action directives/protocols for each responder task.

TABLE 3-3

IMPORTANT PRINCIPLES REGARDING THE DEVELOPMENT AND IMPLEMENTATION OF SAFETY POLICY IN INDUSTRY (ADAPTED FROM ORGANIZATION FOR ECONOMIC COOPERATION AND DEVELOPMENT, 1992: GUIDING PRINCIPLES FOR CHEMICAL ACCIDENT PREVENTION, PREPAREDNESS AND RESPONSE, ENVIRONMENTAL MONOGRAPH NO. 51)

Establishment of Safety Policy by Industry

1. Management of hazardous installation has the primary responsibility for preventing accidents involving hazardous substances, and for developing the means to do so.
2. Effective overall management of hazardous installations necessarily includes effective management of safety; there is a clear correlation been safely run installations and well-managed operations. Therefore, safety should be an integral part of the business activities of the enterprise, and adequate resources should be made available for taking the necessary measures to prevent accidents and to pay for the consequences of any accidents which do occur.
3. All installations in an enterprise should aim to reach the ultimate goal of "zero incident," and resources must be targeted towards this goal.
4. Management should not become complacent if there have not been any accidents at an installation over a period of time; continuous efforts are needed to maintain safety.
5. Each enterprise should establish a corporate safety culture.
6. Each enterprise should have a clear and meaningful statement of its Safety Policy, agreed, promulgated and applied at the highest levels of the enterprise, reflecting the corporate safety culture and incorporating the "zero incident" goal as well as the safety objectives established by public authorities.
7. The development and implementation by an enterprise of policies and practices relating to accident prevention and preparedness should be coordinated and integrated with its activities relating to occupational safety, health and environmental protection as part of the enterprise's total risk management program.
8. The responsibility for day-to-day management of safety should be in the hands of line management at individual installations.
9. All employees have a continuing role and responsibility in the prevention of accidents by carrying out their jobs in a safe manner, and by contributing actively to the development and implementation of safety policies and practices. Employees at all levels, including manager, should be motivated and educated to recognize safety as a top priority and its continuing improvement as a main corporate aim.
10. Producers of hazardous substances have a responsibility to promote the safe management of substances they produce throughout the total life cycle of the substances, from their design through production and use to their final disposal or elimination, consistent with the principle of "product stewardship." Such producers should make special efforts to help prevent accidents during the handling and use of a hazardous substance by downstream users.
11. Enterprises selling hazardous substances should actively try to determine whether their customers have adequate facilities and know-how to handle the substances (including, as appropriate, processing, use and disposal of the substances). If such determination cannot be achieved, judgment has to be exercised to decide whether to accept such customers. If customers are found to be incapable of safely handling the hazardous substances, the seller of the substances should assist the customer in obtaining this capability or else not accept such customers.
12. Smaller enterprises with limited resources should examine the need for assistance on safety matters from external consultants, professional trade associations and public authorities as well as from suppliers. Suppliers of hazardous substances should be supportive by ensuring that people are available to provide advice in order to achieve an appropriate level of safety.
13. Larger enterprises and/or trade associations should offer assistance to small and medium-sized companies in meeting safety objectives.
14. Enterprises and trade associations should take action strongly to encourage enterprises which act less responsibly to meet the appropriate safety objectives.

TABLE 3-4

IMPORTANT PRINCIPLES REGARDING THE DEVELOPMENT AND IMPLEMENTATION OF SAFETY POLICY IN PUBLIC AGENCIES (ADAPTED FROM ORGANIZATION FOR ECONOMIC COOPERATION AND DEVELOPMENT, 1992: GUIDING PRINCIPLES FOR CHEMICAL ACCIDENT PREVENTION, PREPAREDNESS AND RESPONSE, ENVIRONMENTAL MONOGRAPH NO. 51)

Establishment of Safety Objectives & a Control Framework by Public Authorities

1. Public authorities should ensure that appropriate safety objectives are established as part of a long-term strategy.
2. Public authorities should develop a clear and coherent control framework covering all aspects of accident prevention.
3. Public authorities should have available appropriate staff to carry out their role and responsibilities in the prevention of accidents, and should ensure that the staff is adequately educated and trained.
4. A coordinating mechanism should be established where more than one competent public authority exist, in order to minimize overlapping and conflicting requirements from various public authorities.
5. In establishing safety objectives, as well as the control framework, public authorities should consult with representatives of the other stakeholders, including: relevant public authorities, industry, professional and trade associations, independent experts, trade unions, interest groups, and the public.
6. Public authorities should establish the criteria for identifying those hazardous installations considered to have the potential to cause major accidents.
7. The requirements established by public authorities should be applied fairly and uniformly to ensure that enterprises of all sizes and types, whether national or foreign, are required to meet the same overall safety objectives.
8. The control framework should allow flexibility in the methods used to meet the safety objectives and requirements.
9. The requirements and guidance established by public authorities should stimulate innovation and promote the use of improved safety technology and safety practices. The control requirements should be considered minimum; industry should be encouraged to achieve a higher level of safety than would be achieved by adherence to established standards and guidance alone.
10. The requirements and guidance should be reviewed periodically and, where necessary, amended within a reasonable time to take into account technical progress, additional knowledge and international developments.
11. The control framework should include provisions for the enforcement of requirements, and adequate resources should be available to the public authorities for monitoring and enforcement activities.
12. Public authorities should establish procedures for the notification and reporting to them of certain specified categories of hazardous installations.
13. Public authorities should also establish a system for the submission of detailed information for certain categories of hazardous installations.
14. Public authorities should consider which installations, or modifications to installations, are so potentially hazardous that the installations should not be allowed to operate without the prior and continuing approval of identified public authorities. In these cases, a form of licensing control could be utilized which would require management to submit full details of all relevant aspects of its projected activity to the authority in advance of siting and startup, and periodically thereafter. There should be an opportunity for public input into these licensing decisions.
15. Public authorities should establish a requirement for the reporting of certain incidents by the management of hazardous installations.

Continued

TABLE 3-4—*Continued*

IMPORTANT PRINCIPLES REGARDING THE DEVELOPMENT AND
IMPLEMENTATION OF SAFETY POLICY IN PUBLIC AGENCIES
(ADAPTED FROM ORGANIZATION FOR ECONOMIC COOPERATION
AND DEVELOPMENT, 1992: GUIDING PRINCIPLES FOR CHEMICAL
ACCIDENT PREVENTION, PREPAREDNESS AND RESPONSE,
ENVIRONMENTAL MONOGRAPH NO. 51)

16. In order to assist industry in importing safety at hazardous installations, public authorities should consider whether to undertake such additional activities as: provision of technical assistance, promotion of training programs, encouragement of research, and fostering of public awareness.

17. Public authorities in neighboring countries should exchange information and establish a dialogue concerning installations which, in the event of an accident, have the potential of causing transfrontier damage.

18. National and, where appropriate, regional public authorities should cooperate internationally to improve prevention of accidents involving hazardous substances as well as to improve emergency preparedness and response.

19. Cooperation should be promoted in the preparation of guidance documents across countries, industry groups and international organizations.

20. A worldwide network should be established to promote the sharing among enterprises and countries of information related to the prevention of, preparedness for, and response to accidents involving hazardous substances. This is particularly important as a means of providing access to information for those with less capability with respect to the safe handling of chemicals.

21. Trade associations, local chambers of commerce and other organizations can be a useful means of disseminating chemical accident prevention information to smaller enterprises which might be unaware of the existence of such information.

3. Conduct training of personnel on the basis of task responsibilities, using combinations of various techniques, including (but not limited to):
   - Lectures
   - Demonstrations
   - Group discussions
   - Problem-solving workshops
   - Table-top exercises
   - Field exercises
   - Critiques of simulations
   - Role-playing sessions
   - Multiagency/multifacility exercises

4. Evaluate training on the basis of informational and behavioral objectives as defined by required skill levels (using external as well as internal evaluators).

5. On the basis of training, revise (as necessary) the emergency response plan and/or training methods and techniques.

6. Institute facility/agency-wide practice drills (announced and unannounced), and schedule drills to occur over a range of weather conditions and work-shift schedules.

7. Practice. Train. Drill. Practice again!

STUDY GUIDE

True or False

1. Those who are best prepared to detect an impending crisis in any facility are on-site personnel who have been trained to manage and operate the facility.
2. The essential element of any successful emergency response plan is effective practice and training.
3. Those who have the primary responsibility for ensuring that an emergency response plan is effective are the first responders.
4. It is important to coordinate selected in-plant training exercises with local competent authorities.
5. In-plant audits should be conducted by personnel specifically authorized to ensure that corrective action is immediately taken.
6. Corporate recognition of and commitments to planning objectives that are included in the emergency response plan cannot be used by legal authority as binding commitments between the facility and regulatory authority, employees, or the general public.
7. In the United States, it has become increasingly common that, for purposes of regulatory compliance, specific individuals must be assigned specific responsibilities in written plans regarding health and safety.
8. The number, type, description, and location of all emergency equipment and supplies must be clearly identified, including on- and off-site equipment and supplies.
9. During an actual emergency, the first duty of facility management is to provide appropriate and precise data and information to responders.
10. It is the responsibility of the community emergency responders to determine whether or not it is safe for plant personnel to reenter the facility upon the conclusion of response efforts.
11. Hazard assessment includes not only the assessment of chemical stocks, but also of process byproducts and the products of combustion.
12. In case of any in-plant emergency, all personnel should be instructed to shut down all in-use equipment and power supplies.
13. It is generally accepted that emergency planners should compose essentially prewritten texts that are to be used for the notification of external responders.
14. Depleted emergency containment and other incident response materials must be replaced prior to reentry of personnel after an incident.
15. It is typically not sufficient to rely upon a single mode of evacuation alarm (e.g., sound, flashing light).

Multiple Choice

1. A key characteristic of any effective emergency response plan in industry is
   A. flexibility
   B. compliance with detailed regulatory requirements
   C. concern for minimizing disruption of production schedules

2. Those who are best able to discern a developing emergency in a production facility are
   A. the surrounding community
   B. managers and operators of the facility
   C. governmental officials
   D. community responders
3. The basic purpose of a written emergency response plan is to
   A. document which actions are to be taken
   B. train personnel how to perform their emergency response duties
   C. document the assignment of specific responsibilities and authority
4. In industry, the primary responsibility for a well developed, comprehensive, and effective emergency response plan lies with
   A. owners, operators, and managers of the facility
   B. the in-plant emergency response coordinator
   C. the community emergency response coordinator
   D. the appropriate regulatory agency
5. For any emergency response plan, the most essential ingredient is
   A. the assignment of responsibility for implementation of the plan
   B. coordination with community emergency responders
   C. effective training
   D. secure communication
6. In-plant audits for identifying and correcting developing problems should be conducted by
   A. a contracted emergency response contractor
   B. the local fire chief
   C. those who have the specific authority to implement immediate corrective action
   D. the vice president of operations
7. Whenever the external (community) emergency response authority may take command of an in-plant emergency, the facility emergency response coordinator will ensure that he or she will act only at the command of
   A. the facility operations manager
   B. the external emergency response authority
   C. the facility owner
8. A comprehensive in-plant emergency response plan includes an assessment of not only hazardous materials stock supplies, but also
   A. hazardous byproducts of in-plant processing
   B. hazardous byproducts of the combustion of stock supplies
   C. both A and B
9. The assessment of hazards and risks includes consideration of
   A. on- and off-site target human populations
   B. on- and off-site target environmental resources
   C. both A and B
10. The primary purpose of monitoring the evacuation of a facility is to
    A. ensure compliance with evacuation procedures
    B. identify personnel who may be in need of immediate medical attention
    C. document the efficiency of evacuation

Essays

1. Given the wide variety of kinds of information that have to be considered in the development and implementation of a comprehensive emergency response plan, discuss the types of personnel you would include in an in-plant planning team having responsibility for developing that plan.
2. The average manufacturing plant has from 300 to 3000 stock chemicals (including chemical components of mixtures). Discuss how you would have your planning team format data on the characteristics/hazards associated with these chemical so that it would be most usable to emergency response personnel.
3. Assume that your in-plant team has completed what the team considers to be a comprehensive emergency response plan. What do you then propose to do with it?

Case Study

You are the safety officer of a small pharmaceutical company that specializes in the production of half a dozen pharmaceutical products. Your inventory of stock chemicals is on the order of 1000 individual chemicals. In addition to its production line of finished pharmaceutical products, the company has an active research and development laboratory and a quality control laboratory. The company is located in a zoned-for-business area that is interspersed with private homes, schools, and a wide variety of small service-oriented companies in a community of about 10,000.

1. Explain how you would begin to develop an emergency response plan?
2. What considerations will influence your choice of persons to serve on your team for developing this plan?
3. What special concerns do you think you should give priority to?

PHYSICAL AND CHEMICAL HAZARDS

INTRODUCTION

Hazard assessment (or analysis) includes the identification, evaluation, and mitigation (or control) of hazards.

The process of hazard identification involves the identification of (a) potential sources of hazards and (b) types (categories, classes) of hazards. Sources of hazards are inclusive of a wide range of structures, materials, operations, activities, circumstances, and phenomena, including (but not limited to) such sources as those included in Table 4.1 and Figure 4.1. Types of hazards include specific physical, chemical, and biological agents (see Figure 4.2) that, regardless of their source or the circumstance in which they are encountered, present specific categories of safety and health risk.

Throughout the world, industrial facilities are under increasing regulatory control regarding the types of hazards that typically are associated with particular types of industrial activities, such as the hazards associated with the industrial use of chemicals. Agencies and organizations having responsibility for responding to emergencies also give particular attention to the types of hazards they are likely to encounter in their role as responders. In some instances, the range of hazards likely to be encountered is comprehensive, as in the case of a fire-fighting company; in others, the range may be considerably more narrow, as in the case of a specialized water rescue team. The range of hazards to be associated with either a particular industrial facility or the typical activities of an emergency response organization does not in itself, determine the actual risk presented to employees, response personnel, or the general public. Rather, the wider the range of potential hazards, the more comprehensive must be the planning effort devoted to maximizing both the day-to-day control of those hazards and, in the event of an incident, an efficient and effective emergency response. This is particularly the case at the municipal level, which, after all, is typically the level that is inclusive of all the diverse hazards associated with

TABLE 4-1

TYPICAL EXAMPLES OF COMMUNITY HAZARDS (ADAPTED FROM U.S. FIRE ADMINISTRATION, 1995: TECHNICAL RESCUE PROGRAM DEVELOPMENT MANUAL [FA-159])

Source of Risk	Potential Hazards
Sewers	Confined spaces; toxic gases; oxygen deficiency
Rivers/flood ducts, flood-prone areas	Swift water rescue; calm water rescue; toxic water environments; surface and underwater rescue; ice rescue
Industrial facilities	Hazardous materials; toxic gas emissions; confined spaces; machinery entrapment
Cliffs/gorges/ravines/mountains	Above grade and below grade rescue
Agricultural facilities	Dust explosions; confined spaces, hazardous materials; fertilizers; machinery entrapment
Cesspools/tanks	Toxic gases; oxygen deficiency; confined spaces
New construction	Structural collapse; trench rescue; machinery entrapment
Old buildings	Structural collapse
Wells/caves	Confined spaces; hazardous environments
High-rises	High angle rescue; elevator rescue
Earthquakes/hurricanes/tornadoes	Collapse rescue; extrication; disaster response
Transfer facilities	Hazardous materials; toxic gas emissions; confined spaces, machinery entrapment
Transportation centers	Hazardous materials; toxic gas emissions; confined spaces; machinery entrapment; derailment

different industries, the diverse needs of local response services that must meet all contingencies, and, in addition, all nonindustrial hazards that put the public as well as response personnel at risk.

Given the diversity of hazard sources (see Image 4.1) as well as types of physical, chemical, and biological hazards, there can be no question that the comprehensive identification of hazards, which is the first step toward the effective control of those hazards, must be achieved through a partnership of facility owners and operators, emergency response authorities, and municipal authorities. In the United States, such a partnership is the basic objective of the Federal Emergency and Community Right-to-Know Act of 1986 (EPCRA: SARA Title III).

STRUCTURAL, MATERIAL, AND OPERATIONAL SOURCES OF HAZARDS

Whether at a facility or municipal level, it is useful to approach hazard identification by focusing first on particular structures, materials, and operations that may present specific types of risks in a variety of different circumstances.

Structural elements of any facility (e.g., general layout, floor plans, HVAC and sprinkler system, construction materials, engineering specification) should be examined from the perspective of their becoming not only a source of

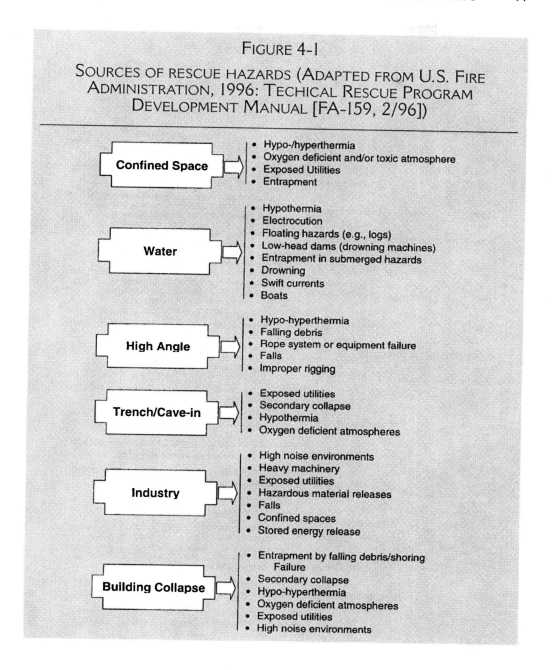

FIGURE 4-1

SOURCES OF RESCUE HAZARDS (ADAPTED FROM U.S. FIRE ADMINISTRATION, 1996: TECHICAL RESCUE PROGRAM DEVELOPMENT MANUAL [FA-159, 2/96])

Confined Space
- Hypo-/hyperthermia
- Oxygen deficient and/or toxic atmosphere
- Exposed Utilities
- Entrapment

Water
- Hypothermia
- Electrocution
- Floating hazards (e.g., logs)
- Low-head dams (drowning machines)
- Entrapment in submerged hazards
- Drowning
- Swift currents
- Boats

High Angle
- Hypo-hyperthermia
- Falling debris
- Rope system or equipment failure
- Falls
- Improper rigging

Trench/Cave-in
- Exposed utilities
- Secondary collapse
- Hypothermia
- Oxygen deficient atmospheres

Industry
- High noise environments
- Heavy machinery
- Exposed utilities
- Hazardous material releases
- Falls
- Confined spaces
- Stored energy release

Building Collapse
- Entrapment by falling debris/shoring Failure
- Secondary collapse
- Hypo-hyperthermia
- Oxygen deficient atmospheres
- Exposed utilities
- High noise environments

hazard to persons who work or live inside the facility (employees, tenants), but also to emergency response personnel who may have to respond to an emergency in that facility. Just how structural elements may contribute to an actual emergency depends, of course, on circumstances, including not only normal circumstances, but also floods, fire, earthquake, and terrorist attack.

For example, if a building is constructed down-slope of unconsolidated (or unprotected) soils, design specifications may or may not be sufficient to prevent building collapse as a result of mud flow. If significant mud flow is not a possibility, heavy rainfall and subsequent flooding may result in an explosion due to the location of nonflood-protected ground-level or subterranean storage

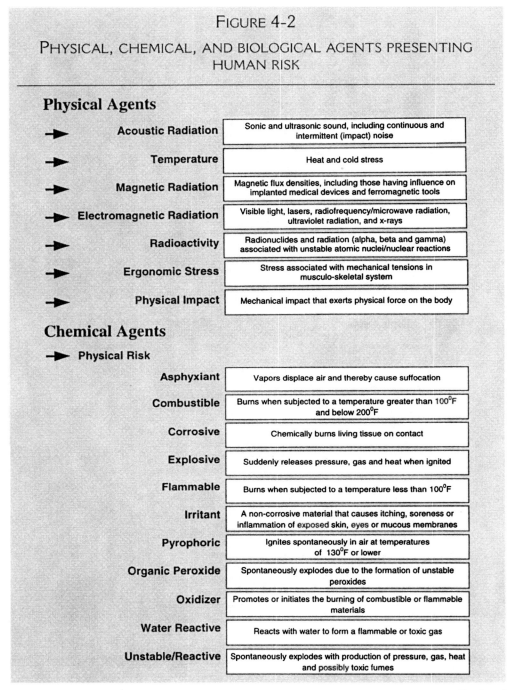

FIGURE 4-2

PHYSICAL, CHEMICAL, AND BIOLOGICAL AGENTS PRESENTING HUMAN RISK

Physical Agents

Acoustic Radiation	Sonic and ultrasonic sound, including continuous and intermittent (impact) noise
Temperature	Heat and cold stress
Magnetic Radiation	Magnetic flux densities, including those having influence on implanted medical devices and ferromagnetic tools
Electromagnetic Radiation	Visible light, lasers, radiofrequency/microwave radiation, ultraviolet radiation, and x-rays
Radioactivity	Radionuclides and radiation (alpha, beta and gamma) associated with unstable atomic nuclei/nuclear reactions
Ergonomic Stress	Stress associated with mechanical tensions in musculo-skeletal system
Physical Impact	Mechanical impact that exerts physical force on the body

Chemical Agents

Physical Risk

Asphyxiant	Vapors displace air and thereby cause suffocation
Combustible	Burns when subjected to a temperature greater than 100^0F and below 200^0F
Corrosive	Chemically burns living tissue on contact
Explosive	Suddenly releases pressure, gas and heat when ignited
Flammable	Burns when subjected to a temperature less than 100^0F
Irritant	A non-corrosive material that causes itching, soreness or inflammation of exposed skin, eyes or mucous membranes
Pyrophoric	Ignites spontaneously in air at temperatures of 130^0F or lower
Organic Peroxide	Spontaneously explodes due to the formation of unstable peroxides
Oxidizer	Promotes or initiates the burning of combustible or flammable materials
Water Reactive	Reacts with water to form a flammable or toxic gas
Unstable/Reactive	Spontaneously explodes with production of pressure, gas, heat and possibly toxic fumes

Continued

areas used for storing water-reactive chemicals. If glass is a substantial component of structural design, that glass presents clear categorical risk to employees during an explosion, as well as to firefighters during fire-fighting operations.

Are stairwells kept at positive atmospheric pressure to ensure their effectiveness as smoke-free evacuation routes during fire? What state-of-the-art design and engineering features are employed to minimize catastrophic releases of

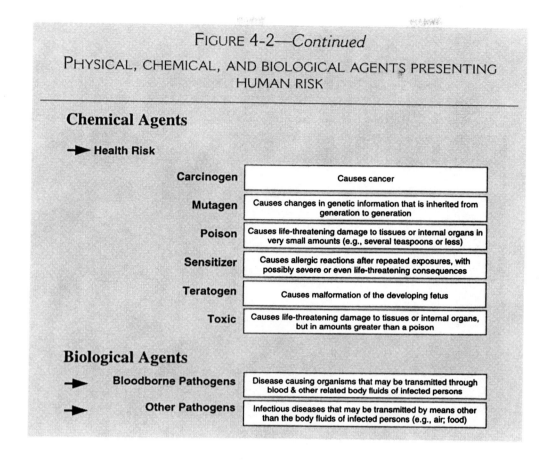

FIGURE 4-2—*Continued*

PHYSICAL, CHEMICAL, AND BIOLOGICAL AGENTS PRESENTING HUMAN RISK

Chemical Agents

➤ Health Risk

Carcinogen	Causes cancer
Mutagen	Causes changes in genetic information that is inherited from generation to generation
Poison	Causes life-threatening damage to tissues or internal organs in very small amounts (e.g., several teaspoons or less)
Sensitizer	Causes allergic reactions after repeated exposures, with possibly severe or even life-threatening consequences
Teratogen	Causes malformation of the developing fetus
Toxic	Causes life-threatening damage to tissues or internal organs, but in amounts greater than a poison

Biological Agents

➤ **Bloodborne Pathogens**	Disease causing organisms that may be transmitted through blood & other related body fluids of infected persons
➤ **Other Pathogens**	Infectious diseases that may be transmitted by means other than the body fluids of infected persons (e.g., air; food)

hazardous fuels (e.g., natural gas, oil) during an earthquake, or of other hazardous chemicals in even normal circumstances? How do design features make maximum use of fire- and toxic-rated materials to minimize the spread of fire and toxic fumes? In congested industrial areas (e.g., seaports; industrial parks; mixed industrial, commercial, and residential areas), how do facility layout and transportation corridors (ground, water, and air) facilitate or prevent rapid access by community response authorities?

Several guidelines are well worth considering in the process of identifying potential hazards related to structures and their design features:

1. It is clearly not sufficient to disregard any potential hazard simply because a particular structural or design element meets the requirements of a particular legal building or construction code. After all, an appropriate code may be lacking or, if legally pertinent, such a code may not be as effective as state-of-the-art practice and technology. The basic rule must be: Regulatory compliance is always a *de minimus* requirement when it comes to health and safety.

2. Structural and design elements should be planned and implemented only after consideration of appropriately defined worst-case circumstances that, at a minimum, include fire, flood, storm, power-outage, earthquake, and terrorist attack.

IMAGE 4-1 (SEE PAGE 96)

WEST GLENWOOD, CO, JUNE 16, 2002: HELITACK AND HOT SHOT CREWS CONTINUE TO EXTINGUISH SPOT FIRES IN THE FORESTS OF NO NAME CREEK ABOVE GLENWOOD SPRINGS

Source: Andrea Booher/FEMA News Photo

3. Owners, operators, and other parties having legal responsibility for a facility should understand that, regardless of any regulatory rule, there can be no excuse for failing to coordinate with and seek the advice of local emergency response authorities regarding the structural design of that facility, including both original plans and any subsequent modifications.

Materials to be considered in any comprehensive identification of potential hazards typically include (beyond structural materials considered earlier) materials serving as feedstock to industrial and commercial processes, by-products of operational processes, and final process wastes.

Increasingly, global attention has begun to focus on what is commonly called the *product cycle*, which is the totality of material and energy transformations that take place over the production, use, and final disposal of manufactured goods. It should therefore be no surprise that the comprehensive identification of hazards is increasingly presumed to be inclusive of those hazards associated not only with process feedstock, process by-products, and process wastes, but also with so-called "finished goods" as well as any material resultant from the use and environmental degradation of those goods. The assessment of new pharmaceutical products in the United States under the National Environmental Policy Act (NEPA), for example, requires a comprehensive evaluation of potential hazards associated with the metabolic decomposition of drugs, which materials may be released to the environment in human feces and urine.

Given the fact that a technologically developed society depends upon the daily industrial and commercial use of roughly 60,000 chemicals, the identification of hazards related to these individual chemicals and to the multitude of their combinations is certainly a daunting and demanding task, but it is a task that is nonetheless required under a constantly expanding number of laws, regulations, and standards implemented not only by national and local governments, but also by international organizations.

It is a mistake for municipal managers and community response services to assume that manufacturing industries are the only potential source of hazardous chemicals. For example, hazardous chemicals may be released into the community during various phases of new development or urban renewal projects, including property management phases prior to construction (e.g., demolition of existing structures that may contain hazardous chemicals as structural components), construction (e.g., on-site storage and preparation of materials and supplies, placement of contaminated fill, fugitive dusts that contain chemical contaminants), and operational and maintenance phases (e.g., runoff from operational site, application of pesticides).

Given the variety of potential sources of hazardous chemicals and materials, it is necessary to collate data and information from different sources, including:

- Chemical inventories of industrial and commercial facilities (including feedstock, process, and waste chemicals)
- Records of previous land use that might be used to indicate chemical contaminants in soils or downstream wetland and lentic muds
- Records of previous incidents involving chemical releases in specific areas (including records maintained by regulatory agencies and/or community response services)
- Data and information that may be available through scientific surveys of local or regional resources (including surveys conducted by corporations and other land buyers to identify potential contamination of on-the-market parcels)

Operations that must be considered in any assessment of potential hazards include all on- and off-site operations (whether industrial or commercial) that involve the transportation, delivery, storage, handling, processing, or disposal of hazardous chemicals and materials.

Although chemicals and materials typically are described as being hazardous by different regulatory agencies using diverse criteria and standards (and having different jurisdictional objectives), it is important to understand that all chemicals are intrinsically hazardous, and that the actual degree of risk presented by any hazard depends on a number of factors, including (but not limited to) the magnitude of exposure (dose), the route by which the chemical enters the body (route of entry), and the sensitivity of the person who is exposed (hypersensitive or hyposensitive). For purposes of identifying hazards that might trigger an emergency response, it is therefore inexcusable to dismiss

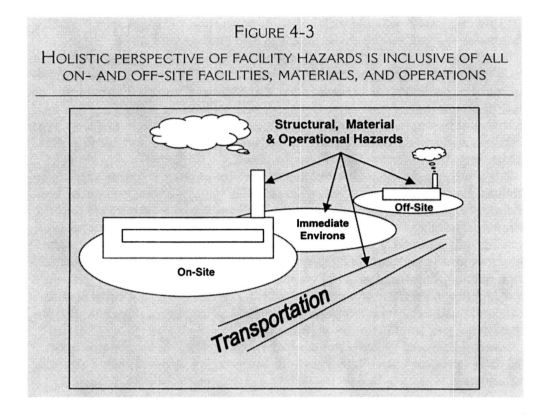

FIGURE 4-3

HOLISTIC PERSPECTIVE OF FACILITY HAZARDS IS INCLUSIVE OF ALL
ON- AND OFF-SITE FACILITIES, MATERIALS, AND OPERATIONS

as unimportant any chemical or material simply because it does not have a regulation-based designation (e.g., a *hazardous material* [under the U.S. Department of Transportation regulations pursuant to the Hazardous Material Transportation Act] or a *hazardous waste* [under the U.S. Environmental Protection Agency regulations pursuant to the Resource conservation and Recovery Act]).

In light of these considerations, the assessment of structural, material, and operational hazards for the purpose of identifying potential hazards that may result in an emergency must be holistic—inclusive of not only industrial and commercial facilities, but also any activity (e.g., land clearing, delivery of feedstock chemicals) or circumstance (e.g., flood) that can trigger or otherwise exacerbate any emergency involving either or all; inclusive also of not only on-site features of any particular facility or activity, but also all off-site subsidiary or complementary activities and services related (directly or indirectly) to that facility, whether in the immediate environs or distantly located (see Figure 4.3).

HAZARD EVALUATION AND MITIGATION

Standard procedures for analyzing hazards (including those discussed in Chapter 2) are listed in Figures 4.4 and 4.5, which summarize basic attributes. It must be emphasized that the indicated procedures are representative only of

FIGURE 4-4

COMPARISON OF COMMON PROCESS SAFETY ASSESSMENT
TECHNIQUES: BASIC TYPE OF ANALYSIS AND USEFULNESS FOR
PRIORITIZING AND MITIGATING HAZARDS

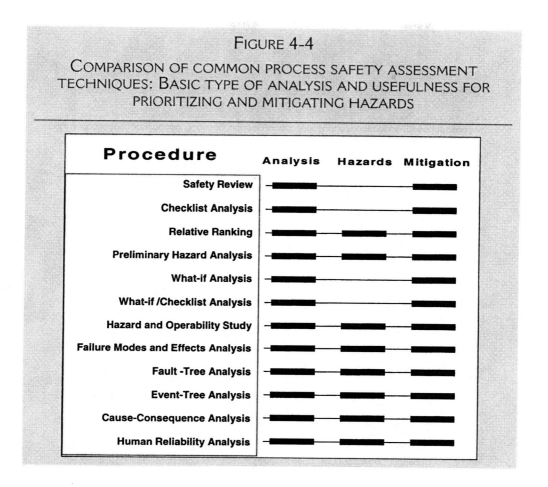

many techniques in common use; new procedures are continually under development. Moreover, because some are more demanding than others (in terms of time, skills of personnel, required data and information bases, and the complexity of protocols), there continues be very active development of commercially available software (especially for Failure Modes and Effects Analysis, and Fault- and Event-Tree Analysis) to facilitate their use.

The early development of most of these procedures was greatly influenced in the United States by the Chemical Process Regulations (29 CFR 1910.119). Such hazard analysis procedures are therefore sometimes referred to as *process hazard analysis procedures*. As this name implies, the primary focus of these procedures is to assess and minimize the potential for a catastrophic release of known hazardous chemicals, such as occurred at the Union Carbide pesticide plant in Bhopal, India.

The achievement of this objective is vital to the proactive phase of emergency response planning, but it must be made clear that analytical techniques used to assess the potential for catastrophic releases in the chemical industry are not sufficient to provide all the types of hazard assessment required by comprehensive emergency response planning. For example, process hazard analysis typically does not define just what the health and safety hazards of a released chemical

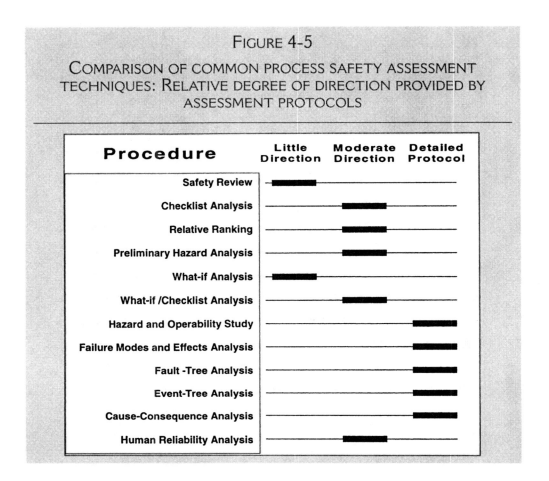

FIGURE 4-5

COMPARISON OF COMMON PROCESS SAFETY ASSESSMENT TECHNIQUES: RELATIVE DEGREE OF DIRECTION PROVIDED BY ASSESSMENT PROTOCOLS

are but, rather, only the sequence of events in the processing of chemicals that may lead to release. They do not define the actual health and safety probabilities attendant to chemical exposures but, rather, the process-related probabilities of equipment or human failure that will lead to human exposure.

Another common approach to hazard assessment, and one that is to be clearly distinguished from a "process hazard analysis," generally is referred to as Job-Task Analysis. This type of analysis (see Figure 4.6), which was originally given major impetus in the United States by hazardous waste regulations under RCRA and, soon thereafter, by workplace health and safety regulations under OSHA, involves the resolution of each workplace job into specifically defined individual tasks. Each task requirement is then assessed with respect to health and safety risks likely to be encountered, with subsequent identification and evaluation of alternative means of reducing risks.

The focus of Job-Task Analysis is clearly on (a) the manner in which the worker is exposed to a hazard, (b) the risk to that specific worker as a consequence of exposure to the hazard, and (c) means for mitigating that risk. This approach to hazard assessment is a basic tool in any modern industrial program of occupational safety and health; it is also particularly relevant to corporate and municipal authorities having primary responsibility for emergency

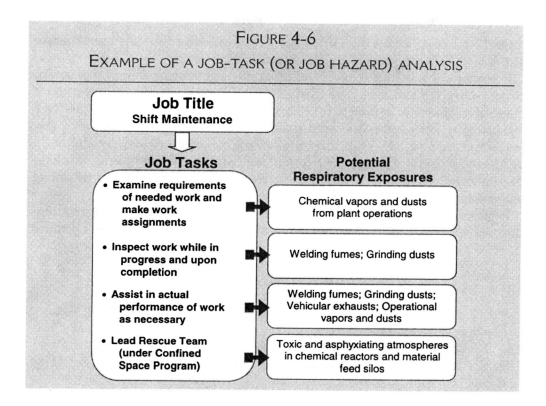

FIGURE 4-6

EXAMPLE OF A JOB-TASK (OR JOB HAZARD) ANALYSIS

Job Title
Shift Maintenance

Job Tasks

- Examine requirements of needed work and make work assignments
- Inspect work while in progress and upon completion
- Assist in actual performance of work as necessary
- Lead Rescue Team (under Confined Space Program)

Potential Respiratory Exposures

Chemical vapors and dusts from plant operations

Welding fumes; Grinding dusts

Welding fumes; Grinding dusts; Vehicular exhausts; Operational vapors and dusts

Toxic and asphyxiating atmospheres in chemical reactors and material feed silos

response planning because it requires that careful attention be given to each phase of hazard assessment (i.e., hazard identification, hazard evaluation, and mitigation) at the level of "human risk" rather than, as with process hazard assessment techniques, at the level of "process risk." This is not to say that process risk assessment is less important to emergency response planning than human risk assessment. Both are necessary; neither alone is sufficient.

Scientific Basis of Chemical Hazard and Risk Assessment

As a potential harm or injury, any hazard associated with a chemical is intrinsic to that chemical and cannot be altered. Any chemical may be associated with several or more hazards.

Each of the roughly 60,000 chemicals in daily commercial use in technologically developed nations may be considered hazardous because, depending on the degree and nature of human exposure to it, each may result in harm to the exposed person. Certainly, some chemicals are more hazardous than others; the more hazardous are most often listed by regulatory agencies as requiring special attention regarding their handling, shipping, and storage.

As shown in Figure 4.2, chemical hazards often are categorized as physical or health hazards. Physical hazards are those that result in physical injury to the exposed person; health hazards are those that result in physiological injury. Specific types of physical or health hazards are indicated by the *hazard class* of a chemical.

Risk results from exposure to hazard. Generally, risk increases with exposure; the greater the exposure, the greater the risk. However, it is important to distinguish between physical and health risks.

Essentially a statistical concept, the risk of experiencing injury or harm to health reflects the fact that individuals within a population demonstrate a range of tolerance with respect to chemical exposure. Some individuals are very sensitive (hypersensitive) to certain chemicals, whereas others are relatively insensitive (hyposensitive). Contrarily, the risk of experiencing physical harm or injury as a result of exposure (for example, to a very strong acid or alkali) does not reflect a range of biological tolerance to chemical burns among humans but, rather to the statistics of accidents.

Figure 4.7 depicts an example of a *dose-effect* (or *dose-response*) relationship between the *dose* of a toxic chemical (express as weight of the chemical

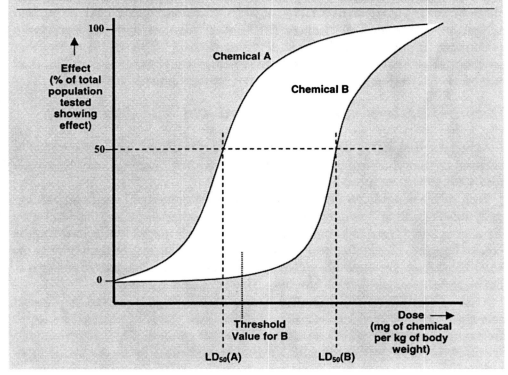

FIGURE 4-7

DOSE–EFFECT (RESPONSE) RELATIONSHIPS FOR TWO CHEMICALS. NOTE THAT SOME EFFECT (I.E., LETHALITY) IS SEEN FOR EVERY INCREMENTAL INCREASE IN DOSE OF CHEMICAL A, WHEREAS INCREASE IN DOSE OF CHEMICAL B RESULTS IN EFFECT ONLY AFTER SOME THRESHOLD VALUE OF DOSE. CHEMICAL A DOES NOT HAVE A THRESHOLD VALUE. THE PERCENTAGE OF POPULATION SHOWING LETHAL EFFECTS MAY ALSO BE READ AS THE PROBABILITY OF DEATH FOR AN INDIVIDUAL (E.G., 50% OF THE POPULATION IS EQUIVALENT TO A 0.5 PROBABILITY FOR ANY INDIVIDUAL)

per body weight; e.g., mg/kg) and the probability of a particular *effect* (e.g., lethality). Dose-effect relationships may be established on the basis of laboratory experiments with animals, and may also be inferred on the basis of epidemiological studies of humans. In laboratory studies, the experimenter controls the exposure of the test organisms to the toxic chemical. In epidemiological studies, exposures are inferred from information about workplace and other exposures of humans to the chemical of interest; that is, exposures are not controlled by the experimenter but by the life-experience of the persons included in the study.

As shown in Figure 4.7, the LD_{50} (Lethal Dose) represents the dose at which 50% of the test organisms (of the same species) exposed to that dose are expected to die. This statistic essentially states that any one organism exposed to that dose has a 50/50 chance (or 0.5 probability) of dying. If some health effect of exposure other than lethality is of interest (e.g., loss of hair, rapid heart beat), an ED_{50} (Effective Dose) can be similarly determined.

LD_{50} and ED_{50} data are based on studies in which the chemical actually is introduced into the organism (e.g., through inhalation or infection); the dose therefore defines that introduced amount. In many instances, the concentration of the toxic chemical in the atmosphere or water in which the test organism lives or functions (i.e., ambient concentration) is known, but the amount actually taken into the organism is unknown. In such cases, LC_{50} (Lethal Concentration) and EC_{50} (Effective Concentration) are used to denote, respectively, the 50/50 probability of lethality or other health effect.

LD_{50}, LC_{50}, ED_{50}, and EC_{50} values are very useful for defining the relative toxicities of different chemicals. For example, Table 4.2 includes LD_{50} values and commonly used categories of relative toxicity. Although these terms are in general use, LD_{50} values do have important limitations when comparing the

TABLE 4-2
COMMONLY USED TERMS AND CRITERIA THAT DESCRIBE RELATIVE TOXICITY

Relative Toxicity	LD_{50} (mg/kg)[1]	Lethal Amount[2]	Examples of Chemicals[3]
Extremely Toxic (Poison)	< 1	< 7 drops	Dioxin Botulinus toxin Tetrodotoxin
Highly Toxic (Poison)	1 - 50	7 drops -1 teaspoon	Hydrogen cyanide Nickel oxide Arsenic trioxide
Very Toxic	50 - 500	1 teaspoon -1 ounce	Methylene chloride Phenol DDT
Moderately Toxic	500 - 5000	1 ounce -1 pint	Benzene Chloroform Chromium chloride
Slightly Toxic	> 5000	> 1 pint	Acetone Ethyl alcohol Ferrous sulfate

1. As tested by the oral route in rats
2. Lethal amount for average adult human, based on liquid with density of water
3. As tested by various routes in several animal species

toxicity of two or more chemicals. For example, Figure 4.8 shows the straight-line portions of the dose-effect curves for two different chemicals. Note that, although both chemicals have equal LD_{50}s, increasing the dose of chemical B results in a smaller incremental increase in risk than does increasing the dose of chemical A.

The dose of a chemical received as a result of exposure is of paramount importance with respect to the health hazard of a chemical. However, it can be irrelevant in certain circumstances. For example, once allergic to a particular chemical, a person can experience a life-threatening episode upon even the most minuscule exposure to that chemical (i.e., an allergen). Also, no well-defined relationship exists between the dose of carcinogens (causing cancer), mutagens (causing mutations), and teratogens (causing disruption of fetal development) and the risk of actually experiencing their respective hazards.

FIGURE 4-8

DOSE–EFFECT CURVES FOR TWO CHEMICALS WITH THE SAME LD_{50}. NOTE THAT INCREMENTAL INCREASES IN DOSE RESULT IN GREATER INCREASES IN THE PERCENTAGE OF POPULATION AFFECTED FOR CHEMICAL A THAN FOR CHEMICAL B. AT DOSES LESS THAN THE LD_{50}, CHEMICAL B IS MORE POTENT THAN CHEMICAL A; AT DOSES GREATER THAN THE LD_{50}, CHEMICAL A IS MORE POTENT THAN CHEMICAL B

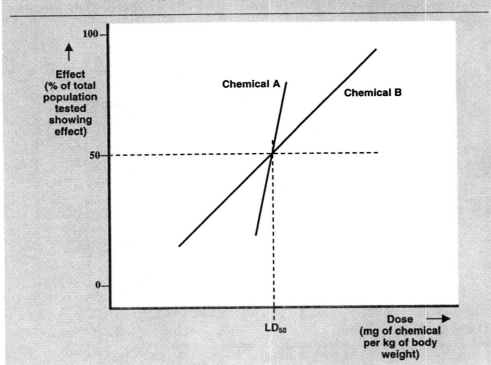

Acute and Chronic Effects

The various effects of chemical exposure may be described as *acute* or *chronic* effects. Acute effects are those that occur very quickly (e.g., minutes, hours, or days) after exposure to the causative chemical agent. Asphyxiants, explosives, pyrophorics, organic peroxides, water reactive and unstable/reactive chemicals, corrosives, and poisons typically produce acute effects. Chronic effects are those that occur only after long periods of time after exposure (e.g., years, decades), including effects of sensitizers and carcinogens. Many lung cancers related to exposure to asbestos or cigarette smoke, for example, develop only several decades after exposure.

Because chronic effects become apparent only over extended periods of time, they are particularly difficult to relate to a specific exposure to a particular chemical. As a result, our current state of knowledge regarding the potential effects of the more than 60,000 chemicals in daily commerce tends to be much more extensive with respect to their acute effects than their chronic effects. This state of affairs, however, reflects only the methodological difficulties involved in the scientific investigation of chronic effects, and does not imply that chronic effects should be of less concern. In fact, the paucity of scientific data regarding the chronic effects of exposure to hazardous chemical is sufficient reason to be particularly wary regarding the potential for chronic effects, especially with regard to periodic exposures experienced by community emergency response service personnel.

The phrase *target organ effects* often is used to specify the particular organs, tissues, cells, and physiologically important systems that are particularly affected by a specific hazardous chemical, regardless of whether the effects are acute or chronic. Target organs typically affected by a wide range of common commercial/industrial chemicals include the skin, eye, mucous membranes, respiratory tract, lungs, liver, kidney, reproductive system, and central nervous system.

Dimensions of Exposure

The term "exposure" denotes some measure of the amount of chemical to which a person may be subjected, either as a dose that enters the body or, as is often the case in emergency response, as a concentration in the ambient atmosphere or water (see Image 4.2). However, the risk of experiencing a health effect is not simply a function of dose or concentration. Other facts that influence risk include age, gender, general state-of-health, lifestyle, and any medications (or drugs) a person may use.

In some instances, one or more of these exposure-related factors may dramatically increase the risk of exposure to a hazardous chemical. For example, smoking cigarettes imparts a certain risk to a person of developing lung cancer; so does exposure to asbestos. However, the combination of these two factors results in a substantially greater risk than is imparted by either factor alone. The interaction of two or more chemicals to multiply the risks associated with any one chemical is an example of the phenomenon of *synergy*.

IMAGE 4-2

AN OSHA EMPLOYEE INSTRUCTS A NEW YORK POLICE OFFICER IN THE USE OF RESPIRATORY PROTECTION AT THE WORLD TRADE CENTER SITE. DURING THE FIRST TWO MONTHS OF THE RECOVERY EFFORT, OSHA DISTRIBUTED ABOUT 110,000 RESPIRATORS, CONDUCTED QUANTITATIVE FIT-TESTING, AND INSTRUCTED WEARERS IN HOW TO USE RESPIRATORS

Source: Donna Miles/OSHA News Photo

Synergistic effects should be of particular concern to emergency response service personnel who, given the nature of their work, are likely to experience periodic exposures to a wide range of chemicals over a long period of time.

Another important factor that can directly influence risk is the particular means (*route of entry*) by which a chemical comes into contact with a person, including:

- Inhalation
- Ingestion (contamination of food and/or hands used to prepare or consume food; nasal drippings that have become contaminated through inhalation)
- Skin or eye contact (where the chemical action is at the surface)
- Absorption (through intact skin, eye, or mucous membranes; no physical lesions are necessary)
- Puncture (or injections)

Not all chemicals can enter the body through all possible routes of entry. Most chemicals do, however, enter the body through two or more routes.

The toxicity of many chemicals is greater or lesser (i.e., in terms of the dose required to produce certain health effects) depending on the specific chemical and its specific route of entry. For example, a toxic chemical that is ingested will often have a higher LD_{50} (i.e., be less toxic on a per-dose basis) than the same chemical injected directly into the bloodstream. Knowing the various routes of entry for different hazardous chemicals is important because, by blocking those routes of entry through the appropriate use of personal protective clothing and equipment (e.g., impervious gloves; respirators), we can effectively prevent exposure and thereby minimize risk.

Environmental Transport and Transformation

Once a chemical enters the environment, it is subject to a variety of mechanisms that transport it from place to place and from one environmental medium to another (e.g., soil to air or air to water). During transport, a chemical may also undergo transformation because of dynamic physical, chemical, and biological processes (e.g., combustion; decomposition). Some environmentally mediated transformations of chemicals can result in the production of a chemical (or chemical by-product) that is more toxic than the original chemical.

The environmental transport and transformation of a chemical often is referred to as the *environmental fate* of that chemical. Computerized multimedia environmental models describe the environmental fate of chemicals and are increasingly available. These models are important for calculating the probable concentrations of different chemical species in different environmental media that may be expected as a result of chemical release to the environment. Computerized models are also particularly useful to emergency response personnel insofar as they have the capacity to predict atmospheric flows of chemical plumes under a variety of ambient conditions. Such models most often are called air (or water) *dispersion models*.

CONSOLIDATION OF RELEVANT TECHNICAL INFORMATION

There is no question that there are vast amounts of information and data regarding hazard assessment that are readily available through governmental agencies (e.g., Nuclear Regulatory Commission, National Institute of Occupational Safety and Heath, the U.S. Fire Administration), private organizations (e.g., CHEMTREC) and electronic network services (e.g., National Pesticide Telecommunications Network), corporate regulatory compliance plans (e.g., Hazard Communication Plan), chemical manufacturers (e.g., Material Safety Data Sheets), and, of course, commercial publications. However, the ready availability of printed or electronic information and data does not in any way guarantee the practical application of that information and data to the manifold needs of effective emergency response planning.

Whether at the national, regional, municipal, or corporate level, relevant data and information must not only be collected but, most importantly, be organized into useful formats and translated into protocols that can be used directly to meet the urgent needs of site-specific emergency response. It is not, after all, the amount of information and data on hand that is important for preventing, planning for, or responding to an emergency; rather, what is important is precisely how specific information and data are consolidated into actual proactive and reactive practice (see Image 4.3).

For example, Figure 4.9 is a summary of information related to the in-plant inventory of chemicals in a small manufacturing company. This information is consolidated into a computerized database (of roughly 1500 chemicals, including feedstock, by-products, and wastes) that serves as the basis of corporate policy, operational procedures, and all personnel training related to occupational health and safety, including both in-house and community emergency response services.

There are many readily available, excellent guides that are very useful as generic tools (see Figure 4.10) for implementing effective emergency prevention, planning, and response programs, but there can be no substitute for those tools specifically fashioned to meet facility-specific, place-specific, and circumstance-specific emergency response needs.

IMAGE 4-3

AN IRON WORKER CUTS UP HUGE PIECES OF TWISTED STEEL FOR REMOVAL FROM THE WORLD TRADE CENTER SITE. OSHA WORKED CLOSELY WITH CONTRACTORS AT THE SITE TO ENSURE WORKERS WERE FAMILIAR WITH POTENTIAL AIR CONTAMINANT HAZARDS, PERSONAL PROTECTIVE EQUIPMENT REQUIREMENTS, AND OVERALL SAFETY RULES

Source: Shawn Moore/OSHA News Photo

FIGURE 4-9

EXAMPLE OF CONTENTS OF A COMPUTERIZED INDUSTRIAL CHEMICAL DATABASE USED TO INFORM AND DIRECT ALL DECISION MAKING REGARDING PERSONNEL HEALTH AND SAFETY PROGRAMS, INCLUDING THE EMERGENCY RESPONSE PROGRAM

Chemical Data Base

- Common synonyms
- Hazards
- Routes of entry
- Target organs
- Acute effects
- Chronic effects
- Health & safety standards
- Symptoms of exposure
- Protective clothing
- Protective equipment
- Disposal requirements
- Chemical compatibility
- Persistence in environment
- Storage requirements
- Proper labeling

- Environmental transport
- Environmental transformations
- Emergency response actions
- Ambient monitoring
- Personal monitoring
- Medical surveillance
- Medical treatment
- Environmental remediation measures
- Combustion products
- Personal Risk factors
- Health and environmental synergisms

STUDY GUIDE

True or False

1. In an industrial/commercial facility, effective control of hazards depends upon a partnership between owners/operators of the facility, emergency response authorities, and municipal authorities.
2. Facility structures, materials, and operations present specific kinds of risks not only to workers, but also to emergency responders.
3. Given the great diversity of industrial/commercial facilities, it is almost impossible to construct an overview of the types of hazards that might be experienced by emergency responders.
4. Hazards associated with structural components of a facility depend upon the specific environmental circumstances that prevail at the time of an actual emergency.

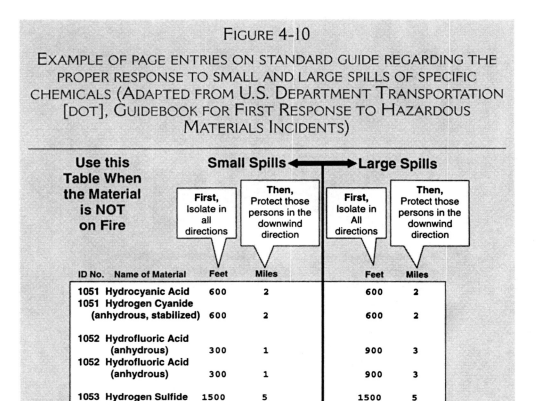

FIGURE 4-10

EXAMPLE OF PAGE ENTRIES ON STANDARD GUIDE REGARDING THE
PROPER RESPONSE TO SMALL AND LARGE SPILLS OF SPECIFIC
CHEMICALS (ADAPTED FROM U.S. DEPARTMENT TRANSPORTATION
[DOT], GUIDEBOOK FOR FIRST RESPONSE TO HAZARDOUS
MATERIALS INCIDENTS)

Use this Table When the Material is NOT on Fire	Small Spills		Large Spills	
	First, Isolate in all directions	Then, Protect those persons in the downwind direction	First, Isolate in All directions	Then, Protect those persons in the downwind direction
ID No. Name of Material	Feet	Miles	Feet	Miles
1051 Hydrocyanic Acid	600	2	600	2
1051 Hydrogen Cyanide (anhydrous, stabilized)	600	2	600	2
1052 Hydrofluoric Acid (anhydrous)	300	1	900	3
1052 Hydrofluoric Acid (anhydrous)	300	1	900	3
1053 Hydrogen Sulfide	1500	5	1500	5

5. Hazardous materials associated with any facility include not only chemical feedstock to industrial processes, but also by-products of operational processes and final process wastes.

6. The product cycle is the totality of material and energy transformations that take place in the production, use, and final disposal of manufactured goods.

7. Any facility audit of hazardous materials must include both on- and off-site operations that involve the transportation, delivery, storage, handling, processing, or disposal of hazardous chemicals and materials.

8. Typically, a single chemical is associated with a single hazard.

9. A hazard is a potential type of harm or injury; a risk is the probability that an individual (or group) in fact will experience that harm or injury.

10. The same hazard will present the same risk to all exposed persons.

11. There is little scientific basis for comparing the relative toxicity of different chemicals.

12. Chronic effects of chemicals are those that occur only after long periods of time (e.g., years) after initial exposure.

13. Human exposure to a chemical hazard is measured only in terms of the amount of chemical to which one is exposed.

14. The way a chemical enters the body (route of entry) can directly affect the degree of risk.

15. The environment can cause changes in chemicals that may be more dangerous to human health than the original chemical.

Multiple Choice

1. An oxidizer is a chemical that
   A. chemically burns living tissue on contact
   B. reacts with water to form a flammable or toxic gas
   C. promotes or initiates the burning of combustible or flammable materials
   D. spontaneously explodes.
2. The dose needed to produce a certain effect is the basic difference between a poison and a
   A. toxic chemical
   B. irritant
   C. pyrophoric substance
3. A threshold value is the dose (or concentration) below which
   A. all effects are lethal
   B. no effects can be detected
   C. 50% of the exposed population will die
4. The dose-response relationship demonstrates that the risk of an adverse effect of a chemical exposure increases with
   A. age of the exposed person
   B. dose of the chemical
   C. frequency of exposure to the chemical
5. One chemical may be more or less potent than another chemical, depending upon their
   A. ED_{50}
   B. LD_{50}
   C. LD_{50} and dose
6. Factors other than the dose of chemical that can significantly influence the effects of exposure to that chemical include
   A. age
   B. gender
   C. general state of health
   D. lifestyle
   E. medications used by the exposed person
   F. all of the above
7. Synergistic effects of chemicals always involve
   A. exposure to two or more chemicals at the same time
   B. exposure to large doses of a sensitizer
   C. exposure to a chemical through inhalation
8. Dispersion models are useful to emergency responders because they predict
   A. the toxicity of chemicals
   B. the hazards to be associated with chemicals
   C. the movement of chemicals in water or in the atmosphere

Essays

1. As an Incident Commander, what basic types of information would you require your responding team to possess before ordering them to enter a facility that has experienced a chemical spill?

2. As an in-plant safety officer, how would you use the information in a "job task analysis" in the development of an emergency response plan (see Chapter 3)?
3. Discuss the significance of the fact that many chemicals do not have a threshold level.3.

Case Study

Assume that you are a local fire chief in a small community that includes several industrial facilities, including a plastic extrusion company, a small pharmaceutical company, and a furniture manufacturer. None of these firms has provided you with any information concerning the potential hazards they may present to the community at large.

1. What information do you want to have concerning each company?
2. In what format do you want that information?
3. What will you do with that information?

BIOHAZARDS

INTRODUCTION

Over the last three decades of the twentieth century, it became increasingly evident that, in addition to those hazards and risks typically faced by emergency response personnel, much more attention had to be given to the hazards and risks posed by the continually expanding profusion of industrial chemicals. Although this concern will likely remain for the foreseeable future of the twenty-first century, it also became evident by the mid 1980s that biological agents of disease had also become of cardinal importance, not only with respect to the health of the general public, but, more specifically, to those who, by the nature of their occupation, experience routine exposure to blood and other body fluids. Most recently, of course, the risk of biological agents has been underscored by the continuing threat of both domestic and foreign terrorism.

Of immediate concern in 1983, of course, was the abrupt, devastating emergence onto the world scene of HIV (Human Immunodeficiency Virus), the viral agent that may lead to fully developed AIDS (Acquired Immune Deficiency Syndrome) in HIV-infected persons. In the United States, the growing recognition of AIDS as being communicable through exposure to the blood and certain body fluids of infected persons prompted federal enactment of the Health Omnibus Programs Extension Act of 1988, Title II (AIDS Amendments of 1988) as a means of giving specific guidance to health workers, public safety workers, and emergency responders.

Even in the then-absence of precise understanding of the etiology of HIV, the U.S. Department of Health and Human Services (Public Health Service, Centers for Disease Control, and National Institute of Occupational Safety and Health) moved quickly (1989) to issue specific guidelines to reduce workplace risk with regard to not only HIV, but also Hepatitis B virus (HBV). The insightful inclusion of HBV along with HIV in these guidelines (U.S. Department of

Health and Human Service, 1989; Guidelines for Prevention of Transmission of Human Immunodeficiency Virus and Hepatitis B Virus to Health-Care and Public-Safety Workers) was based on the following assumptions:

1. Modes of transmission for HBV are similar to those of HIV.
2. Potential for HBV transmission in the workplace is greater than for HIV.
3. There is much greater experience with controlling workplace transmission of HBV.
4. Practices to prevent the transmission of HBV will also minimize the risk of transmission of HIV.

In keeping with this approach to providing practical guidance for minimizing the risk of infection by a number of diverse pathogens that nonetheless demonstrate similarities in modes of transmission, the U.S. Occupational Safety and Health Administration (OSHA) implemented its own "Bloodborne PATHOGEN Standard" (29 CFR 1910.1030) for American industry in 1992.

Although the recognition of blood-borne pathogens as an occupational risk to any personnel having contact with blood and blood-related body fluids of infected persons is now firmly established, blood-borne pathogens do not exhaust the range of biohazards that may present health risks to emergency responders (see Image 5.1).

With the rapidly growing mobility of individuals and high population densities of urban areas throughout the globe, with the ever-developing industrial and commercial use of biotechnology, and, as a consequence of the misuse of previous pharmaceutical technology, the general public continues to be at highly significant risk due to a host of pathogenic and parasitic organisms, with yearly deaths from infectious and parasitic organisms exceeding, worldwide, the deaths due to all other causes (see Figure 5.1) and accounting even today for at least one-third of all deaths. Of growing worldwide concern are the following:

1. *Pathogenic organisms that (a) have been long existent but equally long isolated in remote areas of the world, or (b) are newly evolved through genetic mutations of existing pathogens.*

   Diseases caused by these pathogens, including at least 30 new disease agents identified in the period from 1973 to 1995, are referred to as *emerging diseases* (see Table 5.1).

   As even the most remote areas of the world become more accessible to world travelers, so does the world itself become more accessible to these pathogens. It might be assumed that emergency response personnel are at no greater risk than the general public, but it also must be emphasized that it is becoming increasingly likely that a major catastrophe (e.g., earthquake, high-rise building collapse), wherever it occurs, will elicit the response of specialized emergency teams from throughout the world.

2. *Pathogenic organisms cultured in medical research and other facilities, including facilities in which genetically engineered organisms are produced.*

IMAGE 5-1

AN OSHA INDUSTRIAL HYGIENIST PREPARES AIR SAMPLES FOR PROCESSING AT OSHA'S SALT LAKE TECHNICAL CENTER. THE LABORATORY PROVIDES 24-HOUR-A-DAY SUPPORT TO ANALYZE AIR AND BULK SAMPLES COLLECTED AT DISASTER SITES

Source: Shawn Moore/OSHA News Photo

There have long been laboratories and specialized storage depots housing pathogenic organisms used for research purposes. Emergency responders therefore must be prepared to deal with biohazard risks associated with such facilities. In addition, a growing number of laboratories and commercial industries involve genetic engineering of potential pathogens—a situation that greatly increases the probability of such facilities becoming involved in catastrophic emergencies. Thus the following considerations become of paramount concern to emergency response planning:

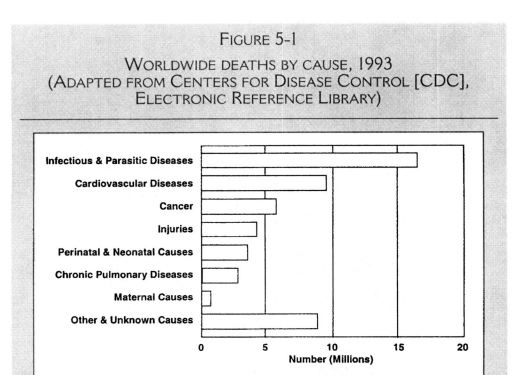

FIGURE 5-1

WORLDWIDE DEATHS BY CAUSE, 1993
(ADAPTED FROM CENTERS FOR DISEASE CONTROL [CDC],
ELECTRONIC REFERENCE LIBRARY)

TABLE 5-1

EXAMPLES OF EMERGENT INFECTIOUS DISEASES RECOGNIZED
SINCE 1973 (ADAPTED FROM CISET, 1997: GLOBAL MICROBIAL
THREATS IN THE 1990S [NATIONAL CENTER FOR INFECTIOUS
DISEASES ELECTRONIC REFERENCE LIBRARY])

Year	Microbe	Type	Disease
1973	Rotavirus	Virus	Major cause of infantile diarrhea
1975	Parvovirus B19	Virus	Aplastic crisis in chronic hemolytic anemia
1976	*Cryptosporidium parvum*	Parasite	Acute and chronic diarrhea
1977	Ebola virus	Virus	Ebola hemorrhagic fever
1977	*Legionella pneumophila*	Bacterium	Legionnaires' disease
1977	Hantaan virus	Virus	Hemorrhagic fever with renal syndrome
1977	*Campylobacter jejuni*	Bacterium	Enterric pathogen
1980	HTLV-1	Virus	T-cell lymphoma-leukemia
1981	Toxic *Staphylococcus aureus*	Bacterium	Toxic shock syndrome
1982	*Escherichia coli* (0157:H7)	Bacterium	Homorrhagic colitis; hemolytic uremic syndrome
1982	HTLV-II	Virus	Hairy cell leukemia
1982	*Borrelia burgdorferi*	Bacterium	Lyme disease
1983	HIV	Virus	AIDS
1983	*Helicobacter pylori*	Bacterium	Peptic ulcer disease
1985	*Enterocytozoon bieneusi*	Parasite	Persistent diarrhea

Continued

TABLE 5-1—*Continued*

EXAMPLES OF EMERGENT INFECTIOUS DISEASES RECOGNIZED
SINCE 1973 (ADAPTED FROM CISET, 1997: GLOBAL MICROBIAL
THREATS IN THE 1990s [NATIONAL CENTER FOR INFECTIOUS
DISEASES ELECTRONIC REFERENCE LIBRARY])

Year	Microbe	Type	Disease
1986	*Cyclospora cayatanensis*	Parasite	Persistent diarrhea
1988	Human herpesvirus-6	Virus	Roseola subitum
1988	Hepatitis E	Virus	Enterically transmitted non-A, non-B hepatitis
1989	*Ehrlichia chafeensis*	Bacterium	Human ehdlichiosis
1989	Hepatitis C	Virus	Parenterally transmitted non-A, non-B hepatitis
1991	Guanarito virus	Virus	Venezuelan hemorrhagic fever
1991	*Encephalitozoon heliem*	Parasite	Conjunctivitis, disseminated disease
1991	New species of Babesia	Parasite	Atypical babesiosis
1992	*Vibrio cholerae* 0139	Bacterium	New strain associated with epidemic cholera
1992	*Bartonella henselae*	Bacterium	Cat-scratch disease; baciliary angiomatosis
1993	Sin nombre virus	Virus	Adult respiratory distress syndrome
1993	Encephalitozoon cuniculi	Parasite	Disseminated disease
1994	Sabia virus	Virus	Brazilian hemorrhagic fever
1995	HHV-8	Virus	Associated with Kaposi sarcoma in AIDS patients

- Given the rapid growth and development of contemporary societies, many facilities in which pathogens are cultured and stored (and/or genetically manipulated) are located in very dense population areas.
- Given recent advances in biotechnology as well as the ready availability of that technology, it is highly unlikely that any legal authority knows precisely what is actually being done anywhere regarding the production or genetic manipulation of pathogens—except in those specific cases were legal sanction or financial support is sought by those doing the work
- Given recent experience with both current technology and domestic as well as foreign terrorism, it is not unimaginable that a pathogen may well become a favorite weapon of choice for any knowledgeable sociopath (note the use of Anthrax "powders" in the United States several years ago).

In light of these considerations, it is suggested to be only prudent for emergency response planning to take careful account of the real possibility that the occupational risks associated with blood-borne pathogens do not define the full range of risks attendant to biohazards.

3. *Well-known pathogens that, due to society's overuse of antibiotics, have evolved into new strains resistant to those antibiotics.*

Diseases caused by well-known pathogens that were once controlled but that have become either resistant to antibiotics or are, for other reasons (e.g., changes in human behavior, development of natural resources;, changes in public policy), ascendant are referred to as *reemerging diseases* (see Table 5.2).

Again, microbial resistance to antibiotics is a risk presented to the general public, with seemingly no special relevance to the emergency responder. However, during emergency response activities involving close personal contact with victims, it is only prudent (as in the case of HIV) that emergency response personnel take appropriate measures to protect themselves from any infection (see Image 5.2).

TABLE 5-2

SOME FACTORS LEADING TO THE ONGOING REEMERGENCE OF INFECTIOUS DISEASES

Disease or Agent	Factors in Re-emergence
Viral Infections	
Rabies	Breakdown in public health measures; changes in land use; travel
Dengue/dengue hemorrhagic fever	Transportation, travel and migration, urbanization
Yellow Fever	Favorable conditions for mosquito vector
Parasitic Infections	
Malaria	Drug and insecticide resistance; civil strife; lack of economic resources
Shistosomiasis	Dam construction, improved irrigation, and ecological changes favoring the snail host
Neurocysticercosis	Immigration
Acanthamebiasis	Introduction of soft contact lenses
Visceral leishmaniasis	War, population displacement, immigration, habitat changes favorable to the insect vector and increase in immunocompromised human hosts
Toxoplasmosis	Increase in immunocompromised human hosts
Giardiasis	Increased use of child-care facilities
Echinococcosis	Ecological changes that affect the habitats of the intemediate (animal) hosts
Bacterial Infections	
Group A Streptococcus	Uncertain
Trench fever	Breakdown in public health measures
Plague	Economic development; land use
Diphtheria	Interruption of immunization program due to political changes
Tuberculosis	Human demographics and behavior; industry and technology; international commerce and travel; breakdown of public health measures; microbial adaptation
Pertussis	Refusal to vaccinate in some parts of the world because of the belief that injections or vaccines are not safe
Salmonella	Industry and technology; human demographics and behavior; microbial adaptation; food changes
Pneumococcus	Human demographics; microbial adaptation; International travel and commerce; misuse and overuse of antibiotics
Cholera	Travel; new strain; reduced water chlorination

IMAGE 5-2

NEW YORK, NY, SEPTEMBER 22, 2001: RESCUE OPERATIONS
CONTINUE FAR INTO THE NIGHT AT THE WORLD TRADE
CENTER. AN IMPORTANT CONSIDERATION IN ANY RESCUE
EFFORT IS CONTAMINATION OF RESCUE WORKERS WITH BLOOD
AND BODY FLUIDS

Source: Andrea Booher/FEMA News Photo

BLOOD-BORNE PATHOGENS

Blood-borne pathogens are those pathogenic organisms that may be found in certain body fluids of infected persons. They may be transmitted to noninfected persons through their contact with contaminated body fluids, including:

- Human blood, blood components, and products made from human blood
- Semen (male reproductive secretion)
- Vaginal secretions (female reproductive secretions)
- Cerebrospinal fluid (associated with brain and spinal cord)
- Synovial fluid (associated with membranes in bone joints)
- Pleural fluid (associated with lung)
- Pericardial fluid (associated with chest cavity)
- Peritoneal fluid (associated with abdominal cavity)
- Amniotic fluid (associated with membranous sack covering fetus)
- Saliva (only in dental procedures where there is a high probability of blood becoming mixed with saliva)
- Any other body fluid that is visibly contaminated with blood
- All body fluids in situations where it is difficult to differentiate between body fluids

The two blood-borne pathogens of primary concern are the hepatitis virus (HV), specifically types B and C (HBV and HCV, respectively), and the so-called AIDS virus (HIV), which is really a number of distinctly different genetic strains of the same virus.

HBV (see Figure 5.2) is of particular concern as an occupational hazard not only because it causes a long-term disabling liver disease possibly leading to cirrhosis and even liver cancer, but also because of its efficient transmission from one person to another following contact with infected blood and body

FIGURE 5-2

BASIC FACTS ABOUT HEPATITIS B (ADAPTED FROM INFORMATION FROM NATIONAL CENTER FOR INFECTIOUS DISEASE [NCID], ELECTRONIC REFERENCE LIBRARY)

Fact Sheet	Hepatitis B
Clinical Features	Jaundice, fatigue, abdominal pain, loss of appetite, intermittent nausea, vomiting
Etiologic Agent	Hepatitis B virus
Incidence	140,000 - 320,000 infections/year in United States
Sequellae	Of symptomatic infections, 8,400 - 19,000 hospitalizations/year and 140 - 320 deaths/year Of all infections, 8,000 - 32,000 chronic infections/year, and 5,000 - 6,000 deaths/year from chronic liver disease including primary liver cancer
Prevalence	Estimated 1 - 1.25 million chronically infected Americans
Costs	Estimated $ 700 million (1991 dollars)/years (medical and work loss)
Transmission	Bloodborne; sexual; perinatal
Risk Groups	+ Injection drug user + Sexual/household contacts + Sexually active heterosexuals of infected persons + Homosexual men + Infants born of infected + Infants/children of immigrants mothers from disease-endemic areas + Health care workers + Low socioeconomic level + Hemodialysis patients
Trends	Incident increased though 1985 and then declined 55% through 1993 because of wider use of vaccine among adults, modification of high-risk behaviors, and possibly a decrease in the number of susceptible persons. Since 1993, increases observed among the three major risk groups: sexually active heterosexual, homosexual men, and injection drug users.
Prevention	+ Hepatitis B vaccine available since 1982 + Screening pregnant women and treatment of infants born to infected women + Routine vaccination of infants and 11-12 year olds + Catch-up vaccination of high-risk groups of all ages + Screening of blood or tissue donors

fluids. HBV has caused (and continues to cause) more cases of occupationally linked infectious disease than any other blood-borne pathogen. In fact, the probability of infection by HBV is on the order of 100,000 times greater than the probability of infection by HIV.

HBV infection may require an extended period of incubation and become manifest in highly diverse symptoms (see Figure 5.3), with many infected persons becoming long-term carriers and therefore potential sources of new infection.

In the United States, it has been estimated that on the order of 300,000 persons, including 9,000 healthcare workers, become infected with HBV every

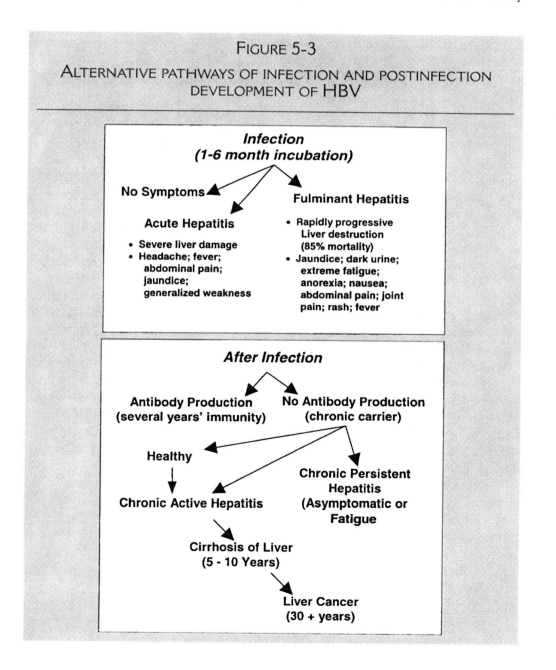

FIGURE 5-3

ALTERNATIVE PATHWAYS OF INFECTION AND POSTINFECTION DEVELOPMENT OF HBV

**Infection
(1-6 month incubation)**

No Symptoms

Acute Hepatitis
- Severe liver damage
- Headache; fever; abdominal pain; jaundice; generalized weakness

Fulminant Hepatitis
- Rapidly progressive Liver destruction (85% mortality)
- Jaundice; dark urine; extreme fatigue; anorexia; nausea; abdominal pain; joint pain; rash; fever

After Infection

Antibody Production
(several years' immunity)

No Antibody Production
(chronic carrier)

Healthy

Chronic Active Hepatitis

Chronic Persistent Hepatitis
(Asymptomatic or Fatigue

Cirrhosis of Liver
(5 - 10 Years)

Liver Cancer
(30 + years)

year. Worldwide, about 300 million persons are chronic carriers of HBV. In southeast Asia and tropical Africa, chronic carriers represent at least 10% of the population; in North America and most of western Europe, this group is less than 1%.

Historically, primary attention was given to HBV as the primary occupationally linked hepatitis virus, and HCV (a non-A, non-B strain), which is also transmitted through blood and other body fluids, was considered to present relatively little risk in the workplace. However, HCV has now been demonstrated to present potentially significant workplace risk, with upward of 40% of hepatitis infections previously attributed to HBV now possibly attributable to HCV.

HIV contravenes the body's capacity to resist a variety of life-threatening infections. HIV infection may also lead to severe weight loss, fatigue, neurological disorders, and certain cancers, including cancer of the skin or other connective tissue (sarcoma) and cancer of the lymph nodes or lymph tissues (lymphoma).

First discovered in 1979, AIDS (see Figure 5.4) quickly attained the status of a global epidemic, with estimates of actual cases worldwide approaching 600,000 in less than a decade. However, a distinction must be made between HIV infection and the development of AIDS.

Most HIV-infected persons do appear to develop antibodies to HIV within six to 12 weeks of exposure to HIV, and some may show neither outward symptoms nor an analytically detectable antibody response (e.g., by means of an HIV screening test) for even longer periods. Finally, even before the full-blown development of AIDS, which is indicated by essentially the collapse of the immune system and the subsequent development of opportunistic diseases (e.g., pneumonia, fungal diseases of the throat and lung, Kaposi's sarcoma, tuberculosis), an HIV-infected person may develop other symptoms, including severe, involuntary weight loss, chronic diarrhea, constant or intermittent weakness, and extended periods of fever—conditions that may themselves result in death.

Regardless of the progress of specific symptoms, and regardless of the length of time over which infected persons may remain asymptomatic, all HIV-infected persons must be considered capable of transmitting HIV to others. However, it is important to emphasize that *HIV transmission requires intimate contact with contaminated blood and/or other body fluids.* There is no documented evidence of HIV transmission simply through casual and even close physical contact with infected persons.

In addition to HBV, HCV, and HIV, blood-borne pathogens include a variety of highly infectious agents that pose significant risks to workers in various parts of the world. These pathogens (see Figure 5.5) include bacterial, protozoan, and viral species that, through a variety of disease vectors (e.g., mosquitoes, ticks, or lice), ultimately contaminate human blood and other body fluids.

Any effort to minimize the risk of infection by blood-borne diseases must be predicated the following three considerations:

1. Exposure to the blood and body fluids of infected persons always presents a real risk of contracting the disease.

2. Analytical tests devised to detect the presence of infection have inherent limits. In some instances, such limits become manifest in *false negatives*, which are analytical results that indicate a disease is not present when it actually is present. For example, a person who is infected with HIV may nonetheless be completely asymptomatic, with blood showing no detectable levels of HIV antibody for weeks and even months after infection. Negative analytical results are therefore false; they do not prove that infection is absent, nor do they demonstrate that a person is incapable of transmitting the disease to others.

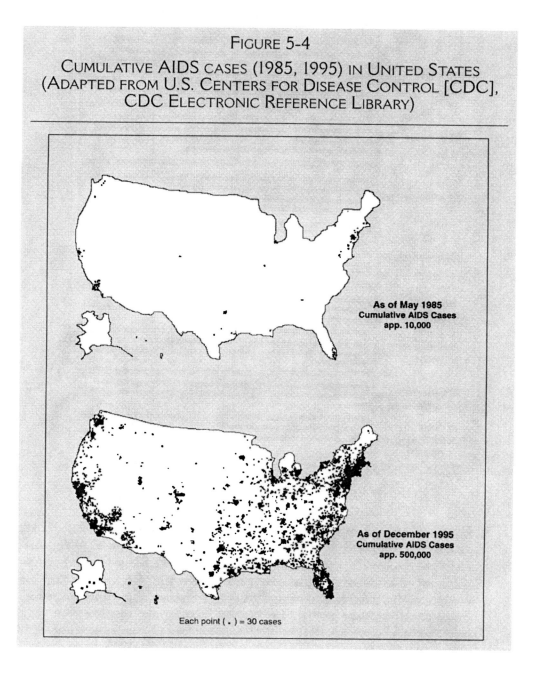

FIGURE 5-4

CUMULATIVE AIDS CASES (1985, 1995) IN UNITED STATES (ADAPTED FROM U.S. CENTERS FOR DISEASE CONTROL [CDC], CDC ELECTRONIC REFERENCE LIBRARY)

As of May 1985
Cumulative AIDS Cases
app. 10,000

As of December 1995
Cumulative AIDS Cases
app. 500,000

Each point (.) = 30 cases

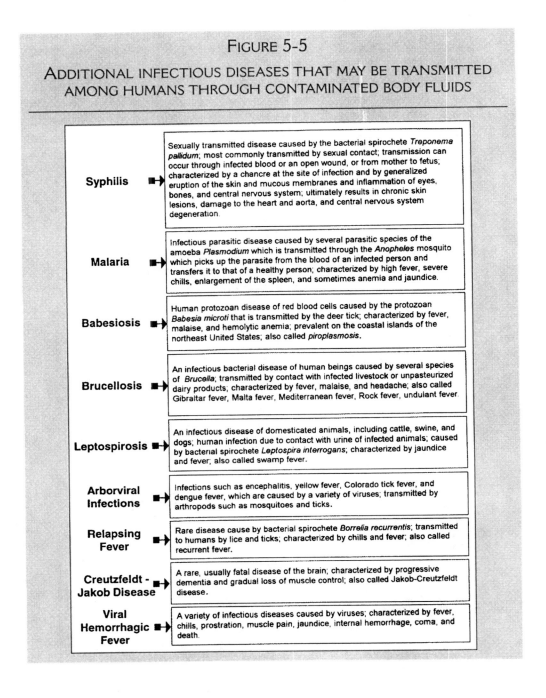

FIGURE 5-5

ADDITIONAL INFECTIOUS DISEASES THAT MAY BE TRANSMITTED AMONG HUMANS THROUGH CONTAMINATED BODY FLUIDS

Syphilis ■➔ Sexually transmitted disease caused by the bacterial spirochete *Treponema pallidum*; most commonly transmitted by sexual contact; transmission can occur through infected blood or an open wound, or from mother to fetus; characterized by a chancre at the site of infection and by generalized eruption of the skin and mucous membranes and inflammation of eyes, bones, and central nervous system; ultimately results in chronic skin lesions, damage to the heart and aorta, and central nervous system degeneration.

Malaria ■➔ Infectious parasitic disease caused by several parasitic species of the amoeba *Plasmodium* which is transmitted through the *Anopheles* mosquito which picks up the parasite from the blood of an infected person and transfers it to that of a healthy person; characterized by high fever, severe chills, enlargement of the spleen, and sometimes anemia and jaundice.

Babesiosis ■➔ Human protozoan disease of red blood cells caused by the protozoan *Babesia microti* that is transmitted by the deer tick; characterized by fever, malaise, and hemolytic anemia; prevalent on the coastal islands of the northeast United States; also called *piroplasmosis*.

Brucellosis ■➔ An infectious bacterial disease of human beings caused by several species of *Brucella*; transmitted by contact with infected livestock or unpasteurized dairy products; characterized by fever, malaise, and headache; also called Gibraltar fever, Malta fever, Mediterranean fever, Rock fever, undulant fever.

Leptospirosis ■➔ An infectious disease of domesticated animals, including cattle, swine, and dogs; human infection due to contact with urine of infected animals; caused by bacterial spirochete *Leptospira interrogans*; characterized by jaundice and fever; also called swamp fever.

Arborviral Infections ■➔ Infections such as encephalitis, yellow fever, Colorado tick fever, and dengue fever, which are caused by a variety of viruses; transmitted by arthropods such as mosquitoes and ticks.

Relapsing Fever ■➔ Rare disease cause by bacterial spirochete *Borrelia recurrentis*; transmitted to humans by lice and ticks; characterized by chills and fever; also called recurrent fever.

Creutzfeldt - Jakob Disease ■➔ A rare, usually fatal disease of the brain; characterized by progressive dementia and gradual loss of muscle control; also called Jakob-Creutzfeldt disease.

Viral Hemorrhagic Fever ■➔ A variety of infectious diseases caused by viruses; characterized by fever, chills, prostration, muscle pain, jaundice, internal hemorrhage, coma, and death.

3. Given the range in variation of human response to infection (from grossly symptomatic to completely asymptomatic response), given different periods of latency typically associated with the signs and symptoms of blood-borne diseases, and given the inherent limits of analytical procedures performed to detect disease (e.g., false negative, outright laboratory error), no person can safely assume that any human blood or related body fluid is safely free of contamination with infectious agents.

TABLE 5-3

GUIDELINES FOR USE OF PROTECTIVE CLOTHING AND EQUIPMENT TO MANAGE RISK OF INFECTION DUE TO CONTAMINATED BODY FLUIDS (ADAPTED FROM U.S. FIRE ADMINISTRATION, 1992: GUIDE TO DEVELOPING AND MANAGING AN EMERGENCY SERVICE INFECTION CONTROL PROGRAM [FA-112])

Task or Activity	Disposable Gloves	Gown	Mask	Protective Eyewear
Bleeding Control (with spurting blood)	Yes	Yes	Yes	Yes
Bleeding Control (with minimal bleeding)	Yes	No	No	No
Emergency Childbirth	Yes	Yes	Yes	Yes
Blood Drawing	Yes	No	No	No
Starting an Intravenous (IV) Line	Yes	Yes	Yes	Yes
Endotracheal Intubation, Esophageal Obturator Use	Yes	No	No	No
Oral/Nasal Suctioning (manual cleaning airway)	Yes	No	No	No
Handling & Cleaning Contaminated Instruments	Yes	No	No	No
Measuring Blood Pressure	No	No	No	No
Measuring Temperature	No	No	No	No
Giving an Injection	No	No	No	No

It is particularly important that these considerations be specifically integrated with job-task analysis (see Chapter 4) so that individual job-tasks that might involve exposure of emergency response personnel to blood or body fluids can be identified and proper protective equipment (see Table 5.3) issued to responsible personnel.

UNIVERSAL PRECAUTIONS

Universal precautions are procedures specifically designed to control the risk of infection by blood-borne pathogens in a wide range of different work-related circumstances. These precautions involve vaccination, engineering controls, work practice controls, and the use of personal protective equipment and clothing (see Image 5.3).

Vaccination

In the United States, personnel who might become exposed to HBV in the performance of their work must be offered immunization against HBV. According to 29 CFR 1910.1030, vaccination must be offered to at-risk personnel within 10 working days of initial job assignment and at no cost to the employee. Other provisions of the OSHA regulations include the following.

IMAGE 5-3

SAN JUAN, PR, NOVEMBER 22, 1996: METRO-DADE TASK FORCE
PERFORMING SEARCH AND RESCUE OPERATIONS INSIDE THE
HUMBERTO VIDAL BUILDING FOLLOWING A GAS MAINLINE
EXPLOSION. TASK FORCE MEMBERS SEARCH THROUGH THE RUBBLE
FOR SURVIVORS. IN ANY URBAN CENTER EXPLOSION,
CONSIDERATION MUST BE GIVEN TO THE POTENTIAL RELEASE OF
BIOHAZARDS FROM AFFECTED BIOLOGICAL RESEARCH AND
BIOTECHNOLOGY CENTERS

Source: Roman Bas/FEMA News Photo

- The vaccination is to be offered at a reasonable time and place and under the supervision of a licensed physician or a healthcare professional licensed to give HBV vaccinations.
- An employee is not required to have a vaccination if (a) the employee has previously receive the complete HBV vaccination series, or (b) tests show

the employee is immune to HBV, or (c) the vaccine is contraindicated for medical reasons.

- An employee is not required to participate in a prescreening program as a prerequisite to receiving the HBV vaccination.
- An employee may refuse to receive the HBV vaccination or, having initially refused, may subsequently decide to receive it.

Work Practice Controls

Work practice controls are those policies and procedures designed to minimize the risk of infection during the performance of routine tasks. Four basic types of work practices are relevant in any situation (including emergency response) where exposure to blood-borne pathogens is possible.

1. *General Work Practices* (apply across the range of work-related tasks):
   - Eating, drinking, smoking, applying cosmetics or lip balm, and wearing contact lenses should be prohibited.
   - Food and beverages should not be stored in cabinets, refrigerators, freezers, or on counters except where such facilities are specifically designated and restricted to the storing or handling of food and beverages.
   - Any procedure involving blood, body fluids, body parts, or potentially infectious materials should be performed to minimize splashing, spraying, or the formation of droplets.
   - Any specimen containing blood, body fluids, or potentially infectious materials should be kept in clearly labeled, leak-proof, closed containers during collection, storage, handling, processing, shipping, and transport.
   - No blood, body fluid or body part should ever be touched or cleaned up without the use of proper protective clothing and equipment.
2. *The Use of "Sharps"* (practices regarding the use and disposal of needles, blades, and other items that may cut or puncture the skin)
   - Needles or other sharps contaminated with human blood or other body fluids should not be bent, broken, sheared, recapped, or removed from their holders.
   - Disposable sharps should be deposited in containers that are puncture-resistant, leak-proof, and color-coded or labeled "Biohazard."
   - Nondisposable sharps should be decontaminated according to written directions.
3. *Accidental Contact* (procedures to be followed after accidental contact with human blood, tissue, or body fluids)
   - Immediately flush eyes with water or wash skin with soap and water.
   - Remove any contaminated clothing immediately and wash any areas of skin that may have been contaminated by fluids soaking through.
   - Obtain medical consultation after contact to determine necessity of follow-up medical treatment or prophylaxis.
4. *Housekeeping* (procedures governing the clean-up of spills of blood, body fluids, and body parts, as well as general housekeeping tasks)

- All blood-soaked rags, papers, and other materials should be placed in biohazard bags, sealed, and disposed of through a biohazard-certified (medical waste) facility.
- Trash receptacles in areas where contamination is likely should be cleaned and decontaminated as soon as possible after contamination.
- All areas contaminated by blood, body fluids, or body parts should be decontaminated.

5. *Personal Protective Clothing and Equipment*
   - Disposable vinyl or latex gloves should be used wherever hand contact with blood-borne pathogens may occur.
   - An emergency packet should be immediately available to emergency responders and other personnel who may become exposed to blood-borne pathogens and should contain (a) disposable vinyl or latex gloves, (b) appropriate disinfectant solution, (c) a supply of absorbent containment material and scoop, (d) biohazard bags, and (e) disposable towels (for stanching copious flows of blood without exposing responder to blood splash).
   - Disposable gloves must not be cleaned or washed for reuse. However, they should be cleaned prior to removal and disinfected following removal or discarded into biohazard bags.
   - No petroleum products (e.g., hand creams) should be used in conjunction with latex gloves because such materials may degrade latex.
   - In no circumstance should mouth-to-mouth resuscitation be performed without the use of protective mouthpieces to prevent contact with potential blood-contaminated saliva.
   - Additional protective clothing should be provided as circumstances may require, including fluid-proof aprons, goggles, shoe covers, and face shields.

EXPOSURE CONTROL PLAN

Under the provisions of 29 CFR 1910.1030, an employer must develop a written *exposure control plan*. The specific objectives of this plan are (a) to designate job-related tasks that present the risk of exposure to blood-borne pathogens, (b) to define the schedule and means for implementing exposure controls, and (c) to establish procedures for the evaluation of exposure incidents, personnel training, and record-keeping.

The regulations provide specific guidance regarding those work-related activities that may result in exposure to blood-borne pathogens. Emergency response operations potentially include all of these activities:

- Activities that result in direct exposure of all personnel to blood-borne pathogens (e.g., emergency medical service personnel)
- Activities that result in direct exposure of some personnel to blood-borne pathogens (e.g., rescue personnel)

- Individual tasks and procedures or groups of closely related tasks and procedures in which some or all employers may experience exposure to blood-borne pathogens (e.g., fire brigade)

Among the various procedures to be implemented regarding the control of exposure to blood-borne pathogens, particular attention must be given to oversight and enforcement. It cannot be overemphasized that the protection of emergency responders who might become exposed to blood-borne pathogens means protection from infections that can easily spread to responders' families and to the community at large. This broad social responsibility for the control of disease means that compliance with workplace polices and procedures designed to control severely disabling and even life-threatening disease must be rigorously enforced without exception throughout the emergency response team.

Special attention must also be given to those procedures regarding the evaluation of an incident of exposure, especially the methodical and detailed assessment of any related failures with regard to the identification of specific tasks and personnel at risk (e.g., responder involved in the extrication of victims), the adequacy of work practice control (e.g., personal protective clothing, disinfection of clothing and equipment), and the adequacy of personnel training. Each postexposure incident evaluation should include specific recommendations for revising the exposure control plan as well as precise schedules for implementing those revisions and monitoring their effectiveness.

OTHER PATHOGENS

A large number of diseases may be transmitted to emergency response personnel through means other than contact with the blood or blood-related fluids of infected persons (see Table 5.4). In addition to direct contact with the body and clothing of infected victims, such nonblood-borne diseases may be transmitted via contact with feces, nasal secretions, sputum, sweat, tears, urine, and vomitus. However, emergency responders need not have direct contact with a victim's body, secretions, or clothing to experience the risk of infection. For example, an underwater or swift water rescue effort may require submersion into lakes, ponds, and rivers that contain viable pathogenic organisms deposited there by sewage. Also, the debris of collapsed structures may become contaminated in the immediate area of victim entrapment. Rain and wind can also transfer contaminated body substances to response personnel who are otherwise removed from infected victims.

A basic strategy of minimizing the risk of disease transmission by contact with infected body substances is referred to most commonly as *body substance isolation* (BSI). Whereas universal precautions are based on the assumption that all blood and certain body fluids should be considered potentially infectious for blood-borne pathogens, BSI is based on the assumption that all body fluids and substances are potentially infectious (see Figure 5.6). BSI therefore requires careful attention to proactive, infection-preventive measures, including:

TABLE 5-4

SUMMARY INFORMATION REGARDING NONBLOODBORNE
INFECTIONS DISEASES (ADAPTED FROM U.S. FIRE
ADMINISTRATION, 1992: GUIDE TO DEVELOPING
AND MANAGING AN EMERGENCY SERVICE INFECTION
CONTROL PROGRAM [FA-112])

Disease	Mode of Transmission	Availability of Vaccine	Signs and Symptoms
Chickenpox	Respiratory secretions and contact with moist blisters	No	Fever; rash; skin blisters
Diarrhea	Fecal/Oral contamination	No	Loose, watery stools
German Measles (Rubella)	Respiratory droplets and contact with respiratory secretions	Yes	Fever; rash
Hepatitis A (Infectious Hepatitis)	Fecal/Oral contamination	No	Fever; loss of appetite; jaundice; fatigue
Herpes Simplex (Cold Sores)	Contact of mucous membranes with moist lesions; fingers are at particular risk for becoming infected	No	Skin lesions located around the mount area
Other non-A, non-B Hepatitis	Several viruses with different modes of transmission	No	Fever; headache; fatigue; jaundice
Herpes Zoster (Shingles)	Contact with moist lesions	No	Skin lesions
Influenza	Airborne	Yes	Fever; fatigue; loss of appetite; nausea; headache
Lice	Close head-to-head contact; both body and pubic lice require intimate contact (usually sexual) or sharing of intimate clothing	No	Severe itching and scratching, often with secondary infection; scalp and hairy portions of body may be affected; eggs of head lice attach to hairs as small, round, gray lumps
Measles	Respiratory droplets and contact with nasal or throat secretions; highly communicable	Yes	Fever; rash; bronchitis
Meningitis			
• Meningococcal	Contact with respiratory secretions	No	Fever; severe headache; stiff neck; sore throat
• Haemophilus influenza (usually in very young children)	Contact with respiratory secretions	No	Same
• Viral Meningitis	Fecal/Oral contamination	No	Same
Mononucleosis	Contact with respiratory secretions or saliva, such as with mouth-to-mouth resuscitation	No	Fever; sore throat; fatigue
Mumps (Infectious Parotitis)	Respiratory droplets and contact with saliva	Yes	Fever; rash

Continued

TABLE 5-4—*Continued*

SUMMARY INFORMATION REGARDING NONBLOODBORNE INFECTIONS DISEASES (ADAPTED FROM U.S. FIRE ADMINISTRATION, 1992: GUIDE TO DEVELOPING AND MANAGING AN EMERGENCY SERVICE INFECTION CONTROL PROGRAM [FA-112])

Disease	Mode of Transmission	Availability of Vaccine	Signs and Symptoms
Salmonellosis	Foodborne	No	Sudden onset of fever, abdominal pain, diarrhea, nausea, and frequent vomiting
Scabies	Close body contact	No	Itching; tiny linear burrows or "tracks"; blisters, particularly around finger, wrists, elbows, and skin folds
Tuberculosis *(Pulmonary)*	Airborne	No	Fever; night sweats; weight loss; cough
Whooping Cough *(Pertussis)*	Airborne; direct contact with oral secretions	Yes	Violent cough at night; whooping sound when cough subsides

FIGURE 5-6

BASIC RULES FOR MANAGING RISKS RELATED TO INFECTIOUS DISEASES (ADAPTED FROM U.S. FIRE ADMINISTRATION, 1992: GUIDE TO DEVELOPING AND MANAGING AN EMERGENCY SERVICE INFECTION CONTROL PROGRAM [FA-112])

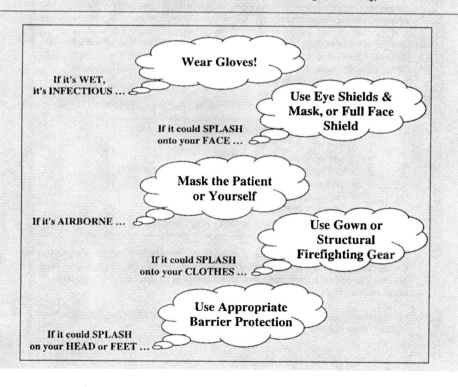

- Proper personal hygiene
- Immunization programs
- Decontamination procedures
- Proper procedures for the handling and disposal of waste

Written SOPs should be prepared (see Figure 5.7) for all procedures that implement these infection-preventive measures.

FIGURE 5-7

EXAMPLE OF STANDARD OPERATING PROCEDURE FOR PERFORMING TASKS UNDER UNIVERSAL AND BODY SUBSTANCE PRECAUTIONS (ADAPTED FROM U.S. FIRE ADMINISTRATION, 1992: GUIDE TO DEVELOPING AND MANAGING AN EMERGENCY SERVICE INFECTION CONTROL PROGRAM [FA-112])

Infection Control Standard Operating Procedures

SOP # IC 5
Scene Operations

---

1. The blood, body fluids, and tissues of all patients are considered potentially infectious, and Universal Precautions/Body Substance Isolation procedures will be used for all patient contact.

2. The choice of personal protective equipment is specified in SOP # IC 4. Personnel will be encouraged to use maximal rather than minimal PPE for each situation.

3. While complete control of the emergency scene is not possible, scene operations as much as possible will attempt to limit splashing, spraying, or aerosolization of body fluids.

4. The minimum number of personnel required to complete the task safely will be used for all on-scene operations. Members not immediately needed will remain a safe distance from operations where communicable disease exposure is possible or anticipated.

5. Handwashing is the most important infection control procedure.

Personnel WILL wash hands
• After each patient contact
• After handling potentially infectious materials
• After cleaning or decontaminating equipment
• After using the bathroom
• Before eating
• Before and after handling or preparing food

6. Handwashing with soap and water will be performed for 10 to 15 seconds. If soap and water are not available at the scene, a waterless handwash may be used, provided that a soap and water wash is performed immediately upon return to quarters or hospital.

7. Eating, drinking, smoking, handling contact lenses, or applying cosmetics or lip balm is prohibited at the scene of operations.

Continued

FIGURE 5-7—*Continued*

EXAMPLE OF STANDARD OPERATING PROCEDURE FOR PERFORMING TASKS UNDER UNIVERSAL AND BODY SUBSTANCE PRECAUTIONS (ADAPTED FROM U.S. FIRE ADMINISTRATION, 1992: GUIDE TO DEVELOPING AND MANAGING AN EMERGENCY SERVICE INFECTION CONTROL PROGRAM [FA-112])

SOP # IC 5
......Continued

8. Used needles and other sharps shall be disposed of in approved sharps containers. Needles will not be recapped, resheathed, bent, broken, or separated from disposable syringes. *The most common occupational blood exposure occurs when needles are recapped.*

9. Sharps containers will be easily accessible on-scene.

10. Disposable resuscitation equipment will be used whenever possible. For CPR, the order of preference is:

    - Disposable bag-valve mask
    - Demand valve resuscitator with disposable mask
    - Disposable pocket mask with one-way valve
    - Mouth-to-mouth resuscitation

11. Mouth-to-mouth resuscitation will be performed only as last resort if no other equipment is available. All members will be issued pocket masks with one-way valves to minimize the need for mouth-to-mouth resuscitation. Disposable resuscitation equipment will be kept readily available during on-scene operations.

12. Patients with suspected airborne communicable diseases will be transported wearing a face mask or particulate respirator whenever possible. Ambulance windows will be open and the ventilation system turned on full whenever possible.

13. Personal protective equipment will be removed after leaving the work area, and as soon as possible if contaminated. After use, all PPE will be placed in leakproof bags, color coded and marked as a biohazard, and transported back to the station for proper disposal.

14. On-scene public relations will be handled by the Department Public Information Officer, if available; if not, the senior line office will assume this function. The public should be reassured that infection control PPE is used as matter of routine for the protection of all members and the victims that they treat. The use of PPE does not imply that a given victim may have a communicable disease.

15. No medical information will be released on-scene.

16. At conclusion of on-scene operations, all potentially contaminated patient care equipment will be removed for appropriate disposal or decontamination and reuse.

In many instances, emergency responders are volunteer personnel who respond to incidents by driving to the emergency site in their own vehicles. This is particularly common in community fire brigades in the United States, which respond not only to fires but also to vehicular accidents and other types of local emergencies. In such a situation, and despite the use of protective gloves and other clothing, special attention must be given to on-site procedures for personal cleaning and disinfection in order to minimize the possibility of individual volunteer responders carrying home (either on their own bodies or in their personal vehicles) infectious materials that can subsequently be transmitted to friends and families.

Another aspect of emergency response that is often overlooked for its potential to spread infectious disease (including both blood-borne and other infectious agents) into the community is the use of private wreckers/recycling companies to transport and dispose of crash vehicles and/or other types of structural debris (e.g., building materials) that may contain large amounts of infectious body substances. Because this task typically takes place at the termination of an incident response (when site control reverts to nonemergency response personnel), there is real potential for (a) direct contamination of personnel involved in subsequent salvage and disposal operations involving contaminated vehicles and debris, and (b) contamination of ambient air, water, and soil during those operations, with subsequent risk to the public.

STUDY GUIDE

True or False

1. A key factor related to the incidence of emergent diseases is the accessibility of even the most remote areas to world travelers.
2. Emergent diseases are of particular concern to emergency responders because major catastrophes typically elicit the response of specialized emergency teams from throughout the world.
3. Genetically modified pathogens also present a problem to the general public as well as emergency responders because the technology for genetic manipulation is widespread.
4. Reemerging diseases present no significant risk to emergency responders.
5. Blood-borne pathogens occur only in human blood.
6. Blood-borne pathogens include hepatitis virus (HV), types B and C, and the human immunodeficiency virus (HIV).
7. The cause of the largest number of deaths in the world is cancer.
8. Many facilities in which pathogens are cultured and stored (or genetically manipulated) are located in very dense population centers.
9. Since 1973, over two dozen emergent diseases have been identified.
10. Some factors that influence the reemergence of infectious diseases (viral, bacterial, and parasitic) include urbanization, changing land use, and immigration.
11. The probability of infection by HBV is on the order of 100,000 times greater than the probability of infection by HIV.
12. A person infected with HBV may show absolutely no symptoms of the disease for months.
13. There is no documented evidence of HIV transmission simply through casual and even close physical contact with infected persons.
14. Universal precautions are procedures specifically designed to control the risk of infection by blood-borne pathogens.

Multiple Choice

1. Work practice controls used to minimize the risk of infections include
   A. general work practices
   B. use of "sharps"
   C. accidental contact
   D. housekeeping
   E. personal protective clothing and equipment
   F. all of the above

2. Body substance isolation (BSI) is a set of procedures based on the assumption that
   A. all blood and certain body fluids should be considered potentially infectious
   B. all body fluids and substances are potentially infectious
   C. both A and B

3. BSI includes proactive infection-preventive measures such as
   A. personal hygiene
   B. immunization programs
   C. decontamination procedures
   D. proper handling and disposal of waste
   E. all of the above

4. A person who is HIV infected
   A. always shows symptoms of the disease within one month
   B. must be considered capable of transmitting HIV to others
   C. cannot transmit HIV to others if the person does not show symptoms within five years

5. Any effort to minimize the risk of infection by blood-borne diseases must be predicated by the following considerations:
   A. exposure to the blood and body fluids of infected persons always presents a real risk
   B. analytical tests have inherent limits
   C. it can never be assumed that any human blood or related body fluid is free of infectious agents
   D. all of the above

6. No blood, body fluid, or body part should ever be touched or cleaned up
   A. unless it is properly tagged as to location and possible identification
   B. without the use of proper protective clothing and equipment
   C. without specific direction by the Incident Commander

7. Nonblood-borne diseases may be transmitted via contact with
   A. feces and urine
   B. nasal secretions and sputum
   C. sweat and tear
   D. vomitus
   E. all of the above

8. With respect to volunteer (including off-duty) responding personnel, particular attention should be given to
   A. the possibility of "carry-home contamination"

B. the status of their inoculation record
C. any aspect of their general health that might make them more susceptible to infectious disease

Essays

1. Review the Standard Operating Procedure for universal and body substance precautions. Identify and discuss particular problems you can envision in implementing this procedure on-site of an emergency.
2. Review Figure 5.6, which is another form of SOP. For each of the five "If it" conditions specified, identify which diseases identified in the text, tables, and figures in this chapter could be transmitted to respondent personnel in the absence of appropriate protective clothing and equipment.

Case Study

Contact a local community responder (e.g., firefighter, emergency medical technician, or ambulance driver) and collect information on the SOPs used by these services to minimize the potential spread of infectious disease due to an emergency event.

1. Comment on the level of detail in such SOPs.
2. Comment on the comprehensiveness of such SOPs.
3. Discuss your opinions regarding the level of preparedness in these organizations.

MEDICAL SURVEILLANCE

INTRODUCTION

Medical surveillance of personnel has become an intrinsic activity in any modern workplace. Various interrelated factors are responsible, including (a) extensive public awareness of health and safety risks, which is engendered by the explosion of telecommunication technology; (b) the increased exposure of corporations, corporate executive, and stockholders to potential liability regarding the exposure of both employees and the general public to workplace chemicals; (c) continually expansive regulatory requirements regarding employee health at all levels of government; (d) the rapid development of a global economy in which the protection of human health is rapidly becoming a basic precept of highly competitive marketing ploys; (e) trade union concerns for the health and well-being of members; and (f) corporate insurance under-writers.

Although the nature and extent of medical surveillance in the workplace are variable with legal jurisdiction and type of industry, the broad dimensions of contemporary workplace medical surveillance are clearly established and pertain directly to any medical surveillance program established for emergency responders (see Image 6.1).

SURVEILLANCE OBJECTIVES AND CONCERNS

Typically, medical surveillance may be subdivided into four basic categories or types of surveillance: preemployment screening, periodic operational monitoring, episodic monitoring, and employment-termination examination.

IMAGE 6-1

NEW YORK, NY, SEPTEMBER 21, 2001: TWO URBAN SEARCH AND RESCUE CREW MEMBERS TAKE A BREAK TO EAT AND REST DURING THE CLEAN UP OPERATIONS AT THE WORLD TRADE CENTER

Source: Michael Rieger/FEMA News Photo

Preemployment Screening

Preemployment screening generally encompasses three objectives:

- To determine the fitness of an employee to perform assigned work
- To identify any health conditions that might exacerbate the effects of work-related hazards
- To establish a baseline health profile that can be used to measure the effects of both short- and long-term exposure to work-related hazards

Although each of these objectives is essential to the protection of workers, each is increasingly the subject of concern regarding a potential abasement of workers rights, especially in light of the possible use of sophisticated clinical and genetic analyses to deny or otherwise restrict employment on the basis of potential healthcare costs likely to be borne by the employer.

It has also become clear that the increasingly widespread use over the past two decades of the "temporary employee" (who typically is ineligible for healthcare and other work-related benefits) may reflect corporate intent to disclaim any long-term financial responsibility for worker health rather than simply to improve cost efficiency by reducing in-house staffs devoted to employment recruitment and training. There can be little doubt that the use of preemployment screening as a means of disenfranchising the employee-at-risk rather than as a means of protecting that employee will long continue to be the focus of legal, political, and social scrutiny and debate, especially with regard to emergency response personnel, a category that includes (a) first responders, who may often be temporary employees or, if full-time employees, personnel whose primary exposure to hazards is defined by normal (i.e., nonemergency) workplace conditions, as well as (b) full-time emergency response personnel. In the latter case, the stringency of preemployment health standards (applied to new hires) could well become of legal relevance in any legal proceeding based on purported bias in hiring practices.

Medical surveillance undertaken to determine fitness for work must be predicated on a precise understanding of the total range of health and safety hazards associated with individual work assignments (including both emergency-related and nonemergency-related tasks) as well as pertinent regulatory requirements (e.g., medical examination for use of respirator). Although primarily defined by job requirements, fitness for work must also be determined on the basis of preexisting health conditions or limitations of the worker—a determination that may often be at odds with the desires of the worker and/or the employer. The employer is well advised that the willingness of a worker to undertake risks contrary to professional medical advice generally does not necessarily abrogate the employer's responsibility for the health and safety of that worker. This fact underscores the importance of implementing a medical surveillance program that complies not only with the requirements of pertinent health and safety regulations but also with the constraints imposed by legal counsel.

In some instances, specific guidance is provided by regulatory authority or by professional organizations. For example, the standard promulgated by the National Fire Prevention Association (NFPA No. 1582; Standard on Medical Requirements for Fire Fighters) establishes medical requirements for both candidate as well as operational firefighters (including age-dependent medical evaluations) regarding:

- Vital signs
- Dermatological system
- Ears, eyes, nose, mouth, and throat
- Cardiovascular system
- Respiratory system
- Gastrointestinal system
- Genitourinary system
- Endocrine and metabolic systems
- Musculoskeletal system
- Neurological system

- Audiometric capacity
- Visual acuity and peripheral vision
- Pulmonary function
- Laboratory analyses
- Diagnostic imaging
- Electrocardiography

In addition to these requirements, various fire departments also use a range of tests to evaluate the physical fitness of candidate and engaged personnel, including (a) determination of body fat, (b) assessment of flexibility, (c) evaluation of aerobic power, and (d) assessment of physical strength.

Of critical importance in any medical surveillance program is the establishment of baseline health profiles of at-risk personnel. The comparison of these profiles with the results of subsequent surveillance is the basic means for detecting changes in health that may be related to routine and nonroutine exposures and stress. It is therefore essential that the medical examination performed in preemployment screening include those measurements of vital signs, vision and hearing measurements, lung function tests, and other clinical biochemical analyses that are directly relevant to work-related exposure and stress. The selection of specific tests and analyses and the type of data and information required must be made only with the professional advice of a medical authority who is provided with all details regarding potential routine and emergency exposures. The medical surveillance program must, therefore, be understood to be specific to the emergency response service; no guideline can be provided for identifying the total range of specific tests and analyses included in a surveillance program that is universally appropriate throughout the emergency response community.

Periodic Operational Monitoring

The sole objective of periodic operational monitoring is the early detection of adverse health effects of routine exposure to hazardous agents and situations. As discussed earlier, periodic operational monitoring must be integrally linked with the baseline profiles established during preemployment screening. In the design of an operational monitoring program, particular attention should be given to the following issues.

1. Because of the wide diversity in types of hazardous agents and situations, the variable progression of different kinds of health impairments and conditions, and the range of work-related exposures, it is highly unlikely that a monitoring schedule appropriate for the early detection of one kind of health condition will be appropriate for the detection of another kind. For example, depending upon specific work-related exposure, an annual schedule for blood testing to detect liver disease may not be appropriate for chest X-rays, which in fact may cause lung injury if used too frequently.

2. Differences noted between baseline profiles and subsequent operational monitoring do not necessarily indicate an actual disease or debilitation;

even if a disease or debilitation is detected, it is not necessarily due to work-related exposure. All medical monitoring data and information are subject to normal variation; abnormal results that may indicate disease or debilitation may reflect home and recreational exposures as well as workplace exposures to hazardous agents; abnormal results, in fact, may not indicate any specific exposure but, rather, simply reflect overall systemic changes in body function.

For example, the alkaline phosphatase test is a very sensitive test that, simply because of its sensitivity, can give widely fluctuating results; it is therefore most often a rather nonspecific indicator of possible liver impairment. The less sensitive gamma glutamyl transpeptidase test is less influenced by extraneous factors, but elevations in this enzyme typically must be on the order of twice the normal amount to trigger clinical concern over liver damage.

The design of an operational monitoring program must therefore be undertaken with a clear understanding of statistical and other criteria of significance that medical professionals must use when interpreting monitoring results. It is strongly recommended that personnel included in a medical surveillance program be provided documentation regarding such criteria.

3. A properly designed medical surveillance program should include a detailed *action plan* that precisely describes steps to be taken whenever operational monitoring results in the detection of a medically significant condition, including follow-up medical examinations, tests, and treatments. The action plan should also provide for the implementation of a comprehensive review of response operations and procedures that may have contributed to the detected health impairment, and which, therefore, should be corrected.

Episodic Monitoring

Episodic medical monitoring includes any nonroutine medical monitoring or surveillance activity undertaken in response to a specific incident, condition, or circumstance (see Image 6.2), such as exposure to a specific chemical (e.g., styrene monomer), or responder complaints of unusual health symptoms (e.g., persistent headaches or nausea). Although episodic monitoring is specifically addressed in particular U.S. regulations (e.g., 29 CFR 1910.1450 [Laboratory Standard]; 29 CFR 1910.120 [Hazardous Waste Operations and Emergency Response]), it is appropriately included in any comprehensive medical surveillance program, regardless of regulatory jurisdiction.

Provisions for episodic medical monitoring must be predicated on several considerations:

1. Although the episode that triggers nonroutine medical surveillance may often be described in terms of objective criteria, such as an uncontrolled exposure to a specific chemical, subjective criteria may alone be sufficient and even critical. Even in the absence of any objectively manifest evidence of exposure, the fact that personnel think they may have

IMAGE 6-2

BOUND BROOK, NJ MEMBERS OF THE RARITAN FIRST AID SQUAD ASSIST A RESIDENT FOLLOWING THE FLOODING IN BOUND BROOK

Source: Andrea Booher/FEMA News Photo

suffered a nonroutine exposure is sufficient cause for medical surveillance and consultation over and above that provided through regularly scheduled operational monitoring. Even though a safety officer may be apt—sometimes, with good reason—to consider an individual complaint the product of an overactive imagination or the purposeful contrivance of a problem employee, safety officers are reminded that individual personnel may be particularly sensitive to a hazardous agent. Should a complaint be ignored simply because it is a singular complaint, it is possible that a real health threat will be ignored, with not only dire consequence to the individual, but also serious legal and financial ramifications for both the safety officer and the response service organization.

2. Whatever the cause or circumstance of the episode, medical authority must be provided with relevant data and information. In many instances, standard forms are used to provide medical professionals the appropriate information (see Figure 6.1). In all instances, it is necessary that

preliminary liaison be established between the safety officer and medical personnel so that the latter have direct access to baseline information that may become relevant to any subsequent episode. Such baseline information should at a minimum include an inventory of relevant chemical agents that, for each listed chemical, identifies hazards, target organs, and routes of entry. It is also recommended that combustion, water-reactive, and other by-products be identified for each chemical included in the inventory, along with the hazards and target organs associated with each chemical.

FIGURE 6-1

EXAMPLE OF A FORM THAT PROVIDES AN ATTENDING PHYSICIAN OR OTHER MEDICAL PROFESSIONAL WITH CRITICAL INFORMATION REGARDING PERSONNEL EXPOSURE TO A HAZARDOUS AGENT

Global Emergency Response Services	**Personnel Exposure Determination**

Employee Identification

Name

Department

Reason for Implementing Determination of Exposure

- ☐ Monitoring Data
- ☐ Observed Spill or Release of Chemical
- ☐ Odor, Taste or Other Sensory Perception of Chemical
- ☐ Procedural/Operational Potential (e.g., open vessel; failure of ventilation)
- ☐ Signs or Symptoms of Chemical Exposure

Name(s) of Chemicals or Chemical Constituents and Relevant OSHA TLV

Chemical Name	OSHA TLV

Available Monitoring Data

Chemical Name	Date	Value/Unit

Description of Incident or Circumstance

Date	Signature of Safety Officer

3. Episodic events that trigger medical surveillance include those predicated by the recognition of health symptoms. It is therefore essential that all personnel receive thorough training in the range of symptoms that may be associated with response-exposure to hazardous agents and understand the importance of reporting such symptoms to the safety officer. Symptoms (see Table 6.1) associated with work-related exposure to hazardous agents typically cannot be differentiated from symptoms associated with nonwork-related exposure or from various health conditions or infections totally unrelated to the work environment. However, the safety officer must understand that the only competent authority for determining the significance of any health symptom is the physician. It is the responsibility of the physician to evaluate symptoms and to determine the relevance of those symptoms to operational exposures; it is the responsibility of the safety officer (and the emergency service

TABLE 6-1

COMMON SYMPTOMS THAT MAY INDICATE EXPOSURE TO HAZARDOUS CHEMICALS (ADAPTED FROM MATERIALS PROVIDED BY DR. DONALD G. ERICKSON)

- Chest pain or discomfort
- Bluish lips or face; extreme paleness
- Persistent coughing or sneezing
- Breathing discomfort; rapid or strained breathing
- Palpitations or fluttering in chest
- Lightheadedness or dizziness; giddiness; fainting
- Headaches (especially persistent, recurrent or progressive)
- Itching or irritation of eye; watering of eye; sensitivity to light
- Visual impairment, including reduced vision, double vision and changes in perception of color
- Loss of physical coordination or dexterity; slurring of speech
- Unusual hair loss
- Bleeding of gums or nose
- Increased sensitivity to noise; changes in hearing acuity; ringing in ears
- Abnormal odor of breath
- Hoarseness
- Fever
- Abnormal sweating or dryness of skin
- Generalized aches and pains; muscle cramping; weakness of a particular muscle
- Prickly sensation in legs, arms, or face
- Prickly or numb sensation in tongue
- Nausea, vomiting, abdominal pain; burning sensation in throat or stomach
- Unusual thirst
- Problems in swallowing; change in taste sensation
- Loss of appetite
- Changes in color of urine
- Unusual skin rashes or swelling; acne-like skin lesions; blisters
- Changes in skin color
- Personality changes
- Abrupt or progressive behavioral changes, including changes in personal grooming; impairment of judgment; aggressiveness; irritability
- Nervousness or restlessness; tremors or shakes
- Lethargy or unusual sleepiness

organization) to ensure that personnel who display health symptoms have immediate access to the physician.

4. As important as symptoms are for triggering medical consultation and surveillance, the limitations of symptoms must be recognized. For example, the health effects of exposure to many hazardous chemicals often require years and decades to develop. In such cases, there may be no readily recognized symptoms for extended periods of time, whereas in others, clear symptoms develop rapidly after exposure to the hazardous agent (see Figure 6.2). In compiling a list of symptoms requiring medical notification, the safety officer must therefore ensure consideration of the range of symptoms associated with both chronic and acute health effects. It is also necessary to identify which particular symptoms require immediate medical response.

The inherent limitations of symptomatology as a trigger to medical consultation require that the safety officer must establish additional criteria for

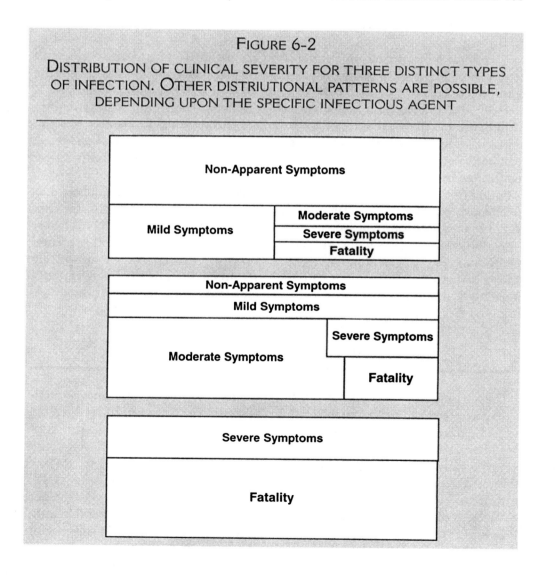

FIGURE 6-2

DISTRIBUTION OF CLINICAL SEVERITY FOR THREE DISTINCT TYPES OF INFECTION. OTHER DISTRITUTIONAL PATTERNS ARE POSSIBLE, DEPENDING UPON THE SPECIFIC INFECTIOUS AGENT

Non-Apparent Symptoms

Mild Symptoms

Moderate Symptoms
Severe Symptoms
Fatality

Non-Apparent Symptoms

Mild Symptoms

Moderate Symptoms

Severe Symptoms

Fatality

Severe Symptoms

Fatality

activating episodic medical monitoring. Examples of such criteria include (but are not limited to):

- Any failure or aberrant function in exposure control devices or procedures during an emergency response incident (e.g., respirator; decontamination procedure)
- An area-wide release of toxic fumes or particles, with the result of potential exposure of nonprotected emergency response personnel
- First-time operational experience with a specific chemical or other hazardous agent
- Notification through post-incident review that personnel might have been exposed to hazardous agents not previously identified or recognized
- Special conditions or circumstances encountered during emergency response operations that might have resulted in unforeseen exposures (e.g., temperature inversions, rain, discovery of purposely hidden hazardous wastes)

Termination Examination

The objective of the termination examination is to complete the total health profile of terminated personnel over the full period of employment. Specific requirements may be defined by pertinent regulations (e.g., in the U.S., 29 CFR 1910.120) or, more commonly, by insurance underwriters, but the termination examination must be based on preemployment screening, operational, and episodic monitoring data and information available to date, as well as on incident-related exposures or health symptoms experienced between the last medical examination and the termination examination.

In addition to taking specimens for the purpose of conducting final clinical or biochemical analyses (e.g., urinalysis, blood count enzymes), it is possible that response organizations will increasingly request specimens to be warehoused for potential future analysis by as yet undeveloped or currently experimental methodologies. Such an approach, which is now rarely practiced, will most likely receive increased attention due to mutually reinforcing trends in rapidly expanding analytical technology and in work-related health and safety litigation.

LIAISON WITH MEDICAL AUTHORITY

The various types of information and data generated in the progress of medical consultation and examination may be described in somewhat different terms by different medical practitioners and measured by different methodologies. For example, a physical examination given by any physician typically varies greatly from one physician to another, especially with regard to the physician's focus on a person's overall health as opposed to a focus on health in terms of work-related activity and risk. Whereas measurements of height and weight have some useful meaning with regard to a person's general health, there is generally little if any significant meaning to these parameters with regard to response-related activities (except those related to physical limitations, such as minimal height/weight restrictions imposed by equipment and/or procedures). As to actual methodologies or procedures, preferred methods are not necessarily those that are most precise but, in some circumstance, may be those that can be performed most rapidly. Which test to perform and which method to employ can only be decided by licensed medical authority, and these decisions must be made on a case-by-case basis.

The fact that decisions about the type of data and information required and the best means for obtaining that data and information are within the sole province of the physician does not mean that the safety officer has little or even no responsibility for the design of an effective and comprehensive medical surveillance program. On the contrary, no other responsibility of the safety officer is more demanding or requires more liaison and coordination with external medical authority. Of particular importance are the following considerations:

1. Most safety officers tend to assume that any licensed medical authority is suitable for the design and implementation of a medical surveillance program. This is definitely not the case—not for the industrial workplace, and certainly not for emergency response personnel. Where possible, the selection of medical professionals should be based on (a) professional experience in occupational medicine, (b) direct professional access to medical and analytical specialists and services regarding laboratory analyses and the timely processing of medically relevant data, (c) demonstrated experience in quality control management of all professional medical services, and (d) state-of-art practices.

2. Even when contracting with medical professionals who have extensive experience in occupational medical specialties, the safety officer must understand the importance of providing these professionals with comprehensive baseline data and information on work-related hazards. Such data and information include not only specific information (e.g., about ambient concentrations of hazardous chemicals), but also all information regarding the potential health significance of those chemicals, such as the target organs of the chemicals themselves and of combustion or water-reactive by-products. Although the safety officer might assume that medical professionals have this information, they often do not, which, given the tens of thousands of different chemicals in daily commerce, is understandable. It is also useful for medical service personnel to become aware of specific correlations and/or recommendations regarding types of hazardous exposures, target organs, and standard medical monitoring practices (see Table 6.2) that are increasingly available through such authorities as professional firefighting services and organizations, HAZMAT specialists, and regulatory agencies. Redundancy of information cannot harm; oversight of information that is readily available can be disastrous.

3. Despite the fact that the selection of appropriate medical testing of personnel is the responsibility of the medical professional, it is necessary that the emergency service safety officer thoroughly understand the basis of selection, including (a) the range of different medical tests and procedures that can be performed, (b) alternative methods for performing the

TABLE 6-2

TARGET ORGANS AND MEDICAL MONITORING ASSOCIATED WITH SELECTED SUBSTANCES (ADAPTED FROM NIOSH, USCG, AND EPA, 1985: OCCUPATIONAL SAFETY AND HEALTH GUIDANCE MANUAL FOR HAZARDOUS WASTE SITE ACTIVITIES)

Substance	Target Organs	Medical Monitoring
Aromatic Hydrocarbons	Blood; Bone marrow; Central nervous system; Eyes; Liver; Respiratory System; Skin; Kidney	Occupational/general medical history emphasizing prior exposure to these or other toxic agents; Medical examination with focus on liver, kidney, nervous system, and skin; Complete blood count; Platelet count; Measurement of kidney and liver function
Asbestos	Lungs; Gastrointestinal system	History and physical examination focused on lungs and gastrointestinal system; Stool test for occult blood evaluation; High quality chest X-ray and pulmonary function test
Halogenated Aliphatic Hydrocarbons	Central nervous system; Kidney; Liver; Skin	Occupational/general medical history emphasizing prior exposure to these or other toxic agents; Medical examination with focus on liver, kidney, nervous system, and skin; Laboratory testing for liver and kidney function; carboxyhemoglobin where relevant
Heavy Metals	Blood; Skin Cardiopulmonary system; Kidney; Liver; Lung; Central nervous system; Gastrointestinal system	Occupational/general medical history emphasizing prior exposure to these or other toxic agents; Medical examination with focus on liver, kidney, nervous system, and skin; Complete blood count; Platelet count; Measurement of kidney and liver function
Herbicides	Kidney; Liver; Central nervous system; Skin	History and physical exam focused on the skin and nervous system; Measurement of liver and kidney function; Urinalysis
Organochlorine Insecticides	Kidney; Liver; Central nervous system	History and physical exam focused on the nervous system; Measurement of kidney and liver function; Complete blood count for exposure to chlorocyclohexanes
Organophosphate & Carbamate Insecticides	Central nervous system; Liver; Kidney	Physical exam focused on the nervous system; Red blood cell cholinesterase levels for recent exposure (plasma cholinesterase for acute exposures); Measurement of delayed neurotoxicity and other effects
Polychlorinated Biphenyls (PCBs)	Liver; Skin; Central nervous system (possibly); Respiratory system (possible)	Physical exam focused on the skin and liver; Serum PCB levels; Triglycerides and cholesterol; Measurement of liver function

various tests and procedures, (c) interpretive criteria to be used in evaluating the significance of medical data and information, and (d) limits associated with the use of any medical data or information for the purpose of diagnosing potential health conditions. In this regard, the safety officer is well advised that, as with any contracted service affecting the health and safety of personnel, any potential liability that might result from incompetence or oversight is not necessarily restricted to the contractor, but might also accrue to the emergency response service itself. In short, it is always best to assume that the emergency service is ultimately responsible for accepting and implementing the professional recommendations of its medical service contractors, including the recommendations made by licensed medical practitioners.

4. Prior to committing to any professional medical surveillance service, the safety officer must ensure that medical surveillance reports will be presented in a format that provides for (a) ready comprehension of the significance of medical data and information by responsible emergency response service personnel, and (b) professional documentation regarding any potential need for follow-up action. Summaries of each type of health monitoring data should clearly highlight the significance of findings and present the basis for the interpretation of that significance (see Figure 6.3).

5. The processing and handling of any health-related information must be monitored assiduously to ensure confidentiality. The safety officer is strongly advised to examine in detail those control measures implemented by all relevant medical service personnel (including external examining physicians and medical-testing laboratory personnel) and, where necessary, to demand additional safeguards.

Of particular importance is the need to ensure that physicians do not report any health information about personnel to emergency service personnel that does not directly relate to work-related conditions or fitness for assigned work. The reporting of medical monitoring results and the maintenance of medical records must be conducted in strict conformity with established rules governing confidentiality and should be closely coordinated with emergency service legal counsel, human resource personnel, and the legal counsel of medical contractors.

TYPES OF MEDICAL SURVEILLANCE

Overview of Standard Medical Tests

Medical service contractors that provide essential medical surveillance services to emergency response organizations (and, increasingly, emergency response services themselves) employ different categorical terms for describing analytical or diagnostic tests. Some of these categories are based on long-used terms that reflect medical specialties and/or health service management, such as hematology, clinical chemistry, and urinalysis (see Table 6.3). Other categories are specific to the types of analyses available for diagnosing structural or functional

FIGURE 6-3

EXAMPLE OF A SUMMARY PRESENTATION OF MEDICAL MONITORING DATA REGARDING BLOOD HAD LEVELS AMONG PERSONNEL. SUCH A CONCISE VERBAL AND GRAPHIC PRESENTATION OF MEDICAL SURVEILANCE DATA IS NECESSARY IN ORDER TO ENSURE THAT NONMEDICALLY TRAINED PERSONS CAN UNDERSTAND THE SIGNIFICANCE OF DETAILED MEDICAL SURVEILLANCE FINDINGS AND THE RECOMMENDATIONS OF PHYSICIANS (ADAPTED FROM MATERIALS PROVIDED BY ENVIRONMENTAL MEDICINE RESOURCES, INC.)

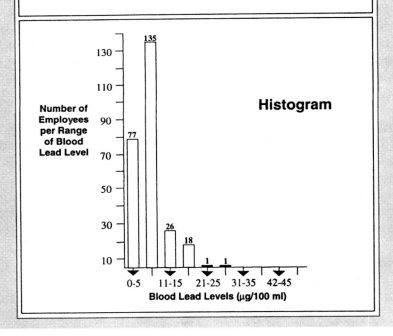

Blood Lead Levels

Inorganic lead is absorbed into the body through the lung and the intestinal tract. Organic lead can be absorbed through the lung, but the skin is the more common route.

In the studied population of personnel, no overt signs of lead intoxication were identified in health history data or by means of the physical examination. The following histogram displays the distribution of the blood lead levels obtained in the 1997 survey of personnel. No level met or exceeded the OSHA standard.

aberrations of specific organs and tissues, such as the liver, kidneys, and blood-forming functions (see Table 6.4).

However categorically described, individual analyses must be selected on the basis of a comprehensive assessment of the types of hazards actually encountered by emergency response personnel.

A detailed job (and task) analysis has long been recognized as the absolutely necessary first step in providing for the proper medical surveillance of emergency response personnel exposed to the smoke, toxic fumes and gases, and airborne

TABLE 6-3

TYPES OF INFORMATION TYPICALLY GENERATED IN A MEDICAL SURVEILLANCE PROGRAM FOR EMERGENCY RESPONSE PERSONNEL

Category	Analyses	
Medical History	-Medical/surgical history	-Allergy history
	-Family history	-Body systems
	-Work-exposure history	
Vital Signs	-Blood Pressure	-Pulse
Respiration	-Respiratory rate	-Pulmonary function
Vision	-Visual acuity	-Depth perception
	-Color vision	-Peripheral vision
Hearing	-Threshold value	
Urinalysis	-Specific gravity	-Albumin -Sugar
	-Blood	-pH -Microscopic examination
Electrocardiogram	-Resting cardiogram	-Stress Test
Radiology	-Chest X-Ray	
Hematology	-White blood cell count	-White cell differential count
	-Red blood cell count	-Platelet count
		-Hemoglobin
	-Corpuscular volume	-Corpuscular hemoglobin
	-Reticulocyte count	
Clinical Chemistries	-Serum glutamic pyruvate transaminase	-Total bilirubin
	-Alkaline phosphatase	-Lactic dehydrogenase
	-Serum glutamic oxaloacetic transaminase	
	-Gamma glutamyl transpeptidase	
	-Blood urea nitrogen	-Creatinine -Serum glucose
	-High density lipoprotein	-Low density lipoprotein
	-Triglyceride	-Sodium -Potassium
		Chloride
Physical Fitness	-Body fat	-Flexibility -Aerobic power
	-Muscular endurance	-Muscular power
	-Muscular strength	-Grip strength

particulates associated with firefighting and the management of hazardous chemical wastes. Since the advent of AIDS and, certainly continuing with our ongoing experience with emerging and reemerging infectious diseases (see Chapter 5), job analysis also increasingly has focused on how specific emergency response tasks can result in the exposure of response personnel to blood-borne pathogens and other biohazards. In the same period, more and more attention has been given to the careful analysis of just how response personnel become subject to the extremely debilitating effects of both physical and psychological stress—two types of risk that have always been attendant to emergency response but that demand closer scrutiny as *ergonomic* and *critical incident stress*.

Ergonomic Stress

Clearly a still developing discipline, ergonomics deals with the causes and consequences of mechanical tensions in the musculoskeletal systems, including those related to vibration, forceful exertion, awkward posture, repetitive

TABLE 6-4

RELEVANCE OF MONITORING TESTS AND ANALYSES TO
FUNCTIONAL HEALTH OF SELECTED ORGAN AND TISSUE
SYSTEMS (ADAPTED FROM NIOSH, USCG, AND EPA, 1985:
OCCUPATIONAL SAFETY AND HEALTH GUIDANCE MANUAL FOR
HAZARDOUS WASTE ACTIVITIES)

Function	Test	Examples of Analyses
Liver:		
General	Blood Tests	Total protein; Albumin; Globulin; Total/direct bilirubin
Obstruction	Enzyme test	Alkaline phosphatase
Cell Injury	Enzyme tests	Gamma glutamyl transpeptidase; Lactic dehydrogenase; Serum glutamic-oxaloacetic transaminase; Serum glutamic-pyruvic transaminase
Kidney:		
General	Blood Tests	Blood urea nitrogen; Creatine; Uric acid
Multiple Systems & Organs:		
General	Urinalysis	Color; Appearance; Specific gravity; pH; Qualitative glucose; Protein; Bile; Acetone; Occult blood; Microscopic examination of centrifuged sediment
Blood-Forming Function:		
General	Blood Tests	Complete blood count with differential and platelet evaluation, including white cell count, red blood count, hemoglobin, hematocrit or packed cell volume, and desired erythrocyte indeces; Reticulocyte count may be appropriate if there is a likelihood of exposure to hemolytic chemicals

and/or prolonged activity, localized bodily impact, and certain environmental conditions (e.g., heat, cold, or noise) (see Image 6.3). Whereas ergonomics tends to focus on mechanical forces operating of muscles, nerves, bones, and tendons, ergonomics also extends into the emotional and other psychological correlates of musculoskeletal dysfunction.

Mechanical stress on muscles, nerves, tendons, and bones can lead to physical injury to joints (e.g., in hand wrist, neck, back, elbow, shoulder, and leg) and surrounding tissue. Most injuries experienced by firefighters and EMTs are due to physical stress on the musculoskeletal system. Typically referred to as *cumulative trauma disorder* (CTD), such an injury may be relatively minor and last for a relatively brief period of time, with primary symptoms expressed as mild discomfort or ache. However, CTD may also become severe, result in acute pain, and may progress even to the point of complete disability.

In formulating a comprehensive medical surveillance program for emergency responders, careful assessment of ergonomic risk factors must be undertaken in stepwise fashion:

IMAGE 6-3

DES MOINES, IOWA, JULY 1993: A TOTAL OF 534 COUNTIES IN NINE STATES WERE DECLARED FOR FEDERAL DISASTER AID. AS A RESULT OF FLOODS, 168,340 PEOPLE REGISTERED FOR FEDERAL ASSISTANCE. MEDICAL SURVEILLANCE OF BOTH VICTIMS AND RESPONSE PERSONNEL IS ESSENTIAL

Source: Andrea Booher/FEMA Photo

1. Identification of specific actions and activities that may result in ergonomic stress, including not only those activities associated with on-site incident operations (see Figure 6.4), but also all activities related to (a) responding to (and returning from) the incident site; (b) post-emergency clean-up, replacement, and refurbishing; and (c) all other nonemergency operations
2. Identification of specific symptoms and syndromes to be associated with different types of ergonomic stress (see Table 6.5), with particular emphasis on target organs and personal risk factors (e.g., level or

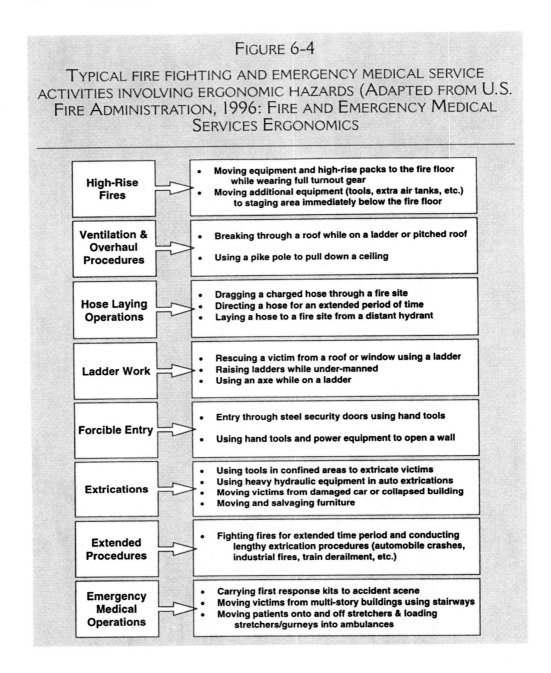

FIGURE 6-4

TYPICAL FIRE FIGHTING AND EMERGENCY MEDICAL SERVICE ACTIVITIES INVOLVING ERGONOMIC HAZARDS (ADAPTED FROM U.S. FIRE ADMINISTRATION, 1996: FIRE AND EMERGENCY MEDICAL SERVICES ERGONOMICS

High-Rise Fires	• Moving equipment and high-rise packs to the fire floor while wearing full turnout gear • Moving additional equipment (tools, extra air tanks, etc.) to staging area immediately below the fire floor
Ventilation & Overhaul Procedures	• Breaking through a roof while on a ladder or pitched roof • Using a pike pole to pull down a ceiling
Hose Laying Operations	• Dragging a charged hose through a fire site • Directing a hose for an extended period of time • Laying a hose to a fire site from a distant hydrant
Ladder Work	• Rescuing a victim from a roof or window using a ladder • Raising ladders while under-manned • Using an axe while on a ladder
Forcible Entry	• Entry through steel security doors using hand tools • Using hand tools and power equipment to open a wall
Extrications	• Using tools in confined areas to extricate victims • Using heavy hydraulic equipment in auto extrications • Moving victims from damaged car or collapsed building • Moving and salvaging furniture
Extended Procedures	• Fighting fires for extended time period and conducting lengthy extrication procedures (automobile crashes, industrial fires, train derailment, etc.)
Emergency Medical Operations	• Carrying first response kits to accident scene • Moving victims from multi-story buildings using stairways • Moving patients onto and off stretchers & loading stretchers/gurneys into ambulances

degree of response activity associated with ergonomic stress, age, physical condition).

3. Establishment of appropriate reporting procedures and personnel training program that promote timely notification of medical authority regarding complaints and symptoms of personnel.

Critical Incident Stress

All persons may be presumed to operate within a range of individual capacity to cope with events and circumstances that may vary from quite normal to

TABLE 6-5

TYPES OF CUMULATIVE TRAUMA DISORDERS (ADAPTED FROM U.S. FIRE ADMINISTRATION, 1996: FIRE AND EMERGENCY MEDICAL SERVICES ERGONOMICS)

Hand & Wrist	Neck & Back
■ Tendinitis: Inflammation of a tendon	■ Tension neck syndrome: Neck soreness, mostly related to static loading or tenseness of neck muscles
■ Synovitis: Inflammation of a tendon sheath	■ Posture strain: Chronic stretching or overuse of neck muscles or related soft tissue
■ Trigger finger: Tendinitis of the finger, typically locking the tendon in its sheath causing a snapping, jerking movement	■ Degenerative disc disease: Chronic degeneration, narrowing, and hardening of a spinal disc, typically with cracking of the disc surface
■ DeQuervain's disease: Tendinitis of the thumb, typically affecting the base of the thumb	■ Herniated disc: Rupturing or bulging out of a spinal disc
■ Ganglion cyst: Synovitis of tendons of the back of the hand causing a bump under the skin	■ Mechanical back syndrome: Degeneration of the spinal facet joints (parts of the vertebrae)
■ Digital neuritis: Inflammation of the nerves in the fingers caused by repeated contact or continuous pressure	■ Ligament sprain: Tearing or straining of a ligament (the fibrous connective tissue that helps support bones)
■ Carpal tunnel syndrome: Compression of the median nerve as it passes through the carpal tunnel	■ Muscle strain: Overstretching or overuse of a muscle
Elbow & Shoulder	
■ Epicondylitis ("tennis elbow"): Tendinitis of the elbow	■ Radial tunnel syndrome: Compression of the radial nerve in the forearm
■ Bursitis: Inflammation of the bursa (small pockets of fluid in the shoulder and elbow that help tendons glide)	■ Thoracic outlet syndrome: Compression of nerves and blood vessels under the collar bone
■ Rotator cuff tendinitis: Tendinitis in the shoulder	
Legs	
■ Subpatellar bursitis ("housemaid's knee"): Inflammation of patellar bursa	■ Shin splints: Microtears and infammation of muscle away from the shin bone
■ Patellar synovitis ("water on the knee": Inflammation of the synovial tissues deep in the knee joint	■ Plantar fascitis: Inflammation of fascia (thick connective tissue) in the arch of the foot
■ Phlebitis: Varicose veins and related blood vessel disorders (from constant standing)	■ Trochanteric bursitis: Inflammation of the bursa at the hip (from constant standing or bearing heaving weight)

extraordinary. Of course, just what is normal and what is extraordinary depends upon the individual—not necessarily as a conscious decision or perception, but simply as an experiential fact. When the individual experiences an incident that essentially overwhelms his or her capacity to cope, that incident may be described as being a *critical incident*; the consequence of a critical incident to the mental and physical well-being of the person who experiences it is known as *critical incident stress.*

Critical incident stress is a common phenomenon in extreme circumstances, such as circumstances involving mass death, coworker suicide, and the injury or death of children. Although such circumstances are encountered frequently by emergency response personnel, the frequency of encounter certainly does not necessarily immunize response personnel against subsequent critical incident stress, nor does it typically alleviate the potential severity of critical incident stress due to many other circumstance of emergency response, including (and perhaps, especially) a prolonged and extremely hazardous rescue effort conducted without success.

Critical incident stress is not only a common but also a quite natural phenomenon that does not in itself indicate any pathological state. People react extraordinarily to extraordinary events, and that immediate distress experience by any person involved in a horrifying incident typically does lessen with the passage of time. However, the actual rate of psychological recovery from the extraordinary to the ordinary is highly variable from person to person (see Figure 6.5), ranging from weeks to months. Unfortunately, there is always a small probability that an individual will not return to a state of fully functional normalcy, that the critical incident stress evolves into a persistent and profound state of *post-traumatic stress disorder* wherein the individual suffers substantial disability.

By the nature of their work, emergency response personnel are at special risk with respect to critical incident stress and post-traumatic stress disorder. After

FIGURE 6-5

TYPICAL PATTERN OF DEVELOPMENT OF CRITICAL INCIDENT STRESS OVER TIME (BASED ON DATA INFORMATION PROVIDED BY J. MITCHELL AND G. BRAY, 1990: EMERGENCY SERVICES STRESS. ENGLEWOOD CLIFFS: PRENTICE-HALL, INC.)

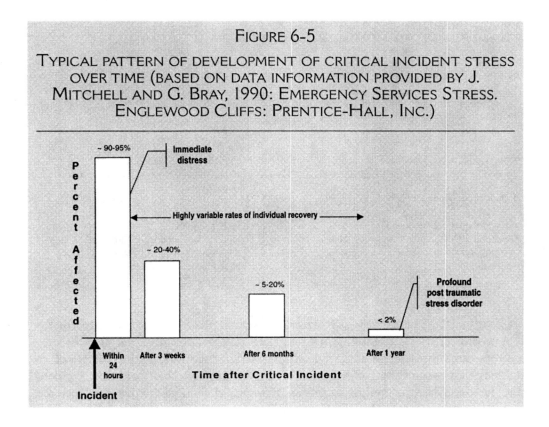

all, each incident presents a wide range of both environmental and psychological sources of stress (see Figure 6.6), one or any combination of which may prove sufficient to overwhelm the individual's capacity to cope.

Signs and symptoms of both critical incident stress and impending post-traumatic stress disorder are recognized as being of the following types:

- Cognitive (individual's state of awareness or capacity to make judgments)
- Emotional (the nature of an individual's feelings)
- Behavioral (objectively observed interactions of an individual with others and the environment)
- Physical (physical and physiological aspects of the body)

Signs and symptoms of critical incident stress in emergency response personnel (see Figure 6.7) should not be viewed as sufficient in and of themselves to initiate immediate intervention by medical authority. However, the emergency service safety officer should take due caution to observe the progress of such signs and symptoms over time. Signs and symptoms typically associated with the impending development of post-traumatic stress syndrome (see Figure 6.8) must trigger immediate intervention by appropriate medical authority.

Of course, in any comprehensive medical surveillance program, proactive approaches to health maintenance must always be given priority. With respect to the dangers of critical incident stress, careful attention must be given to the implementation of programs, procedures, and techniques that help personnel to reduce stress before that stress becomes a significant problem. Many

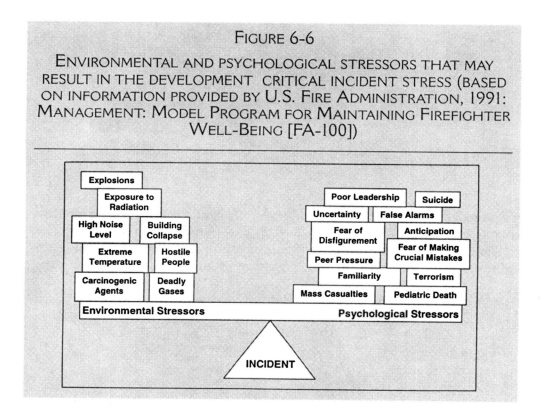

FIGURE 6-6

ENVIRONMENTAL AND PSYCHOLOGICAL STRESSORS THAT MAY RESULT IN THE DEVELOPMENT CRITICAL INCIDENT STRESS (BASED ON INFORMATION PROVIDED BY U.S. FIRE ADMINISTRATION, 1991: MANAGEMENT: MODEL PROGRAM FOR MAINTAINING FIREFIGHTER WELL-BEING [FA-100])

FIGURE 6-7

BEHAVIORAL, PHYSICAL, COGNITIVE, AND EMOTIONAL SYMPTOMS OF CRITICAL INCIDENT STRESS THAT REQUIRE CAUTIONARY MONITORING (BASED ON INFORMATION PROVIDED BY: (A) J. MITCHELL AND G. BRAY, 1990: EMERGENCY SERVICES STRESS. ENGLEWOOD CLIFFS: PRENTICE-HALL, INC.; AND (B) U.S. FIRE ADMINISTRATION, 1991: STRESS MANAGEMENT: MODEL PROGRAM FOR MAINTAINING FIREFIGHTER WELL-BEING [FA-100])

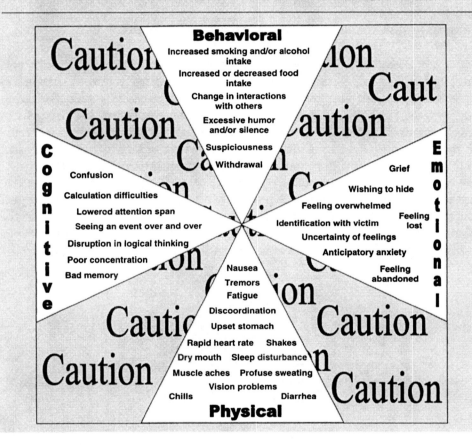

programs have been developed and are readily available through private consultants, governmental agencies (e.g., U.S. Fire Administration), professional organizations, and community services. Such programs typically include several of the following types of stress reduction techniques:

- Centralized relaxation (meditation, selective awareness, brain-wave biofeedback)
- Peripheral relaxation (progressive muscle relaxation, yoga, breath control, biofeedback)
- Time management
- Cognitive reappraisal
- Aerobic physical exercise
- Diversionary techniques (e.g., vacation planning, hobbies, group activities)

FIGURE 6-8

BEHAVIORAL, PHYSICAL, COGNITIVE, AND EMOTIONAL SYMPTOMS OF CRITICAL INCIDENT STRESS THAT REQUIRE IMMEDIATE INTERVENTION (BASED ON INFORMATION PROVIDED BY: (A) J. MITCHELL AND BRAY, 1990: EMERGENCY SERVICES STRESS. ENGLEWOOD CLIFFS: PRENTICE-HALL, INC.; AND (B) U.S. FIRE ADMINISTRATION, 1991: STRESS MANAGEMENT: MODEL PROGRAM FOR MAINTAINING FIREFIGHTER WELL-BEING [FA-100])

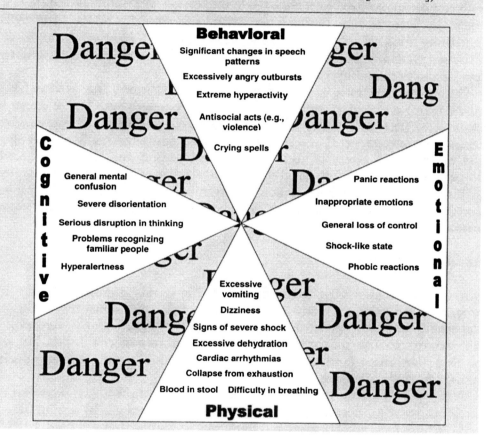

A particularly important approach to stress management, which typically includes both proactive and reactive elements, is the development of a *critical incident stress debriefing* (CISD) *team.*

CISD teams typically are composed of at least one mental health professional and two or more members who may be drawn from emergency response personnel, other emergency support services (e.g., hospital administrator), local clergy, and other community services. Some CISD teams service police, firefighters, and emergency response medical personnel in a multiagency and multijurisdictional setting; some operate within the context of a single organization. The basic functions of the CISD team are to:

- Acquaint personnel with methods for recognizing and reducing both on- and off-the-job stress

- Train personnel in the use of specific techniques to reduce stress in emergency response situations
- Provide assistance to personnel who are experiencing stress
- Serve as a referral service for personnel who need additional support services

Another approach to stress management in emergency response services is the *member assistance program* (MAP), which is also sometimes referred to as the *employee assistance program* (EAP). An EAP or MAP is essentially a referral program (as opposed to a treatment program) designed to assist personnel with respect to personal problems (including drug and alcohol abuse) that can affect job performance. Fire service MAPs (addressed by NFPA 1500-1987) sometimes are used to promote the general health of personnel through such activities as family orientation and educational programs on weight control, stress reduction, and hypertension.

The recent emergence of such programs as CISD teams and MAPs (EAPs) offers many alternative approaches to stress management, but it must be emphasized that such programs typically require significant time and effort to design and implement. Given the very sensitive nature of the issues involved and the potential effect of such programs on mental and emotional health, no organization should undertake to develop either of these programs except with the professional advice of community medical authorities and/or professional organizations having direct experience in both their design and implementation.

Audiometric Testing

Most often audiometric testing is included in standard medical monitoring programs, but there are many instances where this is not done; therefore, special attention must be given to this issue because it is particularly important in emergency response not simply as a matter of the responder's health, but also of his or her safety (and therefore of the safety of all team members) during actual incident operations.

In the United States, regulations include specific requirements regarding the certification of (a) persons performing audiometric testing, (b) testing devices and methodology, and (c) types and frequency of audiograms (29 CFR 1910.95 (g)). Though such requirements may vary from nation to nation, certain aspects of audiometric testing should be emphasized as being of universal concern.

1. The *baseline audiogram* is a hearing test conducted on an employee shortly following exposure to noise above the *action level*. As implied by the term "baseline," the objective of this audiogram is to define the normal hearing capacity of the employee in the absence of any work-related impairment. Subsequent audiograms therefore can be compared with the baseline audiogram to detect work-related hearing impairment and, therefore, the need for appropriate revisions to response service policies and practices regarding hearing protection. Given the importance of the baseline audiogram, it is imperative that it be completed before work-related noise above action levels (see Table 6.6) results in actual hearing

TABLE 6-6

NOISE EXPOSURE LIMITS PROMULGATED BY OSHA AND RECOMMENDED BY AMERICAN CONFERENCE OF GOVERNMENTAL INDUSTRIAL HYGIENISTS

	Duration (hours/day)	Sound Level (dB)
—ACGIH Standards—	16	80
If these levels are exceeded, engineering	8	85
controls will be used to reduce sound	4	90
to acceptable levels or hearing protectors	2	95
will be used	1	100
	1/2	105
	1/4	110
	1/8	115

—OSHA Standard—
Whenever noise exposures equal or exceed an 8-hour time-weighted average (TWA) of 85 dBA, a continuing and effective hearing conservation program shall be instituted

impairment. Also, because noise can cause short-as well as long-term hearing impairment known, respectively, as *temporary threshold shifts* (TTS) and *permanent threshold shifts* (PTS), it is necessary that the affected individual avoid or otherwise be protected from noise prior to examination for a baseline audiogram. In the United States, the standard requires that the baseline audiogram of an employee be established within six months (or, where mobile testing vans are utilized for the purpose, within one year) of the employee's first exposure at or above the action level and that the employee not be exposed to work-related noise for at least 14 hours immediately preceding the test.

2. Subsequent audiograms should be obtained for affected employees over short enough time intervals to ensure the early detection of hearing impairment. In the United States, the standard rule is that audiograms for employees be obtained at least annually. Longer intervals not only lead to an increased risk of permanent threshold shifts in specific individuals, but may also result in progressively expanding hearing risks to the emergency response team.

3. In comparing baseline and subsequent (at least annual) audiograms, the focus is on detecting an impairment of hearing. The action level for such a determination is what is called a *standard threshold shift*, which, in the United States, is a change in hearing of an average of 10dB or more at 2000, 3000, and 4000 Hz in either ear. American OSHA regulations also provide for (but do not mandate) the standardized adjustment of annual audiograms for the aging process in making the determination of a standard threshold shift. The safety officer is advised that the decision as to whether or not to adjust annual audiograms for age should not be left to medical judgment alone because this decision has broad implications that go well beyond the professional purview of any attending physician, including the following.

A. Correcting an annual audiogram for aging is basically equivalent to lessening any observed difference between baseline and annual audiograms, and this, in turn, results in the removal of a possibly desirable safety margin. Of course, by not correcting for age, the safety officer may cause an age-related hearing loss to be falsely attributed to workplace exposure.

B. Even where there may be good reason to exercise this option, serious consideration should be given to the relevance of the database used to perform the age adjustment. For example, are the data (even if provided by regulatory authority) biased to particular national, cultural, or other social (including gender) groups? Is the database current? What is the extent of professional consensus regarding both the utility and the limitations of the database?

4. Where the comparison of an employee's baseline and annual audiogram reveals a standard threshold shift, it is the obligation of the employer to take immediate responsive action. The first action, of course, is informing (in writing) the affected person. The second action is implementing appropriate correction. Determining what constitutes appropriate corrective action is certainly highly problematic because it encompasses not only regulatory mandates, but other considerations as well. Whether corrective action can be taken without continuing to place other responding team members at risk due to potential failure in communication with hearing-impaired personnel is certainly of major concern in any emergency response service.

If a standard threshold shift is attributed to or aggravated by the noise attendant to work-related noise, American OSHA regulations are clear about several required corrective actions, including:

- Other personnel who work under similar work-related conditions as those affected by a standard threshold shift and who do not use hearing protectors will be fitted with hearing protectors, trained in their use and care, and required to use them.
- Other personnel who work in similar work-related conditions as those affected by a standard threshold shift and who do use hearing protectors will be refitted and retrained in the use of hearing protectors and, if necessary, provided with protectors offering greater noise attenuation.
- The person affected by the work-related standard threshold shift will be referred to a clinical audiological evaluation or otological examination.

PROGRAMMATIC REVIEW

Once implemented, a medical surveillance program must be viewed as an essential lifeline for emergency response personnel and therefore should be carefully monitored for effectiveness and efficiency. It is especially important that there be at least an annual review of the entire program, with particular given to the following items:

1. A case-by-case review of any incident involving any aspect of the surveillance program, including episodic exposure to hazardous agents, medical monitoring data that require specific follow-up actions, and discernible trends in the frequency of episodic events or in monitoring data that may signify the need to review operational response procedures, the use of personal protective clothing and equipment, and personnel training requirement.
2. The need to include newly developed medical monitoring tests or to delete or modify other tests (e.g., frequency, methodology) in light of state-of-the-art developments in medical surveillance, changes in response service responsibilities and capabilities, or changes in regulatory requirements or applicable health and safety standards.
3. Performance evaluation of medical service contractors, including attending physicians and analytical laboratories, with particular emphasis on (a) the timeliness, comprehensiveness, and clarity of written reports and recommendations; (b) adequacy of technical and scientific documentation; and (c) satisfaction of response personnel.

It is recommended that, during this annual programmatic review, contracted medical service personnel be requested to (a) present an oral review of findings to date regarding their own quality-control management practices, and (b) discuss their own recommendations regarding any potential changes in the surveillance program.

STUDY GUIDE

True or False

1. One key objective of preemployment screening is to establish a baseline health profile that can be used to measure the effects of both short- and long-term exposure to work-related hazards.
2. Medical surveillance or respondent personnel undertaken to determine fitness for work is based entirely on the type of exposures expected as a result of task assignments.
3. The willingness of a respondent worker to undertake risks contrary to professional medical advice abrogates the employer's responsibility for the health and safety of that worker.
4. In addition to the various operational and medical expertise, the emergency planner should ensure that competent legal authority is involved in the development of a medical surveillance program.
5. A preemployment surveillance should include such tests and assessments that can be used in subsequent health monitoring to identify any changes that may be directly related to emergency-related exposures.

6. The sole objective of periodic operational monitoring is the early detection of adverse health effects of routine exposure to hazardous agents and situations.
7. Differences between baseline profiles and subsequent operational monitoring do not necessarily indicate disease or debilitation due to work-related exposure.
8. A single schedule of health monitoring (e.g., heart, kidney, liver) is sufficient for the detection of all possible adverse effects of exposure to hazardous agents.
9. A properly designed medical surveillance program should include a detailed action plan that precisely describes steps to be taken whenever operational monitoring results in the detection of a medically significant condition.
10. Episodic monitoring is triggered by a specific exposure event.
11. The symptoms of chemical exposure are typically distinct and can easily be distinguished from the symptoms of common diseases and/or health conditions.
12. The distributions of clinical severity for different types of infection are highly variable.
13. Any licensed medical authority is suitable for the design and implementation of a medical surveillance program.
14. A detailed job (or task) analysis is well recognized as the necessary first step in providing for the proper medical surveillance of emergency response personnel.
15. Most injuries experienced by firefighters and emergency medical technicians are due to physical stress on the musculoskeletal system.

Multiple Choice

1. Cumulative Trauma disorder is
   A. a physical condition
   B. a physiological condition
   C. a psychological condition
2. Signs and symptoms of both critical incident stress and impending post-traumatic stress disorder are of the following types:
   A. cognitive
   B. emotional
   C. behavioral
   D. physical
   E. all of the above
3. Critical incident stress may result from
   A. psychological stressors
   B. environmental stressors
   C. both A and B
4. The Critical Incident Stress debriefing (CISD) team
   A. focuses on both proactive and reactive aspects of responders' reactions to a critical incident

    B. may serve different types of emergency response personnel to any particular type of emergency

    C. focuses on the training of personnel in the use of specific techniques for reducing stress in emergency response situations

    D. all of the above

5. A member assistance program (MAP) or an employee assistance program (EAP) is essentially

    A. a referral program designed to assist personnel with respect to personal problems that can affect job performance

    B. a treatment program for personnel

    C. a program that promotes the general health of personnel

    D. A and C

6. A standard threshold shift is determined by

    A. the baseline audiogram

    B. the annual audiogram

    C. the comparison of annual and baseline audiograms

7. The trigger to immediate medical consultation is

    A. the observation of any symptom of exposure to a hazardous agent or situation

    B. operational situations that may suggest potential exposures

    C. both A and B

8. Annual audiograms taken subsequent to the baseline audiogram may be adjusted, taking into account

    A. age of the subject

    B. lifestyle of the subject

    C. severity of recent emergency events

9. The need to include newly developed medical monitoring tests or to delete or modify other tests (e.g., frequency, methodology) may be based on

    A. state-of-the-art developments in medical surveillance

    B. changes in response serve responsibilities and capabilities

    C. changes in regulatory requirements

    D. changes in applicable health and safety standards

    E. all of the above

Essays

1. As a safety officer for either a corporate or municipal organization having emergency response responsibility, how would you ensure that the actual data generated by your surveillance program actually are used to protect the health and safety of your personnel?

2. Given the fact that any comprehensive health surveillance program includes a very wide range of technical and scientific information, how would you, as a safety officer, ensure the adequacy of your surveillance program?

3. How would you as a safety officer begin to collect the information you need to master in order to plan and implement an effective comprehensive medical surveillance program?

Case Study

Contact local public or private organizations/or personnel having emergency response responsibilities (e.g., fire, police, ambulance, hospital, hazardous waste contractor, state emergency management agency) and request information regarding their medical surveillance programs. In light of this information and your reading/discussions of this chapter,

1. How would you assess the comprehensiveness of these programs?
2. How might you evaluate the effectiveness of these programs?
3. What concerns might you have if you were a member of one of these organizations?

PERSONAL PROTECTIVE CLOTHING AND EQUIPMENT

INTRODUCTION

Personal protective clothing (PPC) and equipment (PPE) are selected only after all managerial (sometimes called "administrative") efforts and potential engineering controls have been scrutinized toward the objective of minimizing risk. Managerial efforts include alternative scheduling (to reduce total time of exposure to specific hazards) as well as specific procedures and protocols used to minimize risk. Engineering controls include alternative uses of barriers, area ventilation (e.g., vehicular ventilation systems for ambulances), and spatial isolation of hazards (e.g., biohazard disposal area, on-site decontamination area, hazardous runoff collection system) to confine hazardous materials and/or conditions within areas under management control. PPC and PPE must be viewed as the last available means of controlling responder risk and, therefore, the most sensitive to the consequence of failure. In short, any failure of PPC and/or PPE exposes the responder to immediate and potentially life-threatening risk.

Although protective clothing and equipment are intended specifically to protect the responder against risks otherwise uncontrolled by managerial procedures and engineering controls, PPC and PPE also present additional risks of their own, including impaired vision, mobility, and communication, a well as variable degrees of physical and psychological stress of the user. The objective therefore must be to achieve an effective and assured balance between the risks attendant to the incident and the risks inherent in the use of PPC and PPE, while avoiding both over- and under-protection.

SELECTION OF PPC AND PPE

A wide range of factors must be considered in the selection of PPC and PPE, ranging from general criteria of durability and comfort to highly specific criteria for assessing the capacity of materials to withstand chemical and physical agents and conditions (see Image 7.1). Examples of basic categories of factors to be considered include:

- *Design features* (e.g., sizes and other options, ease of donning, accommodation to use of diverse garment add-ons as well as ancillary equipment, restriction of mobility, visibility in dark, color, weight, comfort for wearer, ease of cleaning and decontamination, ease of field evaluation of functional integrity)
- *Chemical resistance* (e.g., permeation of chemicals through material; discoloration; loss of physical strength and other degradation due to chemical interactions; penetration of liquids, gases, vapors and mists through zippers, seams, closures, seals or material imperfections) (see Image 7.2)
- *Physical quality* (e.g., resistance to wear, tear, puncture, and abrasion; pliability and flexibility under variable environmental conditions; susceptibility to shrinkage; integrity under extremes of temperature; flame resistance; breathability; resistance to decontamination procedures and solutions; resistance to physical shock and vibration; resistance to static electrical charge and electrical current)
- *Vendor-related factors* (e.g., ready availability of replacement parts, cost, special servicing requirements, available customization)

IMAGE 7-1

NEW ORLEANS, LA, AUGUST 31, 2005: A RESIDENT IS TRANSPORTED BY A FEMA URBAN SEARCH AND RESCUE TEAM FROM MISSOURI AWAY FROM HER HOUSE

Source: Jocelyn Augustino/FEMA

IMAGE 7-2

WORKERS IN PROTECTIVE CLOTHING TO GUARD AGAINST CONTAMINATION FROM HAZARDOUS WASTE

Source: CDC

- *Other factors* (e.g., proven effectiveness, use by other similar emergency services, user complaints, professional certification of meeting appropriate engineering standards, documented failures, state-of-the-art technology, maintenance requirements, service life, compatibility with equipment used by other responding services)

No selection of PPC or PPE should be made without full documentation of (a) design and engineering specifications provided by manufacturers, (b) relevant standards, specifications, and guidelines, including those promulgated by governmental agencies (e.g., OSHA, NIOSH, EPA, U.S. Fire Administration) and professional organizations (e.g., National Fire Prevention Association (NFPA), American Conference of Governmental Industrial Hygienists (ACGIH), American Society for Testing and Materials (ASTM)) and (c) available methods and procedures for conducting visual inspections and field assessments (see Table 7.1) of critical parameters of PPC and PPE performance. Much of this information is today readily available via the Internet, especially through Internet links and networks accessible through governmental agencies, including:

- CDC (Center for Disease Control): http://www.cdc.gov/
- EPA (Environmental Protection Agency): http://www.epa.gov/
- U.S. FA (U.S. Fire Administration): http://www.usfa.fema.gov/

TABLE 7-1

RECOMMENDED CHEMICALS FOR FIELD EVALUATION OF THE PERFORMANCE OF PROTECTIVE CLOTHING (ADAPTED FROM U.S. DEPARTMENT OF LABOR, OSHA. OSHA TECHNICAL MANUAL. OSHA ELECTRONIC REFERENCE LIBRARY)

Note	—Chemical—	—Class—
U.S. EPA has developed a portable test kit that allows field qualification of protective clothing materials within one hour using these chemicals.	Acetone	Ketone
	Acetonitrlle	Nitrile
	Ammonia	Strong base gas
	1,3-Butadlene	Olefin gas
	Carbon disulfide	Sulfur-containing organic
	Chlorine	Inorganic gas
	Dichloromethane	Chlorinated hydrocarbon
	Dietheylamine	Amine
	Dimethyl formamide	Amide
	Ethyl acetate	Ester
	Ethylene oxide	Oxygen heterocyclic gas
	Hexane	Aliphatic hydrocarbon
Use of this kit may overcome the absence of specific data and provide additional criteria for selection of appropriate clothing.	Hydrogen chloride	Acid gas
	Methanol	Alcohol
	Methyl chloride	Chlorinated hydrocarbon gas
	Nitrobenzene	Nitrogen-containing organic
	Sodium hydroxide	Inorganic base
	Sulfuric acid	Inorganic acid
	Tetrachloroethylene	Chlorinated hydrocarbon
	Tetrahydrofuran	Oxygen heterocyclic
	Toluene	Aromatic hydrocarbon

- FEMA (Federal Emergency Management Administration): http://www.fema.gov/
- NCID (National Center for Infectious Diseases): http://www.cdc.gov/ncidod/index.htm
- NIOSH (National Institute of Occupational Safety and Health): http://www.niosh.com.my/index.asp
- NRT (National Response Team): http://www.nrt.org/
- OSHA (Occupational Safety and Health Administration): http://www.osha.gov/

PROTECTIVE CLOTHING AND ENSEMBLES

Protective clothing for emergency responders is inclusive of individual items (e.g., bib overalls, helmet, bunker coat) as well as ensembles, which are collections of items that are integrated to meet the needs of a specific condition (e.g., approach suit, fragmentation suit) or constellation of risks (e.g., hazardous chemicals). Each item or ensemble should be considered to offer specific types of protection as well as to impose specific limitations.

Some ensembles are recommended on the basis of generic types of protection required, such as splashes of hazardous chemicals, chemical vapors, and hazardous dusts (see Table 7.2). Other ensembles are recommended on the basis

TABLE 7-2

TYPES OF PROTECTIVE CLOTHING FOR FULL-BODY PROTECTION
(ADAPTED FROM U.S. DEPARTMENT OF LABOR, OSHA. OSHA
TECHNICAL MANUAL. OSHA ELECTRONIC REFERENCE LIBRARY)

Description	Type of Protection	User Considerations
Fully Encapsulating Suit		
One-piece garment; boots and gloves may be integral, attached and replaceable, or separate	Protects against splashes, dust, gases and vapors	Does not allow body heat to escape; may contribute to heat stress in wearer, particularly if worn in conjunction with a closed-circuit SCBA; a cooling garment may be needed; impairs worker mobility, vision, and communication
Nonencapsulating Suit		
Jacket, hood, pants or bib overalls, and one-piece coveralls	Protects against splashes, dust, and other materials, but not against gases and vapors; does not protect parts of head or neck	Do not use where gas-tight or pervasive splashing protection is required; may contribute to heat stress in wearer; tape-seal connections between pant cuffs and boots and between gloves and sleeves
Aprons, Leggings, and Sleeve Protectors		
Fully sleeved and gloved apron; separate coverings for arms and legs; commonly worn over nonencapsulating suit	Provides additional splash protection of chest, forearms, and legs	Whenever possible, should be used over a nonencapsulating suit to minimize potential heat stress; useful for sampling, labeling, and analysis operations; should be used only when there is a low probability of total body contact with contaminats
Firefighters' Protective Clothing		
Gloves, helmet, running or bunker coat, running or bunker pants	Protects against heat, hot water, and some particles; does not protect against gases and vapors, or chemical permeation or degradation. NFPA Standard No. 1971 specifies that a garment consists of an outer shell, an inner liner and a vapor barrier with a minimum water penetration of 25 lbs/in² to prevent passage of hot water	Decontamination is difficult; should not be worn in areas where protection against gases, vapors, chemical splashes or permeation is required

Continued

TABLE 7-2—Continued

TYPES OF PROTECTIVE CLOTHING FOR FULL-BODY PROTECTION (ADAPTED FROM U.S. DEPARTMENT OF LABOR, OSHA. OSHA TECHNICAL MANUAL. OSHA ELECTRONIC REFERENCE LIBRARY)

Description	Type of Protection	User Considerations
Proximity Garment (Approach Suit)		
One- or two-piece overgarment with boot covers, gloves and hood of aluminized nylon or cotton fabric; normally worn over other protective clothing, firefighters' bunker gear, or flame-retardent coveralls	Protects against splashes, dust, gases, and vapors	Does not allow body heat to escape; may contribute to heat stress in wearer, particularly if worn in conjunction with a closed-circuit SCBA; a cooling garmet may be needed; impairs worker mobility, vision, and communication
Blast and Fragmentation Suit		
Blast and fragmentation vests and clothing, bomb blankets, and bomb carriers	Provides some protection against very small detonations; bomb blankets and baskets can help redirect a blast	Does not provide for hearing protection
Radiation-Contamination Protective Suit		
Various types of protective clothing contamination of the body by radioactive particles	Protects against alpha and beta particles; against gamma radiation	Designed to prevent skin contamination; if radiation is detected on site, consult an experienced radiation expert and evacuate personnel until the radiation hazard has been evaluated
Flame/Fire Retardant Coveralls		
Normally worn as an undergarment	Provides protection from flash fires	Adds bulk and may exacerbate heat stress problems and impair mobility

of level of protection required under certain emergency conditions, such as degree of protection required for skin, eyes, and respiratory system when working with hazardous chemicals (see Table 7.3). Still other ensembles are defined essentially by the type of emergency-related mission undertaken, such as search and rescue missions under a wide range of environmental factors.

For example, urban search and rescue missions are classified as involving three distinct sets of conditions: technical rescue, swift water rescue (see Image 7.3), and contaminated water diving. Technical rescues are typically land-based rescues where the principal hazards are physical, such as those encountered in the collapse of structures; the protective turnout clothing usually worn for fire-fighting is too bulky or heavy for responders who must extricate victims from

TABLE 7-3

TYPES OF PROTECTIVE ENSEMBLES (ADAPTED FROM NIOSH, USCG, AND EPA, 1985. OCCUPATIONAL SAFETY AND HEALTH GUIDANCE MANUAL FOR HAZARDOUS WASTE SITE ACTIVITIES)

Level of Protection and Equipment	Overview of Protection	Conditions for Use and Limitations
A	The highest available level of respiratory, skin, and eye protection	▪ The chemical substance has been identified and requires the highest level of protection for skin, eyes, and the respiratory system based on either:
Recommended: ▪ Pressure-demand, full-facepiece SCBA or pressure-demand supplied air respirator with escape SCBA ▪ Fully encapsulating, chemical resistant suit		1. Measured (or potential for) high concentration of atmospheric vapors, gases, or particulates, or 2. Site operations and work functions involving a high potential for splash, immersion, or exposure to unexpected vapors, gases, or particulates of materials that are harmful to skin or capable of being absorbed through the intact skin.
▪ Inner chemical-resistant gloves ▪ Chemical resistant safety boots/shoes ▪ Two-way radio		▪ Substances with a high degree of hazard to the skin are known or suspected to be present, and skin contact is possible. ▪ Operations must be conducted in confined, poorly ventilated areas until the absence of conditions requiring Level A protection is determined.
Optional: ▪ Cooling unit ▪ Coveralls ▪ Long cotton underwear ▪ Hard hat ▪ Disposable gloves and boot covers		▪ Fully encapsulating suit materials must be compatible with the substances involved.
B	The same level of respiratory protection but less skin protection than Level A.	▪ The type and atmospheric concentration of substances have been identified and require a high level of respiratory protection, but less skin protection. This involves atmospheres

Continued

TABLE 7-3—*Continued*

TYPES OF PROTECTIVE ENSEMBLES (ADAPTED FROM NIOSH, USCG, AND EPA, 1985. OCCUPATIONAL SAFETY AND HEALTH GUIDANCE MANUAL FOR HAZARDOUS WASTE SITE ACTIVITIES)

Level of Protection and Equipment	Overview of Protection	Conditions for Use and Limitations
Recommended: ■ Pressure-demand, full facepiece SCBA or pressure-demand supplied air respirator with escape SCBA		– with IDLH concentrations of specific substances that do not represent a severe skin hazard, or
■ Chemical-resistant clothing	This is the minimum level recommended for initial site entries until the hazards have been further identified.	– that do not meet the criteria for use of air-purifying respirators.
■ Inner and outer chemical-resistant gloves		■ Atmosphere contains less than 19.5 % oxygen. ■ Presence of incompletely identified vapors or gases is indicated by direct-reading organic vapor detection instrument, but vapors and gases are not suspected of containing high levels of chemicals harmful to skin or capable of being absorbed through intact skin.
■ Chemical resistant safety boots/shoes ■ Hard Hat ■ Two-way Radio		■ Use only when highly unlikely that the work will generate either high concentrations of vapors, gases, or particulates or splashes of material will affect exposed skin.
Optional: ■ Coveralls ■ Disposable boot covers ■ Face shield ■ Long cotton underwear		

Continued

TABLE 7-3—*Continued*

TYPES OF PROTECTIVE ENSEMBLES (ADAPTED FROM NIOSH, USCG, AND EPA, 1985. OCCUPATIONAL SAFETY AND HEALTH GUIDANCE MANUAL FOR HAZARDOUS WASTE SITE ACTIVITIES)

Level of Protection and Equipment	Overview of Protection	Conditions for Use and Limitations
C	The same level of skin protection as Level B, but a lower level of respiratory protection	▪ Atmospheric contaminants, liquid splashes, or other direct contact will not adversely affect any exposed skin.
Recommended: ▪ Full-facepiece, air purifying, canister equipped respirator ▪ Chemical resistant clothing ▪ Inner and outer chemical resistant gloves ▪ Chemical resistant safety boots/shoes ▪ Hard hat ▪ Two-way radio **Optional:** ▪ Coveralls ▪ Disposable boot covers ▪ Face shield ▪ Escape mask ▪ Long cotton underwear		▪ The types of air contaminants have been identified, concentrations measured, and a canister is available that can remove the contaminant. ▪ All criteria for the use of air- purifying respirators are met. ▪ Atmospheric concentration of chemicals must not exceed IDLH levels. ▪ The atmosphere must contain at least 19.5% oxygen.
D **Recommended:** ▪ Coveralls ▪ Safety boots/shoes ▪ Safety glasses or chemical splash goggles ▪ Hard hat **Optional:** ▪ Gloves ▪ Escape mask ▪ Face shield	No respiratory protection; minimal skin protection	▪ The atmosphere contains no known hazard. ▪ Work functions preclude splashes, immersion, or the potential for unexpected inhalation of or contact with hazardous levels of any chemicals. ▪ This level should not be worn in the Exclusion Zone. ▪ The atmosphere must contain at least 19.5 % oxygen.

IMAGE 7-3

LOS ANGELES COUNTY, CA, NOVEMBER 20, 2003: FIREFIGHTERS
WITH L.A. COUNTY'S WATER RESCUE TEAM PRACTICE THEIR
SKILLS AT A COMPANY DRILL

Source: Jason Pack/FEMA News Photo

collapsed debris. Protective clothing and ensembles used in such circumstances (see Table 7.4) must be flame resistant, but also lighter and permissive of greater physical flexibility than that accorded by typical firefighting gear. Principal criteria used for selecting PPC and PPE for technical rescue include:

- Protection from physical hazards (abrasion, tears, cuts, and punctures)
- High degree of visibility (including light and dark conditions)
- Thermal and physical comfort, fit, and mobility
- Protection from airborne particulates
- Limited flame and heat protection
- Limited chemical flash fire protection
- Limited electrical exposure protection
- Minimal chemical protection
- Minimal protection from biological fluids

Swift water rescue primarily involves the risk of drowning not only through the press of water, but also through entanglement, as well as hypothermia.

TABLE 7-4

TECHNICAL RESCUE ENSEMBLE (ADAPTED FROM U.S. FIRE ADMINISTRATION, 1993: PROTECTIVE CLOTHING AND EQUIPMENT NEEDS OF EMERGENCY RESPONDERS FOR URBAN SEARCH AND RESCUE MISSIONS [FA-136])

Protective Garment

Should cover the wearer's upper and lower torso, legs, and arms; may be one piece coveralls or two piece coat and trousers; two piece garments should have sufficient overlap for mid-torso protection

Materials should resist tearing, snagging, and abrasion due to physical environment

Should be reinforced at elbows and knees; seam and closure strength should be equal to strength of material

Should provide high visibility in the dark

Material should be breathable and comfortable to wear for extended periods of time

Materials should resist ignition when contacted by flame

When exposed to convective and radiant heat, materials should prevent transmission of heat that could burn the wearer's skin. In hot environments, materials should not shrink

Should maintain measured size when repeatedly cleaned

Materials should resist static charge accumulation

Supplemental liners should be provided which prevent chemical penetration of common fire scene chemicals and biologically contaminated liquids

Protective Hood

Should cover the wearer's head and neck with the exception of those areas of the face which may be covered by a SCBA or air purifying respirator

Material should be breathable and comfortable to wear for extended periods of time

Material should resist ignition when contacted by flame. When exposed to convective and radiant heat, materials should prevent transmission of heat that could burn the wearer's skin. In hot environments, material should not shrink

Should maintain measured size when repeatedly cleaned

Material should resist static charge accumulation

Protective Gloves

Should cover the wearer's hands to one inch above the wrist and include a wristlet which prevents entry of foreign objects into the glove

Materials should resist tearing, cutting, punctures, or abrasion due to the physical environment

Materials should be breathable and comfortable to wear for extended periods of time

Should offer adequate dexterity and grip to handle tools and machinery

Materials should resist ignition when contacted by flame. When exposed to convective and radiant heat, should prevent transmission of heat that could burn the wearer's skin. In hot environments, materials should not shrink

Materials should insulate wearer from electrical currents

Supplemental gloves should be provided which prevent chemical penetration of common fire scene chemicals and biologically contaminated fluids

Protective Boots

Should cover wearer's foot from the bottom of the foot to a point eight inches above the foot bottom

Should not have any exposed metal parts. Should include a ladder shank and non-metallic toe protective cap

Upper materials should resist abrasion, cutting, or puncture due to the physical environment. Soles should resist abrasion and puncture

Materials should resist ignition when contacted by flame. When exposed to convective and radiant heat, materials should prevent transmission of heat that could burn the wearer's skin. In hot environments, materials should not shrink

Should maintain water-tight integrity following repeated flexing

Should insulate the wearer from electrical currents

Materials should prevent chemical penetration of common fire scene chemicals and biologically contaminated fluids

Continued

TABLE 7-4—*Continued*

TECHNICAL RESCUE ENSEMBLE (ADAPTED FROM U.S. FIRE ADMINISTRATION, 1993: PROTECTIVE CLOTHING AND EQUIPMENT NEEDS OF EMERGENCY RESPONDERS FOR URBAN SEARCH AND RESCUE MISSIONS [FA-136])

Protective Helmet
Should cover the top of the wearer's head. Should resist impact on top and sides from falling objects
Straps should keep helmet in place when impacted
When exposed to convective and radiant heat, materials should prevent transmission of heat that could burn the wearer's skin. In hot environments, materials should not shrink

Goggles
Should prevent impact of foreign objects to the eyes. Should keep particulates from reaching eyes
Materials should resist ignition when contacted by flame. When exposed to convective and radiant heat, materials should prevent transmission of heat that could burn the wearer's eyes. In hot environments, materials should not shrink

Air Purifying Respirator
Should be NIOSH certified. Should keep fine particulates from entering wearer's respiratory system

Ear Protectors
Should meet ANSI requirements for ear and hearing protection

Appropriate criteria for selecting PPC and PPE for swift water rescue (see Table 7.5) include:

- Flotation (buoyancy)
- Insulation from cold water exposure
- Protection from physical hazard (abrasion, tears, cuts, and punctures)
- High degree of visibility (in light and dark)
- Physical and thermal comfort, fit, and mobility
- Limited chemical protection
- Limited protection from biological fluids

Although diving in contaminated water involves some of the same hazards as those encountered in swift water rescue, primary attention must be given to protecting personnel from exposure to biological and chemical contaminants. Criteria used for selecting PPC and PPE for contaminated water rescue (see Table 7.6) include:

- Integrity of breathing air supply
- Integrity of overall system to water penetration
- Insulation from cold water exposure
- Protection from physical hazards (abrasion, tears, cuts, and punctures)
- Physical and thermal comfort, fit, and mobility
- Protection from chemicals and biological fluids

TABLE 7-5

SWIFT WATER RESCUE ENSEMBLE (ADAPTED FROM U.S. FIRE ADMINISTRATION, 1993: PROTECTIVE CLOTHING AND EQUIPMENT NEEDS OF EMERGENCY RESPONDERS FOR URBAN SEARCH AND RESCUE MISSIONS [FA-136])

Personal Floatation Device
Should meet U.S. Coast Guard requirement for Type III or Type V
Should include hardware for attaching lifeline
Should be corrosion resistant and have sufficient strength to withstand swift water
 forces

Protective Dry Suit
Should cover wearer's upper and lower torso, arms and legs. Should be easily and
 quickly donned
Should prevent water penetration to parts of body covered (should include wrist,
 foot, and neck seals)
Materials should provide insulation from cold water exposure for at least one hour
Materials should resist tearing, snagging, and abrasion due to physical
 environment
Should be reinforced at elbows and knees. Seam and closure strength should be
 equal to strength of material
Should not shrink after cleaning or contact with warm water
Should provide high visibility in the dark
Materials should be breathable and comfortable to wear for extended periods of time
Materials should prevent penetration of diluted chemicals and biological
 contaminants
Materials should not retain contaminants following clean water rinsing

Protective Gloves
Should be 5-fingered design and cover wearer's hands to one inch above the wrist
Should be available in at least 3-sizes
Should limit water penetration to hands and provide insulation from cold water
 exposure for at least one hour
Materials should resist tearing, cutting, and punctures due to physical
 environment
Materials should be breathable and comfortable to wear for extended periods of time
Should offer adequate dexterity and grip to tie knots and operate a knife
Should not slip off wearer's hand if inner glove is worn
Retention straps should not become loosened by use. Metal parts should not
 corrode or rust
Materials should prevent penetration of diluted chemical and biological
 contaminants. Should not retain contaminants following clean water rinsing

Personal Booties
Should cover wearer's feet to one inch above the ankle
Should limit water penetration to feet and provide insulation from cold water
 exposure for at least one hour
Materials should resist tearing, cutting, puncture, and abrasion due to physical
 environment
Soles should be slip-resistant and provide good traction under wet condition
Should accommodate swimming fins
Retention straps should not become loosened in use. Metal parts should not
 corrode or rust
Materials should prevent penetration of diluted chemicals and biological
 contaminants
Should not retain contaminants following clean water rinsing

Swimming Fins
Should allow wearer to walk normally

Continued

TABLE 7-5—Continued

SWIFT WATER RESCUE ENSEMBLE (ADAPTED FROM U.S. FIRE ADMINISTRATION, 1993: PROTECTIVE CLOTHING AND EQUIPMENT NEEDS OF EMERGENCY RESPONDERS FOR URBAN SEARCH AND RESCUE MISSIONS [FA-136])

Helmet
Should cover top of wearer's head. Should resist impact on top and sides from floating or stationary objects
Should be ventilated to allow passage of water. Should not have brim or other surfaces suceptable to swift water forces
Metal parts should not corode or rust

Knife
Should be single edged. Should remain in sheath when inverted and shaken

Whistle
Should be non-metallic. Should not include a pall

TABLE 7-6

CONTAMINATED WATER DIVING ENSEMBLE (ADAPTED FROM U.S. FIRE ADMINISTRATION, 1993: PROTECTIVE CLOTHING AND EQUIPMENT NEEDS OF EMERGENCY RESPONDERS FOR URBAN SEARCH AND RESCUE MISSIONS [FA-136])

Protective Dry Suit
Should cover wearer's upper and lower torso, arms, legs, and feet. Should be hooded or have attachable hood
Should prevent water penetration to parts of body covered (should include wrist, foot, and neck seals)
Materials should provide insulation from cold water exposure for at least one hour
Materials should be rugged and strong and resist tearing, snagging, and abrasion due to physical environment
Wrist, ankle, or neck seal materials should be adjustable for sizing
Wrist, ankle, or neck seal materials should resist cuts and punctures
Should be reinforced at elbows and knees. Seam and closure strength should be equal to material strength
Should provide high visibility in dark
Materials should prevent penetration of diluted chemicals and biological contaminants. Materials should not retain contaminants following clean water rinsing

Protective Gloves
Should mate directly to the dry suit
Materials should prevent penetration of diluted chemicals and biological contaminants
Materials should provide insulation from cold water exposure for at least one hour
Materials should resist cuts, punctures, and abrasion due to physical environment
Should offer adequate dexterity and grip to tie knots and operate a knife
Should not retain contaminants following clean water rinsing

Protective Booties
Should be a part of the drysuit (directly attached)
Materials should prevent penetration of diluted chemicals and biological contaminants
Materials should provide insulation from cold water exposure for at least one hour
Materials should resist tearing, cutting, punctures, and abrasion due to physical environment
Sole materials should resist puncture and wear due to abrasion. Soles should be slip resistant
Should not retain contaminants following clean water rinsing

Continued

TABLE 7-6—*Continued*

CONTAMINATED WATER DIVING ENSEMBLE (ADAPTED FROM U.S. FIRE ADMINISTRATION, 1993: PROTECTIVE CLOTHING AND EQUIPMENT NEEDS OF EMERGENCY RESPONDERS FOR URBAN SEARCH AND RESCUE MISSIONS [FA-136])

Swimming Fins
Should be resistant to diluted chemicals

Helmet
Should cover entire head and neck. Should mate directly to the dry suit with safety mechanism to avoid accidental removal
Should be neutrally buoyant in water
Should include non-return valve in breathing system. Should include emergency valve for connecting bail-out system
Should have double exhaust (for demand diving helmets)
Should have defogging mechanism. Should have shatter resistant face piece
Should have equalizing device (to equalize air pressure in ears)
Should include integrated communication system
Should not retain contaminants following clean water rinsing

Full Face Mask
Should enclose eyes, nose, and mouth. Should include integral second stage regulator
In surface-supplied mode, should be used with bail-out block
Should have equalizing device, automatic defogging mechanism, and earphone pockets
Should have low volume, large buckles and wide straps, and modular communications components
Should not retain contaminants following clean water rinsing

Dry Suit Underwear
Should provide insulating performance even when wet. Must not produce lint

Communication System
May be hard-wire or wireless system. Should be constructed of rugged materials
Should have mechanism to attach electronics housing to tank harness or buoyancy compensator
Exposed parts should be able to be decontaminated
Should include back-up systems (line pull signals)
Compressed Air Supply
Should be a low pressure (175–250 psi) compressor or series of high-pressure bottles
Should include emergency air supply for diver (bail-out system)
Air Manifold Box
Should monitor air-pressure to diver. Should regulate high-pressure air to proper pressure for diver
Should provide connection for top-side emergency air supply
Pneumofathometer
Should usually be contained in the air manifold box. Should be accurate to 0.25% of the gauge's full scale
Umbilical
Should be at least 250 ft long. Air supply hose should have minimal length change when pressurized
Air supply hose should be at least 3/8 in. diameter
Air supply hose should have rated pressure at least 50 % higher than maximum pressure required by regulator or maximum diving depth
Should resist kinking. Should include air-tight fittings and strong diving tether
Should have individual umbilical components (hose, communications wire) connected with plastic cable ties
Should include stainless steel ring to attach air supply hose to diver's harness
Weight Belt/Harness
Materials should be compatible with diluted chemicals. Materials should not retain contaminants following clean water rinsing

Continued

TABLE 7-6—*Continued*

CONTAMINATED WATER DIVING ENSEMBLE (ADAPTED FROM U.S.
FIRE ADMINISTRATION, 1993: PROTECTIVE CLOTHING AND
EQUIPMENT NEEDS OF EMERGENCY RESPONDERS FOR URBAN
SEARCH AND RESCUE MISSIONS [FA-136])

Should be plastic or rubber for easier decontamination. Should include minimum
   amount of weight possible
Should be able to be quickly ditched in emergency. Should not include ankle weights

Diving Tether
Should be attached to harness. Should include synthetic line compatible with
   diluted chemicals and strong enough to bear weight of diver
Should not retain contaminants following clean water rinsing

SCUBA System
Materials should be compatible with diluted chemicals. Should include bail-out
   block mounted on diver's harness
Should include bail-out air bottle with first stage regulations. Should not retain
   contaminants following clean water rinsing

Bail-Out System
Should include 5-minute air supply. Should include bail-out bottle, diver's harness,
   first-stage regulator, relief valve, submersible pressure gauge, quick disconnect
   whip/low pressure whip

Knife and Wire Cutter
Should be single edged and sheathed. Should not fall out of sheath when inverted

Having made a preliminary selection of PPC and PPE on the basis of expected hazards, level of protection required, and/or type of response mission, it becomes absolutely vital to ensure that vendor specifications conform to appropriate technical standards (see Table 7.7), including governmentally enforceable standards and professionally recommended standards. These standards are highly changeable due to changes in technology as well as actual field experience; it is therefore necessary to ensure the use of up-to-date standards, with full awareness of potential trends in both materials research and development and schedules of standard-setting proceedings (see Table 7.8).

Heat Stress

A key consideration regarding all PPC is its contribution to the wearer's *heat stress*. Heat stress may be manifest in several distinct symptoms, including:

- *Heat cramps* (caused by profuse sweating with inadequate replacement of electrolytes): muscle spasms and pain in hands, feet, and abdomen
- *Heat exhaustion* (severe effects of dehydration and stress on body organs due to cardiovascular insufficiency): pale, cool, moist skin; heavy sweating; dizziness; nausea; fainting spells
- *Heat stroke* (temperature regulation of body fails, with body temperature rising to critical levels; increasing risk of death; immediate medical attention required): red, hot, dry skin; lack of or reduced perspiration; nausea; dizziness or confusion; strong, rapid pulse; coma

TABLE 7-7

EXAMPLES OF STANDARDS RELATED TO CLOTHING ENSEMBLES AND RESPIRATORS (BASED ON INFORMATION PROVIDED BY U.S. DEPARTMENT OF LABOR, OSHA. OSHA TECHNICAL MANUAL. OSHA ELECTRONIC REFERENCE LIBRARY)

Vapor-Protective Suit (NFPA Standard 1991)
- Provides "gas tight" Integrity
- Intended for response situations where no chemical contact is permissible
- Equivalent to the clothing required for EPA Level A

Liquid Splash-Protective Suit (NFPA Standard 1992)
- Protection against liquid chemicals in the form of splashes, but not against continuous liquid contact or chemical vapors or gases
- Equivalent to the clothing required for EPA Level B
- It is important to note that, by wearing liquid splash-protective clothing, the wearer accepts exposure to chemical vapors or gases because this clothing does not offer gas-tight performance
- The use of duct tape to seal clothing interfaces does not provide the type of encapsulation necessary for protection against vapors or gases

Support Function Protective Garment (NFPA Standard 1993)
- Provide liquid splash protection but offer limited physical protection
- May comprise several separate protective clothing components (i.e., coveralls, hoods, gloves, and boots)
- Intended for use in non-emergency, nonflammable situations where chemical hazards have been completely characterized
- Examples of support functions include proximity to chemical processes, decontamination, hazardous waste clean-up, and training.
- Should not be used in chemical emergency response or in situations where chemical hazards remain uncharacterized
—Cautionary Note Regarding Respirators—
Protective clothing should completely cover both the wearer and his or her breathing apparatus.

In general, respiratory protective equipment is not designed to resist chemical contamination.

Level A protection (vapor-protective suits) require this configuration. Level B ensembles may be configured either with the SCBA on the outside or inside. However, it is strongly recommended that the wearer's respiratory equipment be worn inside the ensemble to prevent its failure and to reduce decontamination problems. Level C ensemble uses cartridge or canister type respirators that are generally worn outside the clothing.

As shown in Figure 7.1, an increase in relative humidity results in increased risk of heat stress at any ambient temperature. Heat stress can be a significant risk when working in encapsulating suits or other protective clothing, and it is always a risk when undertaking any type of work in direct sunlight or under conditions of high temperature and relative humidity.

There are several strategies that can prove useful for lowering the risk of heat stress. The worker who is acclimatized to working under hot conditions has a lower heart rate and body temperature than a worker who is not acclimatized. Acclimatization most often can be accomplished over a six-day period. During the first day, only 50% of the normal workload and exposure-time is allowed, with 10% more added each day. For a particularly fit individual, acclimatization can be accomplished in two or three days.

TABLE 7-8

NFPA STANDARDS RELATED TO PROTECTIVE CLOTHING AND EQUIPMENT

—Standard—	—Title—
NFPA 1971*	Protective Clothing for Structural Fire Fighting (includes hoods)
NFPA 1972	Helmets for Structural Fire Fighting
NFPA 1973	Gloves for Structural Fire Fighting
NFPA 1974	Protective Footwear for Structural Fire Fighting
NFPA 1975	Station/Work Uniforms for Fire Fighting
NFPA 1976	Protective Clothing for Proximity Fire Fighting
NFPA 1977	Protective Clothing and Equipment for Wildlands Fire Fighting
NFPA 1981	Open–Circuit Self-Contained Breathing Apparatus for the Fire Service
NFPA 1982	Personal Alert Safety Systems (PASS) for Fire Fighters
NFPA 1983	Fire Service Life Safety Rope, Harnesses and Hardware
NFPA 1991	Vapor-Protective Suits for Hazardous Chemical Emergencies
NFPA 1992	Liquid Splash-Protective Suits for Hazardous Chemical Emergencies
NFPA 1993	Support Function Protective Clothing for Hazardous Chemical Operations
NFPA 1999	Protective Clothing for Emergency Medical Operations

*NFPA 1971 incorporates NFPA 1972, 1973 & 1974
NFPA reviews and, as necessary in light of new developments in clothing and material technology and testing protocols, revises its standards on a 5-year cycle.

In an encapsulating suit, acclimated persons sweat more profusely than nonacclimated persons and are therefore more subject to the risk of dehydration. It is therefore necessary to provide for a drinking water program to ensure the proper replacement of water lost through sweating.

Additional approaches for reducing heat stress include the use of the lightest and coolest PPC and PPE possible, along with the use of light-colored clothing that absorbs less heat than dark-colored clothing. Artificially produced shade (e.g., tarp canopy, beach umbrella) can significantly contribute to reducing heat exposure. Finally, cooling vests (often called ice vests) may be worn, though many workers dislike cold so close to the skin.

Cold Stress

In cold environments, a critical factor in the selection of PPC is cold stress, which includes hypothermia and the freezing of flesh. Hypothermia, a potentially catastrophic drop in body temperature with significantly reduced blood flow and rate of metabolism, is of particular concern during cold water diving operations. The freezing of flesh, with potential subsequent development of gangrene, is always a risk attendant to land operations in cold climates, especially when wind velocity acts to increase risk significantly (see Figure 7.2).

Decontamination

Wherever possible and subject to the important constraints of effectiveness and cost, PPC and PPE should be selected to maximize the use of nonreusable

FIGURE 7-1

DETERMINATION OF HEAT STRESS BASED ON AMBIENT TEMPERATURE AND RELATIVE HUMIDITY

		Relative Humidity								
Temperature (°F)		10%	20%	30%	40%	50%	60%	70%	80%	90%
	104	98	104	110	120	132				
	102	97	101	108	117	125				
	100	95	99	105	110	120	132			
	98	93	97	101	106	110	125			
	96	91	95	98	104	108	120	128		
	94	89	93	95	100	105	111	122		
	92	87	90	92	96	100	106	115	122	
	90	85	88	90	92	96	100	106	114	122
	88	82	86	87	89	93	95	100	106	115
	86	80	84	85	87	90	92	96	100	109
	84	78	81	83	85	86	89	91	95	99
	82	77	79	80	81	84	86	89	91	95
	80	75	77	78	79	81	83	85	86	89
	78	72	75	77	78	79	80	81	83	85
	76	70	72	75	76	77	77	77	78	79
	74	68	70	73	74	75	75	75	76	77

Note: Add 10°F when protective clothing worn;
add 10°F when in direct sunlight

80-90: Fatigue possible if exposure is prolonged and there is physical activity

90-105: Heat cramps and heat exhaustion if exposure is prolonged and there is physical activity

105-130: Heat cramps or exhaustion likely; heat stroke possible if exposure prolonged and there is physical activity

Above 130: Heat Stroke Imminent

clothing and equipment in order to minimize the need for decontaminating reusable clothing and equipment—an activity that presents its own risk to personnel and to environmental resources. However, most PPC and PPE items are, in fact, reusable (as well as costly) and therefore must be selected with careful consideration given to means and methods for removing and/or deactivating incident-related contaminants, including dirt and debris as well as chemical and biological agents (see Table 7.9) so as to prevent any loss of integrity or functionality of the PPC and PPE.

Available decontamination methods should be assessed for their compatibility with PPC and PPE materials in a rigorous manner (see Figure 7.3). Appropriate decontamination procedures, cleaning and disinfecting solutions, and associated equipment and materials should then be incorporated into SOPs for both incident-related and normal operations (see Figure 7.4).

FIGURE 7-2

DETERMINATION OF COLD STRESS BASED ON AMBIENT TEMPERATURE AND WIND VELOCITY

		Temperature (°F)											
		45	40	35	30	25	20	15	10	5	0	-5	-10	-15
W	5	43	37	32	27	22	16	11	6	0	-5	-10	-15	-21
i	10	34	28	22	16	10	3	-3	-9	-15	-22	-27	-34	-40
n	15	29	23	16	9	2	-5	-11	-18	-25	-31	-38	-45	-51
d	20	26	19	12	4	-3	-10	-17	-24	-31	-39	-46	-53	-60
S	25	23	16	8	1	-7	-15	-22	-29	-36	-44	-51	-59	-66
p	30	21	13	6	-2	-10	-18	-25	-33	-41	-49	-56	-64	-71
e	35	20	12	4	-4	-12	-20	-27	-35	-43	-52	-58	-67	-75
e														
d	40	19	11	3	-5	-13	-21	-29	-37	-45	-53	-60	-69	-76
(MPH)	45	18	10	2	-6	-14	-22	-30	-38	-46	-54	-62	-70	-78

Wind Speed (MPH)

43 to -22: Little danger for properly clothed person

-24 to -71: Increasing danger; flesh may freeze

-75 to -78: Great danger; flesh may freeze in 30 seconds

TABLE 7-9

DECONTAMINATION METHODS (ADAPTED FROM NIOSH, USCG, AND EPA, 1985: OCCUPATIONAL SAFETY AND HEALTH GUIDANCE MANUAL FOR HAZARDOUS WASTE ACTIVITIES)

Removal _____
- **Contaminant Removal**
  - Water rinse, using pressurized or gravity flow
  - Chemical leaching and extraction
  - Evaporation/vaporization
  - Pressurized air jets
  - Scrubbing/scraping (commonly done using brushes, scrapers, or sponges and water-compatible solvent cleaning solutions)
  - Steam jets

- **Removal of Contaminated surfaces**
  - Disposal of deeply permeated materials (e.g., clothing, floor mats, and seats)
  - Disposal of protective coverings, coatings

Inactivation _____
- **Chemical Detoxification**
  - Halogen stripping
  - Neutralization
  - Oxidation/reduction
  - Thermal degradation

- **Disinfection/Sterilization**
  - Chemical disinfection
  - Dry heat sterilization
  - Gas/vapor sterilization
  - Irradiation
  - Steam sterilization

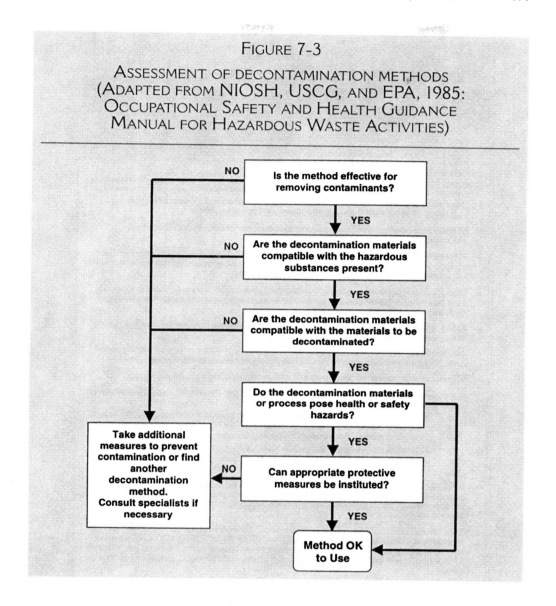

FIGURE 7-3

ASSESSMENT OF DECONTAMINATION METHODS
(ADAPTED FROM NIOSH, USCG, AND EPA, 1985:
OCCUPATIONAL SAFETY AND HEALTH GUIDANCE
MANUAL FOR HAZARDOUS WASTE ACTIVITIES)

Inspection of PPC

In any response organization, written SOPs that give detailed instructions for periodic examination of protective clothing should be readily available and rigidly enforced. Different types and levels of inspections, as well as schedules, are most appropriately developed to cover the lifecycle of the clothing, including (a) receipt of the clothing from the manufacturer; (b) issuance of clothing to personnel; (c) preventive maintenance, prior to and following use (including use during training sessions); (d) receipt of personnel complaints or concerns, governmental or other alerts (e.g., manufacturers, other response agencies) regarding field experience with similar clothing; and (e) upon replacement either in-kind or by substitution with other types, styles, or models.

FIGURE 7-4

EXAMPLE OF SOP INCLUDED IN INFECTION CONTROL PROGRAM (ADAPTED FROM U.S. FIRE ADMINISTRATION, 1992: GUIDE TO DEVELOPING AND MANAGING AN EMERGENCY SERVICE INFECTION CONTROL PROGRAM [FA-112])

Infection Control
Standard Operating Procedures   *SOP # IC6: Post-Response*

- Upon return to quarters, contaminated equipment will be revived and replaced with clean equipment. Supplies of PPE on response vehicles will be replenished.

- Contaminated equipment will be stored only in the decontamination area. Cleaning and decontamination will be performed as soon as practical.

- Disposable equipment and other biohazard waste generated during on-scene operations will be stored in the biohazard disposal area in appropriate leakproof containers. Sharps containers, when full, will be closed and placed in the biohazard disposal area.

- Gloves will be worn for all contact with contaminated equipment or materials. Other PPE will be used depending on splash or spill potential. Heavy-duty utility gloves will be used for cleaning, disinfection, or decontamination of equipment.

- Eating, drinking, smoking, handling contact lenses, or applying cosmetics or lip balm is prohibited during cleaning or decontamination procedures.

- Disinfection will be performed with a department-approved disinfectant or with a 1:100 solution of bleach in water. All disinfectants will be tuberculocidal and EPA approved and registered.

- Any damaged equipment will be cleaned and disinfected before being sent out for repair.

- The manufacturer's guidelines will be used for the cleaning and decontamination of all equipment. Unless otherwise specified:
  - ➤ Durable equipment (backboards, splints, MAST pants) will be washed with hot soapy water, rinsed with clean water, and disinfected with an approved disinfectant or 1:100 bleach solution. Equipment will be allowed to air dry.
  - ➤ Delicate equipment (radios, cardiac monitors, etc.) will be wiped clean of any debris using hot soapy water, wiped with clean water, then wiped with disinfectant or 1:100 bleach solution. Equipment will be allowed to air dry.

- Work surfaces will be decontaminated with an appropriate disinfectant after completion of procedures, and after spillage or contamination with blood or potentially infectious materials. Seats on response vehicles contaminated with body fluids from soiled PPE also will be disinfected upon return to station.

- Contaminated structural firefighting gear (turnout coats/bunker pants) will be cleaned according to manufacturer's recommendations found on attached labels. Normally, this will consist of a wash with hot soapy water followed by a rinse with clean water. Turnout gear will be air-dried. ***Chlorine bleach may impair the fire-retardant properties of structural firefighting gear and will not be used.***

- Contaminated boots will be brush-scrubbed with a hot solution of soapy water, rinsed with clean water, and allowed to air dry.

- Contaminated work clothes (jump suits, T-shirts, uniform pants) will be removed and exchanged for clean clothes. Personnel will shower if body fluids were in contact with skin under work clothes.

- Contaminated work clothes will be laundered at the station using hot water. ***Under no circumstances will contaminated work clothes be laundered at home.***

- Infectious wastes generated during cleaning and decontamination operations will be properly bagged and placed in the biohazard disposal area.

Regardless of the type and schedule of formal inspections, all personnel must understand that it is their responsibility to perform visual checks of PPC prior to, during, and immediately following use, with particular attention given to the following:

- Check coding device (e.g., color, alpha-numeric, name) to ensure that the clothing is properly identified for use in the intended task

- Visually inspect for imperfect seams, nonuniform coatings, tears, malfunctioning closures, improper seals
- Check for pinholes
- Check inside and out for any indications of chemical degradation, including discoloration, swelling, or stiffness
- Flex material and look for cracks, wear, abrasion, or any other sign of deterioration
- Pressurize gloves and hold under water to check for pinholes
- For fully encapsulating suits, check operation of pressure relief valves; inspect fitting of wrist, ankle, and neck seals; check face-shield for cracks and other anomalies

RESPIRATORY PROTECTION

Air Contaminants

Air contaminants include a variety of solid and liquid particles that range greatly in size, from relatively large-size liquid chemical mists (>100 microns) to progressively smaller particles like dusts (e.g., foundry dust and fly ash (1–1000 microns)), fumes and vapors (e.g., metallurgical fumes and oil smoke (0.001–1.0 microns)), bacterial and fungal spores (0.1_1.0 microns), and, finally, gases. The size of an inhaled particle is a key determinant of the depth to which that particle can penetrate into the respiratory tract. Although the depth of penetration is also influenced by the shape of the particle and whether inhalation is primarily through the nose or the mouth, the majority of larger dusts and mists can become deposited along the nasopharyngeal portion of the respiratory tract (above the larynx and including nasal passages), with progressively smaller particles progressing to the upper esophagus, to the tracheobronchial branch of the respiratory tract, and, finally, even to the alveoli of the lung. Particles deposited in nasal passages and within the throat can also ultimately enter the stomach via passage along the esophagus, demonstrating that inhalation, as a route of entry, can be equivalent to ingestion.

Given the range of potential deposition of inhaled particles within both the respiratory and gastrointestinal tract, various organs and tissues become exposed to the diverse health hazards associated with those particles, including such relatively acute afflictions as nasal irritation (e.g., certain chromium dusts), persistent sneezing (e.g., o-chlorobenzylidene malononitrile), and cough (e.g., chlorine), as well as life-threatening acute and chronic afflictions, such as pneumonia (e.g., manganese dusts in lower airways and alveoli), hemorrhage (e.g., boron vapors in alveoli), emphysema (e.g., aluminum abrasives in alveoli), and cancer (e.g., nickel dusts in nasal cavities and the lungs).

Upon being inhaled, various gases, vapors, and mists (e.g., halogenated hydrocarbons, methyl ethyl ketone, methyl methacrylate) can pass directly from the alveoli (or the gastrointestinal tract) into the blood and, depending upon their differential solubilities in body fluids and tissue (e.g., fat), affect other tissues (e.g., bone), organs (e.g., liver), and systems (e.g., central nervous system). Of course, many air contaminants begin to exert their effect immediately upon entry into the blood by triggering an immunological response. For example, many organic dusts, such as cork, malt, and cheese dust, and even those (e.g., pollen) that collect in air conditioners, as well as inorganic dusts (e.g., tungsten carbide, platinum salts, toluene 2,4-diisocyanate, nickel metal), can cause allergenic reactions in hypersensitive persons that can quickly become life threatening. Differential solubilities of air contaminants in body fluids and tissues, as well as their potential as immunological antigens, clearly illustrate that many inhaled contaminants are not simply respiratory hazards, but are in fact hazards to many different organs and tissues.

Action Levels

An *action level* is typically a numerical limit (but it may also be a qualitative situation) that triggers a protective response. In a few cases, action levels may be established by regulation pertaining to specific chemicals, such as benzene or formaldehyde. In most instances, action levels are established by common practice. For example, evacuation from an area containing flammable vapors is most often required whenever ambient concentrations of those vapors attain 10% of the lower explosive limit (LEL). The rationale for any action level is that protection must begin well in advance of an actual life- or health-threatening situation.

In the absence of either an action level or even so much as a standard or guideline established by legal authority (which is by far the most common situation), a criterion for deciding whether or not the measured quantity of an atmosphere requires the use of respiratory protection must be established. Sometimes, given the dearth of action levels and standards as compared with the seemingly limitless number of health and safety standards, the safety officer opts to require respiratory protection regardless of ambient concentration. In normal operational situations (i.e., nonemergency), this practice should be strongly discouraged because the use of any protective clothing or device always produces its own risk. Ideally, the objective should be to balance the risks that derive from the lack of specific protection with the risks (typically of a different kind) that derive from wearing protective equipment. However, during an emergency, it is more often prudent to assume the worst case and prepare accordingly. This usually is done for any or all of several reasons:

- Immediate response action is required even in the absence of specific data and information on ambient air concentrations of chemical or biological contaminants
- Even in situations where ambient air concentrations of chemical contaminants are known, it is always possible that actual concentrations can vary significantly over the spaces to be penetrated by emergency response personnel
- In some situations (e.g., fire, explosion, release of biological pathogens, any other situation resulting in chemical by-products of combustion or chemical reactions), there is no practical alternative but to implement the highest level of respiratory protection.

Types of Respirators

Any reputable manufacturer or supplier of respirators today offers potential clients detailed documentation regarding the broad range of available respirators and the specific uses and limits of each type. In no circumstance should the safety officer purchase any respirator without carefully examining this documentation or consulting with manufacturers' or suppliers' technical staffs.

The basic types of respiratory protection devices (see Figure 7.5) may be briefly described as follows.

Air-Purifying Respirators

These respirators use filter and/or absorbent materials to remove contaminants from inhaled air. *They must not be used in atmospheres that may have either a deficiency (< 19.5%) or an excess (> 22.5%) of oxygen.*

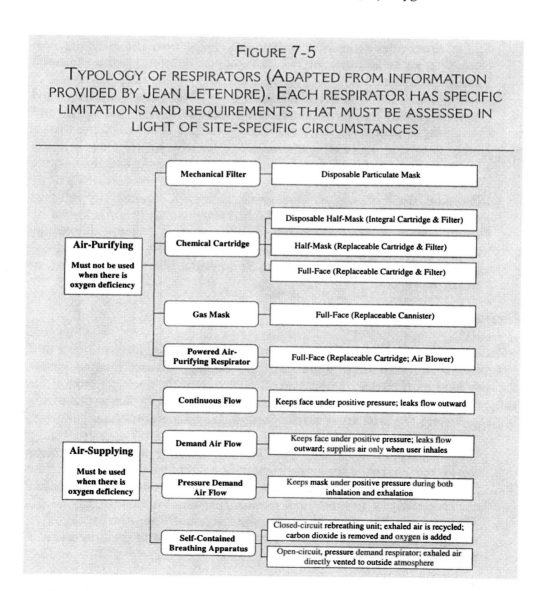

FIGURE 7-5

TYPOLOGY OF RESPIRATORS (ADAPTED FROM INFORMATION PROVIDED BY JEAN LETENDRE). EACH RESPIRATOR HAS SPECIFIC LIMITATIONS AND REQUIREMENTS THAT MUST BE ASSESSED IN LIGHT OF SITE-SPECIFIC CIRCUMSTANCES

1. *Mechanical Filter Respirator*: Removes particles from the air; consists of a simple mesh material that fits over the nose and mouth and is tied with straps of strings behind the neck (some styles include a flexible metal strip that can easily be bent across nose bridge to enforce a more secure seal between face and mask). A comfortable, low-profile, lightweight respirator (often called simply a paper or dust mask) for limited use offers low-cost protection against dust, mist, and fumes, but is not effective for gases, vapors, or nonabsorbable contaminants. Modification of the standard dust mask is the toxic dust mask, having smaller mesh size; no cleaning, disinfection, or spare parts are required. The usual limit for use is set at 10 times the *permissible exposure limit* (PEL).

2. *Chemical Cartridge Respirator*: Either a disposable half-face respirator in which the cartridge is integral to the face-piece or a nondisposable half- or full-face respirator with one or two screw-on chemical cartridges that are specific for particular contaminants. Its use is limited by ambient concentration of contaminants; expended cartridges may be replaced; it is easy to use and needs little cleaning and few if any spare parts. It provides protection against gas, vapor, dust, and mists; its limit usually is set at 10 times the PEL (but may be different) and is indicated on the cartridge.

3. *Gas Mask*: A full-face mask to which is attached a relatively long-lived canister containing absorbent materials that can remove toxic gases and particles; expended canisters may be replaced. It offers a greater capacity of removing high ambient concentrations of contaminants than cartridge respirators; limits for ambient concentrations are specified on the canister

4. *Powered Air-Purifying Respirator*: A helmeted, hooded, or full-face mask containing one or more cartridges through which air is forced by an air blower; it is less exhausting for the user than a chemical cartridge or gas mask respirator. The face- or belt-mounted blower is battery powered; limits usually are set at 100 times the PEL (or as specified on the cartridge).

Air-Supplying Respirators

These respirators consist of a helmet, hood, and full- or half-face mask that is provided air through a compressed air cylinder. They are used in atmospheres that may have a deficiency or an excess of oxygen, or if the concentration of the contaminant vapors, gases, or particles may be immediately dangerous to life (IDL) or beyond the capacity of an air-purifying cartridge or canister.

1. *Continuous Flow Respirator*: The face-piece is kept at positive pressure; air flow is outward from the mask, preventing contaminants from entering the face-piece; it supplies clean, breathable air from a source independent of the contaminated air; and the flow of air remains constant.

2. *Demand Air Flow Respirator*: Supplies air only when the user inhales; exhalations are ejected directly to the atmosphere and the flow of air is regulated by pressure valve.

3. *Pressure Demand Air Flow Respirator*: Supplies air when the user inhales or exhales; exhalations are ejected directly to the atmosphere and the flow of air is regulated by pressure valve.
4. *Self-Contained Breather Apparatus (SCBA)*: Provides an independent air supply that is not mixed with the outside atmosphere and that may be either recycled or exhaled directly into the outside atmosphere, and offers the greatest respiratory protection available. SCBA may be of several types, including open- and closed-circuit units (for multipurpose response operations) and so-called "escape-only" SCBA, which can be used for only a period of five to 15 minutes for the purpose of escaping a hazardous atmosphere (see Figure 7.6).

FIGURE 7-6

ADVANTAGES AND DISADVANTAGES OF DIFFERENT TYPES OF SCBA (ADAPTED FROM NIOSH, USCG, AND EPA, 1985: OCCUPATIONAL SAFETY AND HEALTH GUIDANCE MANUAL FOR HAZARDOUS WASTE ACTIVITIES)

Open-Circuit SCBA

Supplies clean air to the wearer from a cylinder; wearer exhales directly to the atmosphere

Advantages	Disadvantages
Operated in a positive-pressure mode, open-circuit SCBAs provide the highest respiratory protection currently available. A warning alarm signals when only 20-25% of the air supply remains	Shorter operating time (30 to 60 minutes) and heavier weight (up to 35 lbs) than a closed-circuit SCBA. The 30 to 60-minute operating time may vary depending on the size of the air tank and the work rate of the individual

Closed-Circuit SCBA

Recycles exhaled gases by removing CO_2 with an alkaline scrubber and replenishing the consumed oxygen with oxygen from a liquid or gaseous source

Advantages	Disadvantages
Longer operating time (up to 4 hours), and lighter weight (21 to 30 lbs) than open-circuit apparatus. A warning alarm signals when only 20 to 25% of the oxygen supply remains. Oxygen supply is depleted before the CO_2 sorbent scrubber supply, thereby protecting the wearer from CO_2 breakthrough	At very cold temperature, scrubber efficiency may be reduced and CO_2 breakthrough may occur. Units retain heat normally exchanged in exhalation and generate heat in the CO_2 scrubbing operations, adding to danger of heat stress. Auxiliary cooling devices may be required. When worn outside an encapsulating suit, breathing bag may be permeated by chemicals, contaminating breathing apparatus and respirable air. Decontamination of breathing bag may be difficult

Note: Positive-pressure closed-circuit SCBAs offer substantially more protection than negative-pressure units, which are not recommended on hazardous waste sites.

Escape-Only SCBA

Supplies clean air from either an air cylinder or from an oxygen-generating chemical

Advantages	Disadvantages
Lightweight (10 pounds or less), low bulk, easy to carry. Available in pressure-demand and continuous-flow modes	Cannot be used for entry. Provides only 5 to 15 minutes of respiratory protection, depending on the model and wearer breathing rate

In the United States, respirator types are subject to 42 CFR 84 regulations (see Figure 7.7). These regulations provide for nine classes of filters (three levels of filter efficiency, each with three categories of resistance to degradation of filter efficiency) for nonpowered particulate respirators. The three levels of filter efficiency are 95, 99, and 99.97%; the three categories of resistance to degradation of filter efficiency are labeled N (not resistant to oil), R (resistant to oil), and P (oil-proof). Notice of NIOSH certification of nonpowered particulate respirators (see Figure 7.8) is available through the electronic reference library of NIOSH (http:www.cdc.gov/niosh/homepage.html).

FIGURE 7-7

FLOW CHART FOR SELECTING 42 CFR 84 PARTICULATE FILTERS (ADAPTED FROM NIOSH, 1997: PARTICULATE RESPIRATOR SELECTION AND USE: DETAILED GUIDELINES [NIOSH ELECTRONIC LIBRARY

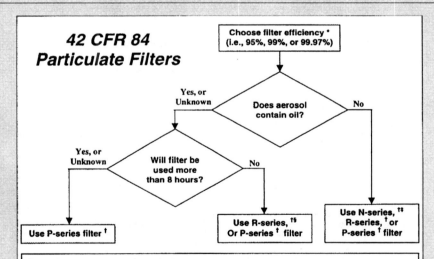

* The higher the filter efficiency, the lower the filter leakage.

† Limited by considerations of hygiene, damage, and breathing resistance.

‡ High (200 mg) filter loading in the certification test is intended to address the potential for filter efficiency degradation by solid or water-based aerosols in the workplace. Accordingly, there is no recommended service time in most workplace settings. However, in dirty workplaces (high aerosol concentrations), service time should only be extended beyond 8 hours of use (continuous or intermittent) by performing an evaluation in specific workplace settings that demonstrates (a) that extended use will not degrade the filter efficiency below the certified efficiency level, or (b) that the total mass loading of the filter is less than 200 mg (100 mg per filter for dual filter respirators).

§ No specific service time limit when oil aerosols are not present. In the presence of oil aerosols, service time may be extended beyond 8 hours of use (continuous or intermittent) by demonstrating (a) that extended use will not degrade the filter efficiency below the certified efficiency level, or (b) that the total mass loading of he filter is less than 200 mg (100 mg per filter for dual-filter. respirators).

FIGURE 7-8
EXAMPLE OF LISTING (PARTIAL) OF PARTICULATE RESPIRATORS
CERTIFIED UNDER 42 CFR 84 (ADAPTED FROM NIOSH, 1997:
NIOSH ELECTRONIC LIBRARY)

FIGURE 7-8
EXAMPLE OF LISTING (PARTIAL) OF PARTICULATE RESPIRATORS
CERTIFIED UNDER 42 CFR 84 (ADAPTED FROM NIOSH, 1997:
NIOSH ELECTRONIC LIBRARY)

Particulate Respirators Certified Under 42 CFR Part 84

NOTICE: Only Non-powered Particulate Respirators are listed here!
Call 1-800-35-NIOSH for listings of other Certified Respirators

On July 10, 1995, a new NIOSH certification program for respirators went into effect. This regulation, 42 CFR Part 84, replaced the long standing regulation 30 CFR Part 11. (Commonly referred to as Part 84 and Part 11, respectively.) The new Part 84 covers all respirator types (self-contained breathing apparatus, air-line respirators, gas and vapor respirators, powered air-purifying respirators, etc.) but only the standards for non-powered, particulate respirators have changed from the provisions of the old Part 11.

The following is a list of ONLY non-powered (negative pressure) particulate respirators that have been tested and certified by NIOSH under provisions of the new Part 84. Many other respirator types, such as particulate filters in combination with gas and vapor cartridges, have been certified under provisions of Part 84 but are not included in this listing. Inquiries about other respirators may be made to 1-800-35-NIOSH. This list is updated periodically as additional non-powered, particulate respirators are certified. The current reference date is provided at the top of the listing.

Note: Multiple listings under the same certification number will be present when that same respirator is marketed by different suppliers under private labels.

Particulate Respirators Certified Under 42 CFR Part 84
Effective Date: July 31, 1997

Approval Number	Supplier/ Phone Number	Respirator Type/ Trade Name	Protection Level/ Series	Exhalation Valve
84A-0001	Better Breathing, Inc.	Filtering Facepiece	N95	Yes
	1-800-638-6275	APR-3-N95-1		
84A-0002	Racal Health and Safety, Inc.	Filtering Facepiece	N95	Yes
	1-800-682-9500	Delta N95		
84A-0003	Racal Health and Safety, Inc.	Filtering Facepiece	N95	No
	1-800-682-9500	Delta N95		
84A-0004	Racal Health and Safety, Inc.	Filtering Facepiece	N100	Yes
	1-800-682-9500	Delta N100		

http://ftp.cdc.gov/niosh   **Continued ....**

General Procedures

A written respiratory protection program is mandatory and must include specific procedures that govern the proper use, maintenance, and replacement of respirators. The following examples of general procedures that apply throughout American industry illustrate the range of issues that must be addressed and adapted to the specific needs and managerial practices of any emergency response organization.

1. Respirators should be selected on the basis of the specific hazards to which individual personnel may be exposed in normal, nonroutine, and emergency response situations. Although the selection of respirators is finally the responsibility of the safety officer, the safety officer will coordinate with all supervisors having responsibility for personnel identified as in need of respiratory protection.

2. Only personnel authorized by the safety officer will utilize respiratory protection. Authorization consists of (a) selection by the safety officer or supervisor on the basis of potential exposure to dangerous atmosphere; (b) appropriate training of personnel in the proper use, maintenance, and limitations of the respirator(s) specified for their use; and (c) completion of medical evaluation and fit-testing requirements.

3. All authorized personnel will be trained in the proper fitting of respirators and taught how to conduct fit-testing. All supervisors of authorized personnel will be fully trained in all aspects regarding the proper use, maintenance, and fitting of respirators.

4. The safety officer will ensure that, whenever possible, training of personnel and supervisors will be conducted by the vendors of respiratory protection devices and will maintain all training records, including the date of training, the names of persons attending the training, the specific subject matter addressed, and the name and affiliation of the trainer.

5. Wherever possible, respirators will be assigned to individual workers for their exclusive use; in such cases, the employee's name will be clearly marked on the respirator.

6. All nondisposable respirators must be cleaned and disinfected after each use and will be stored in a convenient, clean, sanitary, and clearly identified location. Written instructions for the proper cleaning, disinfection, storage, and maintenance of respirators will be included in SOPs posted at each storage location.

7. Respirators will be inspected during cleaning or at least monthly. Worn or deteriorated parts will be replaced. Inspection will include a check on the tightness of connections and the condition of the face-piece, headbands, valves, connecting tubes, and canisters. Rubber or elastomer parts will be inspected for pliability and signs of deterioration. Inspection records will be maintained by supervisors at the respirator storage location.

8. Respirators must be stored to protect against dust, sunlight, heat, extreme cold, excessive moisture, or damaging chemicals. Respirators placed at ready stations will be immediately accessible at all times and should be stored in dedicated and clearly marked compartments. Routinely used respirators, such as dust masks, may be placed in plastic bags for storage. Respirators should not be stored in such places as lockers or tool boxes unless they are in carrying cases or cartons.

9. Respirators should be packed or stored so that the face-piece and exhalation valve will rest in a normal position and function will not be impaired by the elastomer sitting in an abnormal position.

10. All personnel who issue or use canister-type respirators will ensure that canisters purchased or used by them are properly labeled and color coded (in accordance with 29 CFR 1910.134, Table I-1) before they are placed in service, and that labels and colors are properly maintained at all times thereafter.

11. The safety officer will ensure that normal (nonemergency) work areas and operations (e.g., post-incident cleaning and disinfection of contaminated equipment and vehicles) requiring the use of respirators are monitored at least twice a year to ensure proper respiratory protection. The safety officer will maintain written records that document the date of monitoring, the chemical monitored, measurement devices, concentrations, conversion factors, and mathematical transformations of data, as well as any actions undertaken as a result of the monitoring effort.

12. In the case of SCBA, air may be supplied to the respirator from cylinders or air compressors only if in compliance with 29 CFR 1910.134(d).

13. In areas where the wearer of a respiratory protection device could, upon failure of that device, be overcome by a toxic chemical or oxygen deficiency or super-abundance, at least one additional person will be present. Communication will be maintained between both or all individuals present. Planning will be such that one individual will be unaffected by any likely incident and have the proper rescue equipment to effect rescue.

14. Personnel using air-line respirators in atmospheres immediately dangerous to life or health will be equipped with safety harnesses and safety lines for lifting or removing persons from those hazardous atmospheres.

15. In no circumstance will any personnel using a respiratory protective device wear eye contact lenses in any atmosphere that may be chemically contaminated.

Cleaning and Disinfecting Respirators

Respirators should be cleaned and disinfected after each use. Cleaning should be accomplished by washing nonfilter components with detergent in warm water using a soft brush, followed by a thorough rinsing in clean water. If possible, detergents containing a biocide should be used. In no circumstance should any organic solvent (e.g., acetone, benzene) be used, as such solvents typically will damage the rubber face-piece. Alternatively, following detergent washing, a disinfecting rinse can be used such as:

- Hypochlorite disinfecting solution (2 tablespoons of chlorine bleach per gallon of water)
- Iodine disinfecting solution (1 teaspoon of tincture of iodine per gallon of water)

It should be understood that the efficacy of any chemical disinfectant depends upon the time of actual contact between the disinfecting chemical and the target microbe (bacterium, virus, fungus). To ensure contact between disinfectant and microbe, dirt and other substances that can physically interpose between disinfectant and microbe must be completely removed. Once dirt and other interfering substances are removed, allow at least a two-minute immersion of respirator components in the disinfecting solution.

Inspection of Respirators

All respirators should be visually inspected before and after each use, during cleaning and disinfection, and at least monthly (and preferably weekly). Written SOPs should identify signs and symptoms of needed maintenance and replacement, including the following (based on NIOSH recommendations):

1. Disposable Respirator
   - Holes in filter (replace respirator)
   - Poor elasticity or deterioration of straps (replace straps or respirator)
   - Deterioration or excessive deformation of metal nose clip (replace respirator)
2. Air-Purifying Respirator
   - *Rubber face-piece*: excessive dirt (clean thoroughly); cracks, tears, or holes (replace face-piece); permanent distortion (replace face-piece); cracked, stretched, or loose fitting lenses (replace lenses or face-piece)
   - *Head-straps*: breaks or tears (replace head-strap); loss of elasticity (replace head-strap); broken or malfunctioning buckles or attachments (replace items); excessive wearing of head harness (replace head-strap)
   - *Inhalation Valve, Exhalation Valve*: detergent residue, dust particles, or dirt on valve or valve seat (clean); cracks, tears, or distortion of valve material or valve seat (replace); missing or defective valve cover (replace valve cover)
   - *Filter Element(s)*: improper filter and approval designation (replace); missing or worn gaskets (replace gaskets); worn threads (replace filter and/or face-piece); cracks or dents in filter housing (replace filter); deterioration of gas mask canister harness (replace harness); service life indicator (determine proper indicator from manufacturer)
   - *Corrugated Breathing Tube (gas mask)*: cracks or holes (replace tube); missing or loose hose clamps (replace clamps); broken or missing end connectors (replace connectors)
3. Air-Supplying Respirator
   - *Face-piece, head-straps, valves, and breathing tube*: same as for air-purifying respirators, earlier)
   - *Hood, helmet, blouse, or full suit (as applicable)*: rips and worn seams (repair or replace); headgear suspension (adjust properly for wearer); cracks or breaks in face-shield (replace face-shield); damaged or improper fit of protective screen (replace screen)

4. Air Supply System
   - Breathing air quality
   - Breaks or kinks in air supply hose and end fitting attachments (replace)
   - Tightness of connections (adjust; replace as necessary)
   - Proper setting of regulators and valves (see manufacturer's recommendations)
   - Correct operation of air-purifying elements and carbon monoxide or high-temperature alarms (see manufacturer's recommendations)
5. Self-Contained Breathing Apparatus (SCBA)
   - See manufacturer's recommendations

Key Issues in Respirator Maintenance Program

On the basis of its analysis of the cause of deaths of firefighters, NIOSH has strongly emphasized the critical importance of the various standards to be used to establish a policy of providing and operating the highest possible levels of safety and health for all firefighters, including standards directly pertinent to respiratory protection:

- NFPA 1404 specifies the minimum requirements for a fire service respiratory protection program.
- NFPA 100 specifies (1) the minimum requirements for a fire department's occupational safety and health program, and (2) the safety procedures for members involved in rescue, fire suppression, and related activities.
- NFPA 1561 defines the essential elements of an incident management system and other relevant NFPA standards, including NFPA 1971 (clothing), NFPA 1972 (helmets), NFPA 1973 (gloves), NFPA 1974 (footwear), and NFPA 1981 (SCBA).

HEARING PROTECTORS

Work-related noise is inclusive of two basic categories of sound: *impulsive sound*, which is sound that varies more than 40 dB per 0.5 second; and *non-impulsive sound*, which includes so-called continuous and intermittent (i.e., varies less than 40 dB per 0.5 second) sound. The standard action level for the general workplace is an eight-hour *time weighted average* (TWA) of 85 dB.

There are four basic types of ear protectors, each having certain advantages and limitations, especially with regard to personal comfort level:

1. *Enclosure*: helmet type protection, typically providing attenuation of 35 dB at $f < 1000Hz$ and 50 dB at $f > 1000Hz$. Although highly effective for attenuating sound conducted through air, it is not very effective for attenuating sound conducted though one's own body; is relatively

bulky and uncomfortable for most work situations; and generally is used where both hearing and head protection are required.

2. *Aural Insert (earplug)*: most commonly used type of protector in general industry. It is inserted into the ear to plug the ear canal and provides attenuation of up to 25 dB at $f < 1000$ Hz and 35 dB at $f > 1000$ Hz, with common attenuation of 5–15 dB and 15–25 dB, respectively. Three different types are commonly available:

   - *Formable earplug*: designed to be discarded after one-time use, it is made of expandable foams, glass fiber, wax-impregnated cotton, Swedish wool, or mineral-down, and is available in single size only. The degree of attenuation depends on snugness of fit.
   - *Custom molded earplug*: designed to fit an individual's ear; changes in ear canal and drying of the mold material can detract from effectiveness.
   - *Premolded earplug*: made of soft silicone, rubber, or plastic, it fits generic shapes of ear canals, with various modifications for particular situations, including modifications for differential attenuations at different frequencies and for various combinations of continuous or impact noise. Some models are developed for specific occupational groups.

3. *Canal Cap (semi- and supra-aural)*: used to seal the external opening of the ear canal (as opposed to plugging the ear canal);, it is held in place by a band or other head suspension device. The range of attenuation is comparable to that of an earplug; ideal for intermittent use.

4. *Earmuff (circum-aural device)*: domed cup covering the entire external ear, it can provide up to 35 dB attenuation at $f < 1000$ Hz and 45 dB at $F > 1000$ Hz, but often reaching only 10–12 dB. It may be uncomfortable due to slight pressure applied to the side of the head.

Any hearing protector presents a special risk to the wearer during incident operations because of its interference with vital communication among team members (e.g., voice command to evacuate, or other sound signal), and the wearer's awareness of any developing ambient hazard (e.g., noise associated with impending structural collapse). No hearing protector should be worn by any emergency response personnel except when specific provision is made (e.g., integrated earmuff and radio receiver) to ensure that the wearer will not thereby be subject to undue risk.

STUDY GUIDE

True or False

1. PPC and PPE should be viewed as the last available means of controlling responder risk.
2. In selecting protective clothing and equipment, the objective must be to achieve an effective and assured balance between the risks attendant to the incident and the risks inherent in the use of PPC and PPE.

3. Currently, there are relatively few standards for PPC and PPE that are promulgated by regulatory authorities.
4. Protective ensembles are recommended solely on the basis of generic types of protection (e.g., chemical splashes, hazardous dusts).
5. Technical rescues are typically land-based rescues where the principal hazards are physical.
6. Technical rescue ensembles generally give minimal protection from contact with biological fluids.
7. Swift water rescue ensembles maximize insulation from cold water and protection from physical hazard (e.g., abrasion, tears, cuts).
8. Contaminated water rescue ensembles maximize protection from exposure to biological and chemical contaminants.
9. Basic ensembles for personnel at hazardous waste sites are graded with respect to levels of both respiratory protection and skin/eye protection.
10. Regulatory and professional standards regarding protective clothing and equipment are changeful because of continually changing technology and on-going field evaluations of their efficacy.
11. Heat cramps are the least dangerous symptoms associated with heat stress.
12. An increase in relative humidity results in decreased risk of heat stress at any ambient temperature.
13. Regardless of the protective clothing worn, there is always a risk of heat stress when conducting response activities in direct sunlight or under conditions of high temperature and relative humidity.
14. It is possible to reduce the risk of heat stress by acclimatizing personnel to hot conditions over a period of days.
15. Cold stress presents the risks of hypothermia and freezing of flesh.

Multiple Choice

1. Chemical resistance of PPC and PPE is described in terms of
   A. permeation of chemicals through materials
   B. discoloration
   C. loss of physical strength
   D. physical seepage through seams and seals
   E. all of the above
2. No selection of PPC or PPE should be made without full documentation of
   A. manufacturer's design and engineering specifications
   B. relevant standards provided by governmental and professional authorities
   C. available methods and procedures for conducting visual inspections and field assessments
   D. all of the above
3. Pale, cool, and moist skin, heavy sweating, dizziness, nausea, and fainting spells are symptoms of
   A. heat cramps
   B. heat exhaustion
   C. heat stroke

4. Effective strategies for reducing the risk of heat stress include
   A. provide for acclimatization
   B. institute drinking water program
   C. use lightest and coolest and light-colored clothing
   D. provide artificial shade
   E. provide "ice vests"
   F. all of the above
5. In order to minimize the risk of cold stress at an emergency site, it is necessary to know the ambient temperature and
   A. latitude of the site
   B. wind velocity
   C. relative humidity
6. Decontamination of PPC and PPE may involve inactivation by
   A. halogen stripping
   B. neutralization
   C. oxidation/reduction
   D. thermal degradation
   E. all of the above
7. Air-purifying respirators must never be used in an atmosphere that may have
   A. a deficiency ($\leq 19.5\%$) of oxygen
   B. an excess ($\geq 22.5\%$) of oxygen
   C. under either condition A or B
8. At a temperature of 88°F and a relative humidity of 70%, the greatest risk is
   A. heat cramps
   B. heat cramps and heat exhaustion
   C. heat stroke
9. A continuous flow respirator must be used when
   A. there is an oxygen deficiency
   B. there is evidence of chemical aerosols
   C. there is evidence of hazardous biological agents
10. Only personnel authorized by the safety officer will utilize respiratory protection; this authorization consists of
    A. selection on the basis of potential exposure to a dangerous atmosphere
    B. appropriate training of selected personnel in proper use, maintenance, and limitations of the respirator used
    C. completion of medical evaluation and fit testing requirements
    D. all of the above

Essays

1. Review the basic requirements of a respiratory protection program. Identify the key analytical tasks that must be performed in order to meet these requirements.
2. Consider the information provided regarding hearing protectors. Comment on your appreciation of the problems associated with the WTC twin tower collapse related to communications.

3. Review the chapter content on respiratory protection. Comment on your appreciation of the problems associated with the WTC twin tower collapse related to respiratory protection.

Case Study

As a first responder in your community, you have responsibility to take charge of any fire incident at a local college. This college experiences a fire in a laboratory building housing both chemical and biological (including microbiological) laboratories.

1. How should your PPC and PPE program have accounted for the types of risks your personnel might experience?
2. How might you have ensured that your program would be both effective and safe for your personnel?
3. Given the risks that your respondent personnel might experience, how would you go about integrating that information with a plan for dealing with potential risks to the at-large community?

HAZARD AND RISK REDUCTION STRATEGIES

INTRODUCTION

In both proactive and reactive phases of emergency response planning, hazard and risk reduction are central objectives. Hazard reduction strategies include any attempt to minimize the potential harm or injury associated with any substance, situation, or condition (see Image 8.1); risk reduction strategies focus not on the source of potential hazards but, rather, on the exposure of persons to those hazards. As coequal efforts to minimize the impact of emergencies, hazard and risk reduction are today most often described as two aspects of the process of *mitigation*.

The concept of mitigation is, in fact, very broad, inclusive of both proactive and reactive actions taken by both response and nonresponse personnel and by governmental as well as private organizations. With specific regard to emergency response personnel and organizations, the practical application of mitigation is directed toward the following objectives:

1. *Minimize the number of incidents requiring the implementation of emergency response.* Although this objective is very much the bottom-line goal of all emergency response planning, emergency responders themselves can actually do little to advance this objective, except by the advice and training they make available to private corporations and the public at large regarding the control of hazards and protection against exposure. Historically, this type of service (e.g., consulting and training) most often has been viewed as essentially ancillary to the main function of a response organization; however, given the paramount importance of a proactive ethos, there is good reason to argue that consulting and training services are at least as important as actual response services.

   The primary effort in regard to this mitigation objective (beyond consulting and training) must be made by personnel who have primary

IMAGE 8-1

BILOXI, MISS., SEPTEMBER 3, 2005: DAMAGE AND DESTRUCTION TO HOUSES CAUSED BY HURRICANE KATRINA

Source: Mark Wolfe/FEMA

managerial responsibility for potential sources of hazards (e.g., plant manager). Regulatory authorities also play a vital role, of course, by forcing compliance with standards that are recognized as effective means of hazard and risk management—a role, it should be noted, that gives important support to the consulting and training services of response organizations.

2. *Minimize the magnitude of incidents.* The magnitude of an incident maybe described in terms of various criteria, including (but not limited to) (a) geographic extent of hazardous conditions, (b) numbers of persons harmed or killed, (c) loss of property and other resources, (d) duration of the incident, and (e) resources used in response (e.g., personnel, equipment, money).

Certainly these dimensions of an actual incident are influenced by both the proactive and reactive efforts of both response and nonresponse organizations—efforts that, in the midst of an actual incident, become intertwined and hardly distinguishable. For example, the capacity of response personnel to contain a developing emergency in a particular facility is influenced not only by their response preparedness and readiness, but also by the policies of that facility regarding the control of hazardous material inventory and in-plant safeguards, including notification alarms and first-response protocols.

Beyond maintaining their professional response readiness, response organizations are primarily responsible (with regard to this mitigation objective) for minimizing risk to response personnel.

3. *Prevent natural disasters from becoming human-made disasters.* The historic distinction between natural and human-made disasters is today less distinct. Even though humans do not cause earthquakes, humans do choose to build upon earthquake-prone faults; even though humans do not cause torrential storms and floods, humans do choose to build within floodplains. This is not to say that humans can avoid any and all perils of nature by a mere act of volition, but the fact remains that many ordinary natural phenomena essentially become human disasters simply because of specific choices that humans do or do not make—choices that, regrettably, put lives of emergency responders at risk.

   Although the basic responsibility of this type of mitigation must lie with political and municipal planning processes, emergency response organizations can and do play an import role with respect to:

   - Advising corporations and the general public as to proper location of hazardous facilities and operations
   - Providing guidelines for proper design and construction (see Figure 8.1)
   - Ensuring that the risks attendant to the various types of natural hazards are appropriately integrated into response procedures, the selection of personal protective clothing and equipment, and personnel training

4. *Integrate lessons learned from actual incidents and response actions into both proactive and reactive protocols and methods.* A basic tenant of mitigation is that, however high either the level of sophistication of emergency planning and prevention or the level of response preparedness, every emergency incident is a unique opportunity for learning how to plan and how to respond better (see Figure 8.2). This principle applies directly to response and nonresponse organizations with equal relevance.

MITIGATION MEASURES

Being essentially an interactive, community-wide partnership among all organizations having direct or indirect responsibility for managing human risk, mitigation is inclusive of a wide range of measures that contribute in diverse ways to the prevention, control and containment, and minimization of risk. Many of these measures have long been used in both public and private sectors and are variously encapsulated in such phrases as "good engineering practices," "regulatory standards," "pollution control measures," "quality control measures," and (more recently) "green productivity" and "environmental management." As can be inferred from these phrases, the need to mitigate hazardous incidents has resulted not so much in the invention of whole new technologies as it has the marshaling of both existent and developing technologies toward the achievement of a specific common purpose.

Mitigation of Chemical Hazards

Because of the ongoing proliferation of industrial chemicals in densely populated urban areas, renewed attention is being given to specific methods whereby industry can demonstrably reduce both the workplace and community risks associated with hazardous chemicals. It is important to emphasize that whatever an industrial plant does to protect workers and the surrounding community may also provide essential protection to emergency response personnel involved in a facility incident.

Of particular importance are the following techniques, which presume both the redirected use of a current base of knowledge and developing technologies:

- Product reformulation
- Chemical substitution
- Alternative process engineering

FIGURE 8-1

EXAMPLE OF FEMA MITIGATION TIP (FEMA ELECTRONIC REFERENCE LIBRARY)

FEMA - Mitigation
Reducing Risk through Mitigation

Protecting Your Property from Earthquakes

Are You at Risk?
If you aren't sure whether your house is at risk from earthquakes, check with your local building official, city engineer, or planning and zoning administrator. They can tell you whether you are in an earthquake hazard area. Also, they usually can tell you how to protect yourself and your house and property from earthquakes.

What You Can Do
Earthquake protection can involve a variety of changes to your house and property -- changes that can vary in complexity and cost. You may be able to make some types of changes yourself. But complicated or large-scale changes and those that affect the structure of your house or its electrical wiring and plumbing should be carried out only by a professional contractor licensed to work in your state, county, or city. Examples of earthquake protection are anchoring and bracing propane tanks and compressed gas cylinders. These are things that skilled homeowners can probably do on their own.

Anchor and Brace Propane Tanks and Gas Cylinders
During earthquakes, propane tanks can break free of their supporting legs. When a tank falls, there is always a danger of a fire or an explosion. Even when a tank remains on its legs, its supply line can be ruptured. Escaping gas can then cause a fire. Similar problems can occur with smaller, compressed gas cylinders, which are often stored inside a house or garage.
One way to prevent damage to propane tanks and compressed gas cylinders is to anchor and brace them securely. The figure shows how the legs of a propane tank can be braced and anchored. Using a flexible connection on the supply line will help reduce the likelihood of a leak. Compressed gas cylinders, because they have to be periodically replaced, cannot be permanently anchored. But you can use chains to attach them to a wall so that they will remain upright.

Continued

FIGURE 8-1—*Continued*

EXAMPLE OF FEMA MITIGATION TIP (FEMA ELECTRONIC
REFERENCE LIBRARY)

Tips

Keep these points in mind when you anchor and brace propane tanks or compressed gas cylinders:

- Before you alter your propane tank in any way, make sure that the tank is your property and not rented from the propane supplier. Before welding new bracing to the tank legs, you must remove the gas from the tank. You should also check with your propane supplier to find out whether additional precautions are necessary.

- Clear the area around the propane tank to ensure that there are no tall or heavy objects that could fall on the tank or rupture the supply line.

- Keep a wrench near the shutoff valve and make sure the members of your family know how to turn off the supply line if they smell a gas leak. On larger tanks, such as farm tanks, consider installing a seismic shutoff valve that will automatically turn off the gas during an earthquake.

- Provide a flexible connection between the propane tank and the supply line and where the supply line enters the house. But keep in mind that adding a flexible connection to a propane tank line should be done by a licensed contractor, who will ensure that the work is done correctly and according to all applicable codes. This is important for your safety.

- To attach a compressed gas cylinder to a wall, use two lengths of chain around the cylinder -- one just below the top of the cylinder and one just above the bottom. The chains should be attached to eye hooks that are screwed into the wall. In wood-frame walls, the eye hooks must be long enough to penetrate not just the wall but the studs behind it as well. In concrete or masonry block walls, the eye hooks should be installed with expansion anchors or molly bolts.

Estimated Cost

Bracing and anchoring a propane tank will cost about $250. Having flexible connections installed on the tank and at the house will cost about $75. Attaching one gas cylinder to the wall will cost about $50.

Other Sources of Information

Seismic Retrofit Training for Building Contractors and Building Inspectors: Participant Handbook, FEMA, 1995

Reducing the Risks of Nonstructural Earthquake Damage: A Practical Guide, FEMA-74, 1994

Protecting Your Home and Business from Nonstructural Earthquake Damage, FEMA, 1994

To obtain copies of these and other FEMA documents, call FEMA Publications at 1-800-480-2520. Information is also available on the World Wide Web at http//:www.fema.gov.

In many instances, it is possible to reduce or effectively remove a hazard associated with a particular material or process. One of the hazards of a coolant oil, for example, may be toxicity due to a heavy metal constituent (e.g., cadmium). Such an oil may be reformulated to remove the heavy metal constituent and thus remove the hazard of heavy metal toxicity without impairing the usefulness of the coolant. This is an example of *product reformulation*, which is an increasingly important growth industry today precisely because of

FIGURE 8-2

LESSONS LEARNED FROM TRAIN DERAILMENT (ADAPTED FROM FEMA. TECHNICAL RESCUE INCIDENT REPORT: THE DERAILMENT OF THE SUNSET LIMITED; SEPTEMBER 11, 1993; BIG BAYOU CANOT, ALABAMA)

> ➤ Personnel working on or around water must wear appropriate personal flotation devices.

> ➤ Rotate crews and pace activities to avoid premature exhaustion. Don't let them get burned out in the initial stages of an incident — especially if it is likely to be an extended incident.

> ➤ Relieve the initial response personnel and remove them from the immediate operating area as soon as feasible. Critical Incident Stress Management (CISM) is an absolute must! Personnel who were involved in the incident report that their participation in CISM was of significant benefit to them.

> ➤ Rehabilitation, although difficult to establish in the tight operational environment of the scene, was an absolute necessity. Given the high temperatures, constant exposure to the sun, and dangerous working conditions, it was critical to ensure that personnel were rehabilitated on a regular, formal basis. Use a large tarp or parachute cloth to set up a shaded area for rehabilitation.

> ➤ Personnel should be rotated through various jobs on an extended incident to prevent them from being unduly subjected to too much of any particular sight, sound, smell, etc. Ensure that this rotation occurs even if they are being adequately rehabilitated with rest periods, fluids, and nourishment.

> ➤ Personnel must receive ample food and liquids. This will prevent dehydration (which is the most likely source of fatigue) and loss of energy. Care must be taken to ensure that responders do eat and drink as they may have a tendency to ignore sustenance in favor of continuing to work. Personnel must not eat on scene or prior to thoroughly washing their hands.

> ➤ In swampy areas, have ample supplies of mosquito repellent on hand. When working in the sun, make sure that sunblock is available to all personnel.

the concern for human health and safety and the integration of that concern into global marketing strategies.

Another hazard reduction strategy is *chemical substitution*, which, though possibly involving a chemical reformulation of an existing product, primarily focuses on the substitution of a less hazardous chemical for a more hazardous substance. Examples include water-based paint substitutes for oil-based paints, nonchlorine bleaching agents for chlorine-containing bleaching agents, and certain botanical pesticides for synthetic pesticides.

In many situations, neither chemical reformulation nor a simple chemical substitution can be effectively employed. It may therefore become reasonable to consider *alternative process engineering*, as in reengineering a water treatment plant to accomplish disinfection by an ozonation process rather than a chlorination process—an engineering alternative that could significantly reduce both facility and community risk due to catastrophic releases of chlorine caused either by natural phenomena (e.g., earthquake, flood) or by human error. Such an alternative, of course, also would significantly reduce the probability of an incident requiring emergency response services as well as the risk presented to emergency response personnel.

In order to minimize the risk associated with a chemical hazard that cannot itself be reduced, it is necessary to minimize exposure. This is accomplished by implementing exposure-control approaches in the following order:

1. *Management (or executive, or administrative) control*: includes the management of schedules, assignments, and procedures to minimize the frequency and duration of exposure to specific hazards.
2. *Engineering control*: involves the use of space, barriers, and ventilation to limit and isolate exposure.
3. *Personal protective clothing and equipment.*

Both management and engineering control approaches are of particular importance to reducing risk not only at the workplace, but also off-site in the surrounding community.

Specific examples of management and/or engineering control approaches include:

1. *Efficiency Improvement*: the redesign of production processes to improve the efficiency by which hazardous materials are processed; increased efficiency can result in decreased amounts of on-hand hazardous chemicals as well as hazardous waste.
2. *In-Process Recycling*: the rerouting of hazardous materials directly back into a production process, which therefore can result in reduced inventories of feedstock and hazardous waste.
3. *Fugitive Release Control*: the prevention, entrapment, and/or containment of spills, leaks, and air emissions of hazardous substances.
4. *Chemical Inventory Control*: any policy or procedure affecting the purchase and storage of hazardous feedstock chemicals.

The reduction in feedstock chemicals and hazardous waste through efficiency improvement and in-process recycling are particularly relevant to emergency response planning not only because they can reduce the probability of large releases of hazardous chemicals (and, therefore, of incidents requiring response), but also, should an incident occur, the risk of exposure presented to the public and to emergency response personnel.

Fugitive release control, especially those measures taken by industry to implement immediate containment of leaks and spills, can mean the difference between a minor incident that can be handled easily by in-plant operational personnel, and a major incident that requires extensive community response services. With respect to incident response services themselves, fugitive release control must be given special attention in the progress of several operationally related activities (see Image 8.2), including:

- The control of runoff of water and foams used for fighting fires, which may be contaminated with on-site hazardous chemicals and which could contaminate both subsurface and downstream surface water supplies

IMAGE 8-2

NEW YORK, NY, SEPTEMBER 28, 2002: URBAN SEARCH AND RESCUE TEAM MEMBERS REVIEW SITE MAPS AT THE WORLD TRADE CENTER SITE

Source: Andrea Booher/FEMA News Photo

- The control of wind-borne hazardous particles and other contaminated materials released on-site during and after the incident
- The control and containment of incident-related releases (e.g., through tank rupture and other structural failures) of on-site feedstock chemicals, fuels, and hazardous wastes
- The control, containment, and subsequent disposal of waste water and other contaminated materials generated during the decontamination of equipment and clothing used during response operations

Of all the types of management control approaches to the mitigation of risks associated with industrial chemicals, inventory control is probably the most crucial. However, it is also the approach that most often conflicts with standard operating procedures within industrial bureaucracies. This is because the purchasing function in companies, which is intimately connected to finance and production functions, is typically independent of any corporate authority regarding health and safety.

The objective of the purchasing department is to obtain process feedstock, analytical, special purpose, and/or general housekeeping chemicals on schedule with regard to plant operations and at or under cost limitations. It is typically not the objective of purchasing to assess the potential for substitute chemicals having less severe hazards than ordered chemicals, nor to minimize the day-to-day on-site volume of hazardous chemicals. In fact, because the costs of chemicals can be reduced by bulk buying, usual purchasing guidelines generally result in the purchase of volumes in excess of specific operational needs.

The bulk buying of hazardous chemicals results, of course, in the on-site storage of larger volumes of those chemicals than is necessary, with consequent increase in the potential for:

- Major spills or leaks that can result in explosion, fire, uncontrolled reactions, and toxic releases
- The development of an actual emergency due to unstable or reactive chemicals being stored beyond their safe shelf-life (e.g., chemicals that produce organic peroxides)
- The development of a hazardous chemical emergency due to environmental factors (e.g., heat) and/or natural disasters (e.g., flood, earthquake)
- Increased exposure of response and other personnel as well as the general public to hazardous chemicals

The only practical approach toward ensuring that the purchasing function does not lead or contribute to hazardous chemical incidents is to integrate that function into a facility-wide health and safety program. This is most effectively accomplished by inserting the health and safety officer into the chemical procurement decision-making loop, which extends from the determination of operational needs and specifications to the actual purchase of chemicals.

The impact of inventory control on emergency planning and response extends, of course, well beyond the purchase of chemicals. Inventory control also includes protocols that govern the manner in which chemicals actually are stored and handled. Too often, the storage and handling of chemicals is based solely on grounds of production-needs and convenience rather than on an understanding of those physical and chemical characteristics of chemical feedstocks that determine their potential for initiating or exacerbating an emergency.

For example, the vapor density of many commonly used flammable industrial solvents (e.g., ethanol, acetone, benzene) is greater than the density of air (see Table 8.1). The direct consequence of this physical characteristic is that such vapors fall to the bottom of the air column (rather than rising to the top), with potential subsequent risk of an incident due to:

- *Collection of explosive fumes in subfloor conduits used for electrical wiring and junctions*: explosion could occur in these room conduits; also, because subfloor conduits often extend beneath walls, an explosion could occur at fume collection points far removed from the source of fumes and/or extend (through flashback) throughout the plant conduit system
- *Collection at lower floor levels of explosive fumes released on upper floor levels*: this is a particularly dangerous situation because fumes could flow from upper level production areas to lower level areas where cafeterias, smoking-break areas, or other locations where open sparks or flame could ignite the fumes and cause flashback to source areas
- *Misplacement of intake to exhaust ventilation*: in storage areas having exhaust ventilation, the intake to the ventilation system must be at floor level in order to exhaust high-density fumes; misplacement of intake at higher

TABLE 8-1

VAPOR DENSITY (VD) RELATIVE TO AIR FOR COMMON INDUSTRIAL SOLVENTS (VD$_{AIR}$ = 1.0)

Vapor Density for Common Solvents *(relative to air)*	
Methanol	1.1
Ethanol	1.6
Acetone	2.0
Isopropanol	2.1
Butanol	2.5
Tetrahydrofuran	2.5
Petroleum ether	2.5
Benzene	2.8
Furfural	3.3
Isopropyl ether	3.5
Heptane	3.5
n-Butyl ether	4.5

levels, which does not guarantee exhaust of high-density fumes, will lead to a sense of false security; also, external exhaust of such fumes should be high enough to ensure that there is not contact with sparks or flames (e.g., an outside smoking area, or location where there is open-arc machinery)

Regulation-based Mitigation Methods

Regulations typically are viewed by industry simply as constraints imposed by government on business operations, but they are also key sources of information regarding the mitigation of hazards that, if not appropriately addressed, may result in incidents requiring emergency response or otherwise result in conditions that present unnecessary risk to emergency response personnel.

Regardless of the type of hazard or risk addressed by any particular regulation, regulations generally define a wide range of alternative mitigation strategies that may be used to minimize the probability of occurrence and the magnitude of potential incidents, including the following:

- General policies
- Analytical and assessment methods for determining the need for mitigation
- Criteria for selecting appropriate measures from among alternative administrative and engineering mitigation techniques
- Criteria for selecting, testing, and maintaining personal protective clothing and equipment
- Ambient monitoring and medical surveillance techniques
- Personnel training objectives and methods

Of particular importance as sources of information regarding effective mitigation measures for American industry are the following OSHA regulations:

- 29 CFR 1910.95 (*Hearing Conservation*): requires the development of a written program that is inclusive of the ambient monitoring of workplace noise, noise reduction strategies, the selection and proper use of appropriate hearing protections, and the medical surveillance of personnel

- 29 CFR 1910.109 (*Process Safety Management of Explosive and Blasting Agents*): requires the development of written SOPs regarding all aspects of the use of explosive and blasting agents, including marking and labeling, personnel training, and personal protective equipment and clothing
- 29 CFR 1910.119 (*Process Safety Management of Highly Hazardous Chemicals*): for companies that handle any listed chemicals at or above threshold quantities; requires the development of written SOPs regarding the analysis of potential hazards and appropriate use of alarms and other fail-safe systems for preventing and containing chemical releases; includes requirements for a written *management-of-change program*
- 29 CFR 1910.120 (*Hazardous Waste Operations and Emergency Response*): requires the development of a written emergency response plan that will minimize risks to employees engaged in cleanups at uncontrolled hazardous waste sites, in routine operations and corrective actions at RCRA regulated facilities, and in other emergency response activities without regard to location
- 29 CFR 1910.134 (*Respiratory Protection*): requires the development of a written respiratory protection program that is inclusive of the selection, use, and maintenance of respirators as well as personnel training, fit testing, and medical surveillance requirements
- 29 CFR 1910.146 (*Confined Space Entry*): requires the development of written SOPs and facility permits for working within areas having limited openings for human entry and egress and that may present physical and/or chemical hazards; includes specific requirements for response personnel who enter confined spaces for the purpose of rescue.
- 29 CFR 1910.147 (*Energy Control; Lockout/Tagout*): requires the development of written SOPs that direct the servicing of machines and equipment in which unexpected energization or the release of stored energy could cause physical injury to employees performing maintenance tasks
- 29 CFR 1910.252 (*Hotwork*): requires the development of written SOPs and facility permits for performing work that results in the generation of an open flame, spark, or (by any other means) sufficient heat to cause fire or explosion, including such commonly performed work as welding, grinding, drilling, and cutting
- 29 CFR 1910.331-.335 (*Electrical Safety-Related Work Practices*): requires the development of written procedures and training for personnel who, by the nature of their work, can be expected to work with or near an electrical hazard, as well as personnel who may accidentally become exposed to an electrical hazard
- 29 CFR 1910.1030 (*Blood-borne Pathogens*): requires the development of a written exposure control plan to prevent the exposure of personnel to blood-borne pathogens; includes specific requirements regarding the development of SOPs, the use of personal protective clothing and equipment, and personnel training
- 29 CFR 1910.1200 (*Hazard Communication Standard; Right-to-Know*): requires the development of a written program that provides employees

specific information about workplace chemical hazards and the various means used to protect employees from those hazards

- 29 CFR 1910.1450 (*Laboratory Standard*): requires the development of written SOPs regarding the determination of chemical hazards, the use of MSDSs (Material Safety Data Sheets), and personnel training designed to protect laboratory personnel from chemical exposure; essentially a more stringent application of the *Hazard Communication Standard* to laboratories

This list of regulations does not exhaust the OSHA regulations that are relevant to emergency response services and operations, but does give a good representation of OSHA regulations that underscore the comprehensive scope of contemporary hazard management in industry. As well, other regulations promulgated by other agencies (e.g., EPA hazardous waste regulations, DOT hazardous materials regulations, NIOSH respiratory standards, EPA SARA Title III regulations) are equally relevant and should be carefully reviewed by response services not only with respect to potential compliance requirements but, even more importantly, with respect to types of mitigation strategies and methods. It should be noted that many regulations include nonmandatory (as well as mandatory) appendices that typically contain very useful information and guidelines for designing and managing mitigation programs.

Another vitally important reason that community response services should become thoroughly familiar with industrial regulations related to hazard and risk mitigation is that they are thereby better prepared to advise local industrial facilities of minimally acceptable standards and, as appropriate, to work with individual corporations to ensure the implementation of adequate policies and programs that can reduce the probability of in-plant incidents and/or of unnecessary risks to emergency response personnel and to the public at-large during any incident that may nonetheless occur.

Mitigation Methods Employed by Professional Response Services

As important as it is for any response service to keep abreast of technological developments (e.g., chemical substitution, alternative process engineering) and regulatory standards, perhaps the most practical information on effective in-service mitigation methods is available through peer professional organizations, including other response service organizations (e.g., community fire services, local search and rescue services), professional associations (e.g., NFPA), and governmental emergency service agencies (e.g., FEMA, state emergency response agencies). With the advent of the Internet and the ready accessibility of personal computers, instantaneous linkage with such organizations gives good assurance that in-service mitigation protocols, procedures, and equipment are continually improved by actual field experience as well as by new developments in technology.

Information readily available includes criteria, procedures, methods, equipment, and technical assessments having both broad and specialized relevance to any comprehensive in-service mitigation program.

For example, ergonomic criteria developed for the selection of tools by one type of response service (see Figure 8.3) can be readily adapted to the needs of other services, regardless of any dissimilarity of incident-related roles and responsibilities. In the same manner, protocols found to be effective by fire services for selecting personal protective clothing and equipment (see Figure 8.4) can readily be adapted to meet comparable needs in swift-water search and rescue and many other specialized services.

Specialization and compartmentalization are thoroughly pervasive attributes in any type of human endeavor and, unfortunately, typically result in the proliferation of distinct jargon, informational sources, and perhaps more importantly, attitudes that isolate groups that otherwise have a common purpose. With respect to community response services (which certainly do have a common purpose) and, more specifically, to mitigation (which must become a community-wide

---

FIGURE 8-3

ERGONOMIC CRITERIA FOR SELECTING TOOLS (ADAPTED FROM U.S. FIRE ADMINISTRATION, 1996: FIRE AND MEDICAL SERVICES ERGONOMICS)

It is important to find out how the equipment feels when being used before instructing a fire fighter or EMT to use new equipment.
Tools should be tried out and the potential users should understand how they are used and how they feel before being specified.

Hand tools should be selected which are designed so that the wrist can maintain a straight and neutral position.
Whenever possible, tools should be chosen that have been designed to reduce vibration and which do not transmit torque to the hand and arm. Power tools are preferred because they reduce the force and exertion required to perform the task. All devices must be designed for safe operation.

Tool balancers should be used where possible.
The center of gravity must be aligned with the center of the grasping hand; this provides the leverage necessary to keep the tool in alignment. Tools that are continuously held should weigh approximately one to two pounds. Any heavier tools should be counterbalanced.

Handles should contact as much of the hand and fingers as possible.
Handles should be approximately 1.25 – 1.75 inches in diameter. The minimum handle length should be 5 inches. Tools requiring both hands should provide two handles. Handles should be positioned to reduce awkward positions, and have a maximum distance of 2.7 inches. Tools should be equipped with a comfortable grip span between the thumb and forefinger. Vibration dampening materials should be incorporated in or on tool handles. These materials can also be incorporated into gloves. Where possible, use slip-resistance material for tool handles.

Narrow tool handles which concentrate large forces onto small areas of the hand or grips with finger grooves, ridges, or recesses should be avoided.
Short tool handles that press into the palm of the hand and tools that exert force onto the sides or back of the hand should be avoided. Power tools that must be grasped by the motor housing should be avoided. Evaluate pinch points, sharp edges, vibration transmitted to the user, tool exhaust air directed toward the wrist or other parts of the body, and excessive noise.

Safety features should be maintained as originally designed.
Knives and other tools with a cutting edge must be kept sharp. A regularly scheduled tool maintenance program should be established.

FIGURE 8-4

GUIDELINES FOR SELECTING AND PROCURING FIRE-FIGHTING
PROTECTIVE CLOTHING AND EQUIPMENT (ADAPTED FROM U.S.
FIRE ADMINISTRATION, 1993: MINIMUM STANDARDS ON
STRUCTURAL FIRE FIGHTING PROTECTIVE CLOTHING
AND EQUIPMENT: A GUIDE FOR FIRE SERVICE
EDUCATION AND PROCUREMENT [FA-137])

---

1. Identify the specific needs of your department

- Examine current problems or injuries in using protective clothing and equipment. Determine what performance is needed to overcome these limitations.

- Consider current protective clothing and equipment and its compatibility with the items to be purchased.

- Determine if there are specific areas of protection that are required but are not covered in available specifications.

- Decide on clothing sizing and design issues. For example, your department may have a preference for clothing color and trim style. Also, special pockets maybe required to hold radios or other equipment.

2. Write specification that will meet these needs

- Solicit help from other organizations (such as NAFER or SAFER) or departments who may have already developed specifications that you can use or modify.

- Use NFPA standards as the basis for purchase specifications. If deviations from a NFPA standard are made, thoroughly document the reasons for the deviations. This limits department liability.

- If trying to purchase specific manufacturer products, determine those product characteristics that are unique which can be specified.

- Establish rating criteria ahead of time that will allow you to evaluate products to your specification.

- Include provision for returning unsatisfactory products and assessing penalties if bid specifications are not met.

3. Solicit bids for the clothing or equipment needed

- Increase purchasing power by forming collective buys with other departments to obtain larger quantities at volume discounts.

- Require manufacturers to provide one or more samples that can be evaluated. Develop a test plan to evaluate the sample and compare its performance to competing products.

- Specify that manufacturers show evidence of compliance with the appropriate NFPA standard. Have manufacturers supply all data showing compliance of their product in a format that will allow your department to compare all competing products easily.

- Require manufacturers provide complete user instructions and copies of warranties and technical data for examination.

4. Thoroughly evaluate bids

- Carefully check sample clothing against the developed specification.

- Examine information supplied with the products such as instructions, warranties, and technical data. Look for completion of the information and ease of its use.

- Employ an evaluation system to rate products.

- Physically wear or use equipment to ensure that it meets original specifications.

5. Evaluate the performance of clothing or equipment to determine if it does meet the original needs

- Once clothing or equipment has been received, establish SOPs for its use, care, and maintenance.

- Periodically review how clothing or equipment meets fire department needs.

- Revise specifications as needed for new or replacement clothing and equipment. Use the standard as the basis for your specifications.

objective), it is therefore of paramount importance to recognize that all response services, regardless of the special risks engendered by their specific roles, must deal with certain categories of risk that are common to all, including those risks related to the performance levels of personnel, inappropriate protective clothing and equipment, administrative failures, and problems with communication equipment and procedures (see Table 8.2). Risk management techniques designed by one type of service to minimize such risks therefore should be reviewed carefully by other services and, as appropriate, adapted to meet needs that, though similar, may differ to some degree in the operational context of an actual incident.

There are, of course, certain mitigative techniques that must be assiduously employed in precisely the same manner by any response service that may have jurisdictional control of incident response. For example, in a prolonged response effort in an isolated area where potable water for responding personnel is problematic, disinfection of drinking water must be implemented in standard fashion, regardless of type of incident or jurisdictional control. The fact that recommended procedures for disinfecting water may vary from state to

TABLE 8-2

SOURCE OF RISK AND CORRECTIVE ACTIONS (ADAPTED FROM U.S. FIRE ADMINISTRATION, 1996: RISK MANAGEMENT PRACTICES IN THE FIRE SERVICE [FA-166])

Potential Source of Risk	Risk Management Technique
Personnel	
▪ Failure to meet minimum performance requirements..............	▪ Establish minimum performance
▪ Failure to properly train..................	▪ Establish and conduct performance-based training for all personnel
....	▪ Training should conform to relevant OSHA, NFPA, and other standards
▪ Failure to adequately equip personnel..............................	▪ Provide protective equipment that meets NFPA standards
Fire Inspection Practices	
▪ Failure to notify owners of hazards..............................	▪ Require complete records of every inspection
▪ Failure to pursue compliance..............	▪ Consistently issue citations and seek judicial intervention when hazards are imminent
Administration	
▪ Level of service not defined..............	▪ Define level of service for all service deliverables
▪ Incomplete records..................	▪ Document and address all complaints promptly
Communication	
▪ Failure to dispatch promptly..............	▪ Ensure that specific dispatch policies are in place and that performance is monitored
▪ Failure to properly advise callers of potential delay..........................	▪ Establish policy and procedure to address these issues

state (see Figure 8.5) therefore may require that preincident interagency planning and coordination efforts produce consensual disinfection procedures. Similarly, other mitigative actions must be addressed and resolved by preincident consensus, including (a) proper health maintenance of responders regarding the control of insects and other potential disease-vectors and operational nuisances; (b) location and design of temporary morgues, sanitary facilities, and personnel shelter and rehabilitation areas; and (c) local disposal of incident-related debris and other potentially contaminated materials.

Internet links to governmental and professional organizations having specific response expertise are invaluable resources not only for response services, but also for industry as well as the general public. Some of these links (see Figure 8.6) provide immediate access to chemical hazard databases, HAZMAT-related information, and specific alerts regarding both sources and potential mitigation

---

FIGURE 8-5

DIRECTIONS FOR DISINFECTING DRILLED WELLS (ADAPTED FROM CENTERS FOR DISEASE CONTROL AND PREVENTION, NATIONAL CENTER FOR ENVIRONMENTAL HEALTH. FLOOD: A PREVENTION GUIDE TO PROMOTE YOUR PERSONAL HEALTH AND SAFETY, NCEH ELECTRONIC REFERENCE LIBRARY)

Disinfection of Drilled Wells

1. Determine the amount of water in the well by multiplying the gallons per foot (**Table 1**) by the depth of the well in feet. For example, a well with a 6-inch diameter contains 1.5 gallons of water per foot. If the well is 120 feet deep, multiply 1.5 by 120 (1.5 gal/ft x 120 ft = 180 gallons).

2. For each 100 gallons of water in the well, use the amount of chlorine (liquid or granules) indicated in **Table 2**. Mix the total amount of liquid or granules with about 10 gallons of water.

3. Pour the solution into the top of the well before the seal is installed.

4. Connect a hose from a faucet on the discharge side of the pressure tank to the well casing top. Start the pump. Spray the water back into the well and wash the side of the casing for at least 15 minutes.

5. Open every faucet in the system and let the water run until the smell of chlorine can be detected. Then close all the faucets and seal the top of the well.

6. Let stand for several hours, preferably overnight.

7. After you have let the water stand, operate the pump by turning on all faucets, continuing until all odor of chlorine disappears. Adjust the flow of water from faucets or fixtures that discharge into septic tank systems to a low flow to avoid overloading the disposal system.

Table 1	
DIAMETER OF WELL (IN INCHES)	GALLONS PER FOOT OF WATER
3	0.37
4	0.65
5	1.0
6	1.5
8	2.6
10	4.1
12	6.0

Table 2

Laundry Bleach (5.25 % Chlorine) ——➤ 3 cups*

Hypochlorite Granules (70% Chlorine) ➤ 2 ounces**

* 1 Cup = 8-ounce measuring cup
** 1 Ounce = 2 heaping tablespoons of granules

Source: Illinois Department of Public Health
Note: Recommendations may vary from state to state

FIGURE 8-6

KEY INTERNET LINKS AND INFORMATION AVAILABLE THROUGH U.S. NATIONAL RESPONSE TEAM HOMEPAGE (NRT ELECTRONIC LIBRARY)

Preparedness and Response Links to the Internet

Responders' Toolbox:
Chemical, Safety, Health, and Risks

+ Agency for Toxic Substances and Disease Register (ATSDR) – HazDat Databases This Hazardous Substance Release/Health Effects Database provides access to information on the release of hazardous substances from Superfund sites or from emergency events and on the effects of hazardous substances on the health of human populations. (Source: ATSDR/HHS)

+ Agency for Toxic Substances and Disease Registry ToxFAQs Menu - Searches on information of toxic hazardous substances. (Source: ATSDR/HHS)

+ Center for Disease Control Sites Query - Specific facility information for particular sites, such as counties, facility type, etc. (Source: ASTDR)

+ Emergency Response & Research Institute - EmergencyNet News Service - Provides direct links to hazmat-related information such as Hazardous Materials Operations Page, Disaster/Rescue Operations Page, etc. (Source: EmergencyNet)

+ Chemical Safety Alert - "Rupture Hazard of Pressure Vessels" (May 1997) - An EPA/CEPPO Alert document containing information to protect human health and the environment by preventing chemical accidents. This document provides information on the hazards of improperly operated or maintained pressure vessels that could rupture. Currently only available in **Adobe PDF** format. (Source: CEPPO)

+ Chemical Safety Alert - "Fire Hazard from Carbon Absorption Deodorizing Systems" (May 1997) - An EPA/CEPPO Alert document containing information to protect human health and the environment by preventing chemical accidents. This document provides information on activated carbon systems used to adsorb vapors for control of offensive odors. Currently only available in **Adobe PDF** format. (Source: CEPPO)

+ Chemical Safety Alert - "Catastrophic Failure of Storage Tanks" (May 1997) - An EPA/CEPPO Alert document containing information to protect human health and the environment by preventing chemical accidents. This document provides information on the catastrophic failure of aboveground, atmospheric storage tanks. Currently only available in **Adobe PDF** format. (Source: CEPPO)

+ Chemical Safety Alert - "Lightning Hazard to Facilities Handling Flammable Substances" (May 1997) - An EPA/CEPPO Alert document containing information to protect human health and the environment by preventing chemical accidents. This document provides precaution information for industry in case lightning strikes hit equipment and storage or process vessels containing flammable materials. Currently only available in **Adobe PDF** format. (Source: CEPPO)

of various types of hazards. Most links also provide access to international (see Figure 8.7) as well as national sources of information regarding mitigation.

ALL-HAZARD MITIGATION

Environmental and workplace health and safety regulations in the United States have emphasized the importance of assessing the potential impacts of various types of industrial hazardous incidents (e.g., fire, chemical release) and natural disasters (e.g., flood, earthquake) on workplace and community health and safety, and taking effective actions to mitigate the probability and magnitude of

FIGURE 8-7

EXAMPLE OF U.S. INTERNATIONAL TRADE ADMINISTRATION ALERT
ON HANDBOOK RELEVANT TO HAZARD AND RISK MITIGATION
(INTERNATIONAL TRADE ADMINISTRATION, ELECTRONIC
REFERENCE LIBRARY)

International Trade Center Bookstore

Environmental Technologies Export Handbook

Performing Org.: International Trade Administration, Washington, DC.

Price: $28 . Outside the U.S., Canada, and Mexico $56 .
Microfiche price: $14 . Outside the U.S., Canada, and Mexico $28 .
A $4 handling fee is added to each total order, $8 for customers outside of the U.S., Canada, and Mexico.

| Order | Or call the NTIS sales desk at 1-800-553-6847
. *NTIS Order Number:* PB96-137633INO.
144p* Publication Date: Sep 95 .
Document Type: Technical Report

Summary: Table of Contents: Defining Environmental Technology; The Global Environmental Market; Environmental Market Country Summaries (Countries Are Listed Alphabetically); Assistance Programs; Sources of Sales Leads; Sources of Finance; **Major Environmental Trade Associations and Programs; Publications on Environmental Technology Issues;** and Bibliography.

Keywords: * Environmental protection; * International trade; * Market analysis; * Pollution control equipment; Economic development; Exports; Financing; Foreign countries; Investments; Market research; **Mitigation; Pollution abatement;** Remediation; Sales; Technology innovation; Technology transfer; **Waste management**

Primary Subject Category: Environmental Pollution and Control - General (Codes: 68*; 96C*)

To order, call the NTIS sales desk at 1-800-553-6847 This product may also be ordered by fax at (703) 321-8547, or by e-mail at orders@ntis.fedworld.gov. NTIS is located at 5285 Port Royal Road, Springfield, VA 22161.

Visit the International Trade Center Bookstore Search page to locate other related products.
I International Trade Center Bookstore I National Technical Information Service (NTIS) I

---

such impacts. Although governmental agencies have demonstrably moved to integrate response to human-caused and natural disasters into an all-hazard approach to hazard and risk mitigation, much of American industry has failed to do so, persisting in looking at disaster through a regulatory-induced myopia.

The all-hazard approach (which is the premise of the National Response Plan; see Chapter 12) is based on the recognition that any distinction between human-made and nature-made disaster is essentially irrelevant to the risk to human life and health presented by an actual emergency. Just as the risk imposed by a natural disaster such as flood (see Image 8.3) or earthquake can be magnified by subsequent releases of hazardous chemicals or uncontrolled chemical reactions, so the risks imposed by chemical or biological contamination can be magnified by a natural disaster. This recognition of the actual concomitancy of human-made and nature-made risks reflects, perhaps, our growing appreciation

IMAGE 8-3

NEW ORLEANS, LA, AUGUST 31, 2005: FEMA PERSONNEL
DO A SECONDARY SEARCH ON A HOUSE FLOODED
BY HURRICANE KATRINA

Source: Jocelyn Augustino/FEMA

of some of the more dire consequences of the simultaneous urbanization of world society and the rapidly expanding dependency of that society on industrial technology.

There can be no question that industry must continue to improve its mitigation measures with respect to specific types of hazards, such as fire and the release of hazardous materials and wastes (see Table 8.3). However, industry must also significantly increase its effort to evaluate the consequences of facility-related operations and chemical inventories in light of disasters beyond its control, such as hurricane, tornado, flood, winter storm, earthquake, and power outage (see Table 8.4). Of course, another increasingly important source of risk that is typically beyond the operational control of industrial facilities is terrorism. Specifically, industry must consider:

- How such natural disasters as well as terrorism can cause facility-related operations, chemical feedstock, and hazardous wastes to become additional sources of risk to community health and safety
- What managerial and/or engineering means of mitigation can be implemented to reduce those risks
- What quality control measures must be implemented to ensure the effective implementation of selected mitigation measures

TABLE 8-3

BASIC STEPS IN FIRE AND HAZARDOUS MATERIAL INCIDENT MITIGATION (ADAPTED FROM FEMA EMERGENCY MANAGEMENT GUIDE FOR BUSINESS, FEMA ELECTRONIC LIBRARY)

Fire Mitigation

- Meet with fire department to talk about community fire response capabilities; discuss your operations; identify processes and materials that could cause of fuel a fire, or contaminate the environment in a fire
- Have facility inspected for fire hazards; ask about fire codes and regulations
- Request insurance carrier to recommend fire prevention and protection measures; carrier may also offer training
- Distribute fire safety information to employees regarding: how to prevent workplace fire; how to contain a fire; how to evacuate the facility; where to report a fire
- Instruct personnel to use the stairs—not elevators—in a fire; instruct them to crawl on their hands and knees when escaping a hot or smoke-filled area
- Conduct evacuation drills; post maps of evacuation routes in prominent places; keep evacuation routes, including stairways and doorways, clear of debris
- Assign fire wardens for each area to monitor shutdown and evacuation procedures
- Establish procedures for the safe handling and storage of flammable liquids and gases
- Establish procedures to prevent the accumulation of combustible materials
- Provide for the safe disposal of smoking materials
- Establish preventive maintenance schedule to keep equipment operating safely
- Place fire extinguishers in appropriate locations
- Train employees in use of fire extinguishers
- Install smoke detectors; check smoke detectors once a month; change batteries at least once a year
- Establish system for warning personnel of a fire; consider installing a fire alarm with automatic notification to the fire department
- Consider installing a sprinkler system, fire hoses and fire-resistant walls and doors
- Ensure that key personnel are familiar with all fire safety systems
- Identify and mark all utility shutoffs so that electrical power, gas or water can be shut off quickly by fire wardens or responding personnel
- Determine the level of response facility will take if a fire occurs, considering the following options: (a) immediate evacuation of all personnel on alarm, (b) all personnel are trained in fire extinguisher use; personnel in immediate are of a fire attempt to control it; if they cannot, the fire alarm is sounded and all personnel evacuate, (c) only designated personnel are trained in fire extinguisher use, (d) a fire team is trained to fight incipient-stage fires that can be controlled without protective equipment or breathing apparatus; beyond this level of fire, the team evacuates, (e) a fire team is trained and equipped to fight structural fires using protective equipment and breathing apparatus

Hazardous Material Incident Mitigation

- Identify and label all hazardous materials stored, handled, produced and disposed of by facility; follow government regulations that apply to facility and operations; obtain material safety data sheets for all hazardous materials at facility
- Ask local fire department for assistance in developing appropriate response procedures
- Train employees to recognize and report hazardous material spills and releases; train employees in proper handling and storage

Continued

TABLE 8-3—*Continued*

BASIC STEPS IN FIRE AND HAZARDOUS MATERIAL INCIDENT MITIGATION (ADAPTED FROM FEMA EMERGENCY MANAGEMENT GUIDE FOR BUSINESS, FEMA ELECTRONIC LIBRARY)

- Establish a hazardous material response place, including:
  1. procedures to notify management and emergency response organizations of an incident,
  2. procedures to warn employees of an incident, and
  3. procedures for evacuating facility

- Depending on operations, organize and train an emergency response team to confine and control hazardous materials spills in accordance with applicable regulations
- Identify other facilities in area that use hazardous materials; determine when an incident could affect your facility
- Identify highways, railroads and waterways near facility used for the transportation of hazardous materials; determine how a transportation accident near your facility could affect your operations

TABLE 8-4

BASIC STEPS IN MITIGATION WITH RESPECT TO NATURAL DISASTERS (ADAPTED FROM FEMA EMERGENCY MANAGEMENT GUIDE FOR BUSINESS, FEMA ELECTRONIC LIBRARY)

Hurricane & Tornado Mitigation

Hurricane
- Review communication evacuation plans
- Establish facility shutdown procedures; establish warning and evacuation procedures; make plans for assisting employees who may need transportation
- Make plans for communicating with employees' families before and after a hurricane
- Purchase a NOAA Weather Radio with a warning alarm tone and battery backup; listen for hurricane watches and warnings
- Survey facility and take steps to protect outside equipment and structures
- Protect windows; permanent storm shutters offer the best protection; covering windows with 5/8 inch marine plywood is another option
- Consider the need for back-up systems: portable pumps; alternative power sources; battery-powered emergency lighting
- Prepare to relocate records, computers and other items within facility or to another location

Tornado
- Review community tornado warning system
- Purchase a NOAA Weather Radio with a warning alarm tone and battery backup; listen for tornado watches and warnings
- Establish procedures to inform personnel when tornado warnings are posted; consider the need for spotters to be responsible for looking out for approaching storms
- Work with structural engineer or architect to designate shelter areas in facility; ask local emergency management office or National Weather Service office for guidance
- Consider the amount of space needed; adults require about six square feet of space; nursing home and hospital patients require more space

Continued

TABLE 8-4—*Continued*

BASIC STEPS IN MITIGATION WITH RESPECT TO NATURAL DISASTERS (ADAPTED FROM FEMA EMERGENCY MANAGEMENT GUIDE FOR BUSINESS, FEMA ELECTRONIC LIBRARY)

- If an underground protective area is not available, consider: small interior rooms on lowest floor and without windows; hallways on lowest floor away from doors and windows; rooms constructed with reinforced concrete, brick or block with no windows and heavy concrete floor or roof system; protected areas away from doors and windows. *Note: auditoriums, cafeterias and gymnasiums that are covered with a flat, wide-span roof are not considered safe.*
- Make plans for evacuating personnel away from lightweight modular offices or mobile home-size buildings; these structures offer no protection from tornadoes.
- Conduct tornado drills
- Once in a shelter, personnel should protect their heads with their arms and crouch down.

Flood Mitigation

Planning Measures
- Determine location of facility in relation to flood plain and elevation in relation to streams, rivers and dams
- Review community emergency plan and community evacuation routes; know where to find higher ground in case of flood
- Establish warning and evacuation procedures for facility; make plans for assisting employees who may need transportation
- Inspect areas in facility subject to flooding; identify records and equipment that can be moved to a higher location; make plans to move records and equipment in case of flood
- Purchase a NOAA Weather Radio with a warning alarm tone and battery backup; listen for flood watches and warnings
- Consider feasibility of flood-proofing facility

Permanent Flood-proofing Measures
- Fill windows, doors or other openings with water-resistant materials such as concrete blocks or bricks
- Install check valves to prevent water from entering where utility and sewer lines enter facility
- Reinforce walls to resist water pressure; seal walls to prevent or reduce seepage
- Build watertight walls around equipment or work areas within facility that are susceptible to flood damage
- Construct floodwalls or levees outside facility to keep flood waters away
- Elevate facility on walls, columns or compacted fill

Contingent Flood-proofing Measures
- Install watertight barriers called flood shields to prevent passage of water through doors, windows, ventilation shafts or other openings
- Install permanent watertight doors
- Construct movable floodwalls
- Install permanent pumps to remove flood waters

Emergency Flood-proofing Measures
- Build walls with sandbags
- Construct double row of walls with boards and posts to create a crib and fill crib with soil
- Construct a single wall by stacking small beams or planks on top of each other
- Consider need for back-up systems: portable pumps; alternative power sources; battery-powered emergency lighting
- Participate in community flood control projects

Continued

TABLE 8-4—*Continued*

BASIC STEPS IN MITIGATION WITH RESPECT TO NATURAL DISASTERS (ADAPTED FROM FEMA EMERGENCY MANAGEMENT GUIDE FOR BUSINESS, FEMA ELECTRONIC LIBRARY)

Severe Winter Storm & Earthquake Mitigation

Severe Winter Storm
- Listen to NOAA Weather Radio and local radio and television stations for weather information
- Establish procedures for facility shutdown and early release of employees
- Store food, water, blankets, battery-powered radios with extra batteries and other emergency supplies for employees who become stranded at the facility
- Provide a backup power source for critical operations
- Arrange for snow and ice removal from parking lots, walkways, loading docks, etc.

Earthquake
- Assess facility vulnerability to earthquake; ask local governmental agencies for seismic information for area
- Have facility inspected by structural engineer; develop and prioritize strengthening measures, including: adding steel bracing to frames; adding sheer walls to frames; strengthening columns and building foundations; replacing un-reinforced brick filler walls
- Inspect non-structural systems such as air conditioning, communications and pollution control systems; assess potential for damage and prioritize preventive measures
- Inspect facility for any item that could fall, spill, break or move during an earthquake, and take steps to reduce these hazards
- Move large and heavy objects to lower shelves or the floor; hang heavy items away from where people work
- Secure shelves, filing cabinets, tall furniture, desktop equipment, computers, printers, copiers and light fixtures
- Secure fixed equipment and heavy machinery to the floor; larger equipment can be placed on casters and attached to tethers which attach to wall
- Add bracing to suspended ceilings; install safety glass; secure large utility and process piping
- Keep copies of design drawings to be used in assessing facility safety after an earthquake
- Review processes for handling and storing hazardous materials; store incompatible chemicals separately
- Establish post-earthquake evacuation procedures
- Designate areas away from exterior walls and windows where occupants should gather after an earthquake if an evacuation is not necessary
- Conduct earthquake drills; provide personnel with following information: (a) during earthquake, if indoors, stay there; take cover under sturdy furniture or counter; brace against inside wall; protect head and neck, (b) if outdoors, move into the open, away from buildings, street lights and utility wires, (c) after earthquake, stay away from windows, skylights and items that could fall; do not use elevators, (d) use stairways to leave building if it is determined that a building evacuation is necessary

Technological Mitigation

(Technological emergency is any interruption or loss of a utility service, power source, life support systems, information system or equipment needed to keep business in operation)
- Identify all critical operations, including:
  a. utilities, including electric power, gas, water, hydraulics, compressed air, municipal and internal sewer systems, wastewater treatment services
  b. security and alarm systems, elevators, lighting, life support systems, heating, ventilation and air condition systems, electrical distribution system

Continued

TABLE 8-4—*Continued*

BASIC STEPS IN MITIGATION WITH RESPECT TO NATURAL
DISASTERS (ADAPTED FROM FEMA EMERGENCY MANAGEMENT
GUIDE FOR BUSINESS, FEMA ELECTRONIC LIBRARY)

Technological Mitigation

    c. manufacturing equipment, pollution control equipment
    d. communication systems, both data and voice computer networks
    e. transportation systems, including air, highway, railroad and waterway
- Determine the impact of service disruption to each system
- Ensure that key safety and maintenance personnel are thoroughly familiar with all building systems
- Establish procedures for restoring systems
- Determine need for backup systems
- Establish preventive maintenance schedule for all systems and equipment

Shift in Paradigm

Throughout the world, the phrase *shift in paradigm* is used to denote a significant change in perspective (if not methodology) in the conduct of any enterprise, whether in science, business, government, or any other sphere of professional or technical endeavor. Such a change in perspective or methodology begins with a clear recognition that the former way of doing things does not work—that current objectives have outstripped long-established techniques and that, if current objectives are not abruptly abandoned, new methodologies, technologies, and/or new approaches must be constructed and implemented.

The all-hazard approach to mitigation constitutes such a shift in paradigm. At both federal and state levels of government, it cuts across long-established boundaries of jurisdictional authority, forcing not only a significant increase in interagency cooperation, but also a growing partnership among federal and state authorities as well as private sector organizations and corporations. But the shift in paradigm required for effective all-hazard mitigation does exact specific costs, whether paid by governmental agencies or corporations.

For industry, the cost is not measured in jurisdictional authority but, rather, in changing social expectations of moral and legal responsibility. For example, in Southeast Asia, various governments have made substantial efforts to establish as a working principle the dictum: *those who are responsible for causing hazards and risks are precisely the persons who have primary responsibility for controlling those hazards and risks.*

That industry, wherever it might be, should have the primary responsibility for mitigating those hazards and risks posed to the community in which it conducts its operations is, perhaps, an arguable question under locally relevant law. But, that there is a significant worldwide trend toward holding not only corporations but also their owners and officers to such a standard cannot be questioned. In the United States, this trend is well established (if not boldly enunciated) in the plethora of health and safety regulations implemented over the past 20 years; it is also becoming increasingly documented in the case law used in both civil and criminal proceedings.

What, therefore, does an all-hazard approach to hazard and risk mitigation actually portend if not yet signify to industry? Simply, that whatever the specific circumstance, whatever the sequence of specific causes and specific effects, and whatever the level of compliance to specific regulations, corporate owners, officers, and operators (and, perhaps even corporate boards of directors or investors) are increasingly at personal financial and criminal risk if they do not take effective measures to mitigate the hazards and risks they present to the community by way of both their on- and off-site operations.

INCIDENT AND UNIFIED COMMAND

The Incident Command System (ICS) is a single standard incident management system, originally developed (in 1980) in California in response to the threat of wildland fires, that has rapidly become the command system used not only by all U.S. federal agencies having wildland fire management responsibilities, but also by an increasing number of U.S. federal and state agencies as well as operational agencies of other nations that have jurisdictional authority over various types of emergencies, including fires, oil and hazardous chemical releases, earthquakes, storms and other natural disasters, and terrorism. The basic organization of ICS (see Figure 8.8) includes five key functional groups referred to as *sections*.

FIGURE 8-8

OVERVIEW OF STRUCTURE OF INCIDENT COMMAND SYSTEM (ADAPTED FROM NATIONAL INTERAGENCY FIRE CENTER, 1994: INCIDENT COMMAND SYSTEM NATIONAL TRAINING CURRICULUM, MODULE 3 [NFES NO. 2443])

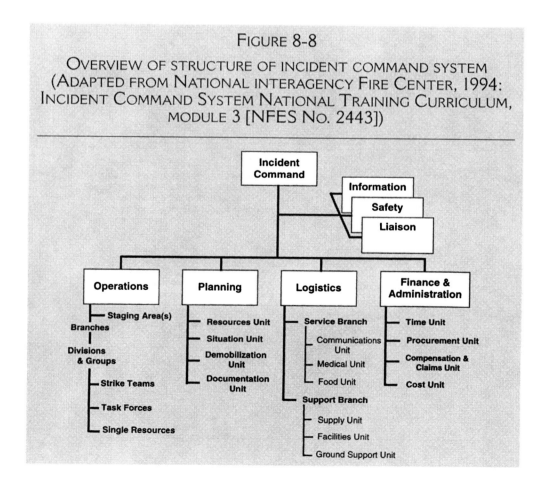

In incidents involving responders from a single jurisdictional authority (i.e., a single-jurisdiction incident), the Incident Commander (IC) is the individual who has final responsibility for the management of response activity. Depending upon various circumstances (see Figure 8.9), Incident Command may be transferred from one individual to another.

In incidents involving more than one agency (i.e., a multijurisdiction or multiagency incident), the command function is assumed by the Unified Command (UC), which may consist of a number of different federal and sate officials as well as other authorities (e.g., emergency coordinator(s) from the industrial facility(-ies)) primarily involved in the incident. However, whether the command function is assumed by the IC or the UC, there is always one person who has final responsibility for the overall response effort.

Much of the flexibility of ICS derives from the various options available to the Incident Commander (and, subsequently, to appropriate incident response managers) regarding the activation of subsidiary components of the ICS management organization. These options are based on the assessment of the ongoing developing nature and extent of the incident. Another aspect of the flexibility of ICS derives from the different modes of coordinating and directing incident response on the basis of (a) multijurisdictional responsibilities, (b) the occurrence of two or more incidents in close proximity, and (c) multiagency and multijurisdictional responsibilities within an extended geographic region.

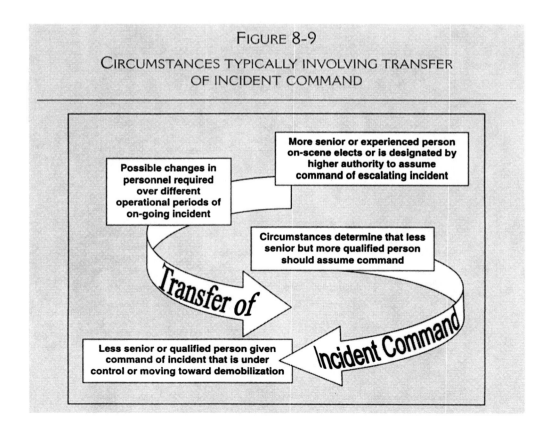

FIGURE 8-9

CIRCUMSTANCES TYPICALLY INVOLVING TRANSFER
OF INCIDENT COMMAND

More senior or experienced person on-scene elects or is designated by higher authority to assume command of escalating incident

Possible changes in personnel required over different operational periods of on-going incident

Circumstances determine that less senior but more qualified person should assume command

Transfer of

Incident Command

Less senior or qualified person given command of incident that is under control or moving toward demobilization

Because of the importance that must be given in the United States to jurisdictional responsibilities of different agencies at federal, regional, state, and local levels, the ICS must be flexible enough to accommodate these differences without sacrificing efficiency and effectiveness at the same time. This is accomplished by extending and adapting ICS to meet the needs of Unified Command, Area Command, Multiagency Coordination Systems, and Emergency Operations Centers (see Figure 8.10). The full range of these adaptations of ICS not only

FIGURE 8-10

COMPARISON OF ALTERNATIVE EMERGENCY MANAGEMENT SYSTEMS UNDER ICS (ADAPTED FROM NATIONAL INTERAGENCY FIRE CENTER, 1994: INCIDENT COMMAND SYSTEM NATIONAL TRAINING CURRICULUM, MODULE 1 [NFES NO. 2468]; MODULE 16 [NFES NO. 2470])

Incident Command System (ICS)	The management system used **to direct all operations at the incident scene.** The Incident Commander (IC) is located at an Incident Command Post (ICP) at the incident scene.
Unified Command (UC)	An application of ICS used when there is more than one agency with incident jurisdiction. **Agencies work together through their designated Incident Commanders at a single ICP to establish a common set of objectives and strategies and a single Incident Action Plan.**
Area Command/ Unified Area Command AC/UAC	Established as necessary **to provide command authority and coordination for two or more incidents in close proximity.** Area Command works directly with Incident Commanders. Area Command becomes Unified Area Command when incidents are multi-jurisdictional. **Area Command may be established at an EOC facility or at some location other than an ICP.**
Multi-Agency Coordination Systems (MACS)	An activity or formal system used **to coordinate resources and support between agencies or jurisdictions.** A MAC Group functions with the MACS. MACS interact with agencies or jurisdictions, not with incidents. MACS are useful for regional situation. **A MACS can be established at a jurisdiction EOC or at a separate facility.**
Emergency Operations Center (EOC)	Also called Expanded Dispatch or Emergency Command and Control Centers. EOCs are used in varying ways at all levels of government and within private industry **to provide coordination, direction, and control during emergencies. EOC facilities can be used to house Area Command and MACS activities as determined by agency or jurisdictional policy.**

ensure proper involvement of diverse responsible authorities in incident response, but they also ensure that the response to a particular incident will not unduly detract from local and regional resources that may be needed in response to multiple incidents.

It must be understood that there is no (nor should there necessarily be) compelling consensus regarding all possible relationships (or even terminology) regarding multiagency coordination and management of incident response. The appropriate relationships among response components (as well as precise definitions) ultimately depend upon the specific procedures in place in particular agencies and organizations at the time place of the incident. The ICS therefore provides a flexible framework of management, rather than a definitive algorithm.

For example, the U.S. National Response Team has promulgated guidelines for adapting the ICS to meet the needs of a Unified Command (Chapter 2; see Figure 2.12). As noted in these guidelines, no attempt is made "to prescribe specifically how a particular organization or individual fits within a given response structure." Despite the lack of prescriptions regarding specific assignments, however, these guidelines do give explicit directions for ensuring that the accommodation of diverse jurisdictional interests of various public agencies and private organizations do not detract from those clear lines of authority, responsibility, and accountability that are firmly established by ICS.

STUDY GUIDE

True or False

1. Emergency responders can help to minimize the number of incidents requiring the implementation of emergency response primarily by the advice and training they make available to private corporations and to the public at large.
2. Mitigation involves the prevention, control and containment, and minimization of risk.
3. The control of runoff of water and foams used for fighting fires, which may be contaminated with on-site hazardous chemicals, is a good example of efficiency improvement.
4. Recommended procedures for disinfecting water may vary from state to state.
5. An "all hazard approach to hazard and risk mitigation" essentially integrates response to human-cause and natural disasters.
6. Chemical substitution is a hazard reduction strategy that involves the substitution of a less hazardous chemical for a more hazardous substance.
7. Reengineering a water treatment plant to accomplish disinfection by ozonation rather than chlorination is a good example of alternative process engineering.
8. The first step for minimizing risk associated with a chemical hazard that cannot itself be reduced is to use engineering controls.

9. In-process recycling involves the rerouting of hazardous materials directly back into a production process, which can result in reduced inventories of feedstock and hazardous waste.
10. Chemical inventory control, which includes any policy or procedure affecting the purchase and storage of hazardous feedstock chemicals, typically requires the direct involvement of the safety officer in the chemical purchasing function.

Multiple Choice

1. The term *mitigation* today means
   A. the reduction of hazard
   B. the reduction of risk
   C. both A and B
2. The mitigation of a chemical hazard may involve
   A. product reformulation
   B. chemical substitution
   C. alternative process engineering
   D. all of the above
3. Rotating personnel through various jobs on an extended incident to prevent them from being unduly subjected to a specific hazard is a good example of exposure-control through
   A. management control
   B. engineering control
   C. both A and B
4. Fugitive release control is
   A. the prevention, entrapment, and/or containment of spills, leaks, and air emissions of hazardous substances
   B. the rerouting of hazardous materials directly back into a production process
   C. the substitution of a less hazardous chemical for a more hazardous material
5. Excessive bulk buying of hazardous chemicals increases the potential for
   A. major spills
   B. the development of an actual emergency due to unstable or reactive chemicals
   C. the development of a hazardous chemical emergency due to environmental factors (e.g., heat) and/or natural disasters (e.g., flood)
   D. increased exposure of response and other personnel as well as the general public to hazardous chemicals
   E. all of the above

Essays

1. Comment on the implication of mitigation at the level of local facilities for community-wide mitigation.
2. Discuss the advantages and disadvantages of the all hazard mitigation approach at the community level.

Case Study

A moderately sized town (population of 20,000) is located on a 25-year flood plain. The town has numerous small manufacturers of small metal-products as well as specialized furniture.

Having community-wide responsibility for public health and safety,

1. What kind of information would you like to have at hand regarding manufacturing chemical feedstock and manufacture processing?
2. What would you do with this information?

DECONTAMINATION

INTRODUCTION

Decontamination is the safe removal or inactivation of any hazardous contaminant that adheres to or is otherwise in contact with personnel, protective clothing, protective equipment, and any other incident-related equipment, vehicles, materials, or debris. Potential contaminants include physical (e.g., asbestos, radioactive substance), chemical (e.g. pesticide), or biological agents (e.g., bacteria, viruses, other parasites). There are several basic objectives to decontamination procedures:

1. *To protect on-site response personnel from direct bodily exposure to contaminants adhering to or absorbed into protective clothing or equipment.* The presumption is that personal protective clothing and equipment are appropriately selected to prevent bodily exposure to contaminants (see Image 9.1). However, unnoticed tears, rips, punctures, and other malfunctions in fact may result in actual exposure (especially during particularly rigorous response activity), in which case timely decontamination can effectively minimize the duration of exposure. Of course, in the absence of malfunction of properly selected PPC and PPE, decontamination prevents personal exposure during donning and doffing procedures.

2. *To prevent the mixing of incompatible or synergistic chemical contaminants derived from different response activities involving exposure of the same personnel and/or equipment to different chemical hazards.* During a particular incident, it may be necessary to use the same PPC, PPE, and other equipment (e.g., hand tools, vehicles, extrication equipment) for a variety of different tasks and also in different locations, with subsequent contamination by different chemicals that may be incompatible. Chemical incompatibilities or synergies in turn may exacerbate

IMAGE 9-1

NEW ORLEANS, LA, AUGUST 31, 2005: FEMA URBAN SEARCH AND RESCUE TASK FORCE BEGIN THEIR MISSION TO ASSIST RESIDENTS AFFECTED BY HURRICANE KATRINA

Source: Jocelyn Augustino/FEMA

existing hazards as well as result in completely different hazards, such as when a relatively innocuous powder becomes highly skin-absorbable and toxic when mixed with water.

3. *To protect off-site personnel (e.g., hospital personnel) from exposure in the process of treating victims, servicing equipment, or handling and transporting incident-related debris or other materials.* Whereas much of the concern regarding decontamination typically focuses on on-site incident-related activities, decontamination procedures play a vital role in the attempt to confine hazardous contaminants to the incident site where they can be better controlled. Transported off-site via victims, response personnel, and response-related equipment and debris, such contaminants readily place off-site response personnel as well as the unsuspecting and unprepared public in danger.

4. *To protect the families of response personnel from "carry-home contamination" (i.e., contamination carried off-site on the body, in clothing, and/or in personal vehicles).* A particularly vulnerable subset of the off-site public at risk due to improper decontamination procedures is composed of the families of response personnel. Even if not personally involved in tasks requiring direct exposure to physical, chemical, or biological contaminants, on-site response personnel can become contaminated indirectly simply by being present at the incident site (see Image 9.2). Also, in many

IMAGE 9-2

NEW ORLEANS, LA, AUGUST 31, 2005: KATRINA STRANDED RESIDENTS ARE BROUGHT TO AN ELEVATED BRIDGE AREA BY BOAT TO AWAIT TRANSPORTATION TO THE MAIN STAGING AREA

Source: Win Henderson/FEMA Photo

situations involving volunteer responders, personnel arrive on-site in personal vehicles and wearing personal clothes. Personal vehicles can become contaminated either by on-site conditions (e.g., wind, rain, runoff) or by the owners themselves, with subsequent risk to family members and friends; contaminated personal clothes can also transport dangerous chemicals directly into the home.

5. *To protect environmental resources (e.g., water, soil, air) and, subsequently, the general public from the incident-related release of contaminants.* Another major risk to the public is the environmental release of contaminants into water, soil, and air not solely as a result of the

incident, but also from emergency response operations themselves. Runoff water from firefighting operations, for example, must not only be contained but collected and finally decontaminated prior to final disposal. Similarly, construction debris from the incident as well as runoff and debris from post-incident decontamination operations (e.g., dirt from vehicle tires, equipment wash water) must be decontaminated prior to release to environmental resources that can serve as hazard transport vectors into the general community.

SCOPE OF DECONTAMINATION PLAN

To meet the objective of decontamination, the decontamination plan must include SOPs pertaining to each of the following:

- Minimization of personnel contact with hazardous substances
- Maximization of responder protection
- Determination of number and layout of decontamination stations
- Determination of decontamination methods and equipment
- Prevention of contamination of clean areas
- Minimization of contact with contaminants during removal of PPC and PPE
- Disposal of materials and equipment that cannot be completely decontaminated and/or that become contaminated as a result of decontamination operations

Although there is typically no time to develop SOPs for a specific incident, generic guidelines and checklists should be developed that can be modified and applied as appropriate on a site-specific basis. However, it should also be understood that the appropriateness of any decontamination method is ultimately determined by the specific nature of the hazardous substance(s) of concern and the on-site conditions encountered during the incident. Therefore, all decontamination plans, guidelines, and checklists must be carefully reviewed and tested against the actual field situation before implementation.

Minimization of Personnel Contact

Typically viewed as the first step in any comprehensive decontamination program, the minimization of personnel contact with hazardous substances (see Figure 9.1) depends upon not only proper clothing, equipment, policies, and procedures, but also (and perhaps most importantly) attitudes. Personnel must understand that there is nothing heroic about exposing themselves (and, thereby, coworkers and family) to hazardous substances; that potentially lethal agents typically are not visible, nor do they necessarily advertise their presence with strong odors; and that, depending upon the nature of the contaminant and its concentration, a contaminant may cause irreversible acute and chronic affects. The only acceptable professional attitude is, therefore, an attitude of seasoned caution that must inform all response behavior and that is encapsulated in the

FIGURE 9-1

EXAMPLES OF STANDARD OPERATING PROCEDURES THAT MINIMIZE CONTACT WITH CHEMICALS (ADAPTED FROM OSHA, OSHA TECHNICAL MANUAL, SECTION 7. OSHA ELECTRONIC LIBRARY)

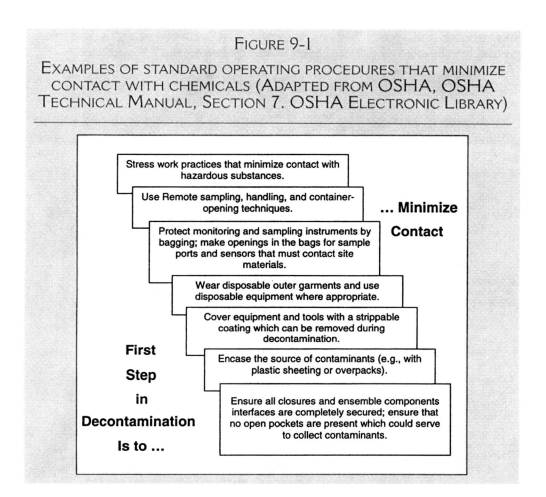

dictum: *The first responsibility of every emergency responder is to protect him- or herself.*

Certain behavioral constraints necessarily follow from this dictum and must be stringently enforced, including:

- Limited access to operational areas, based on functional need, and strict adherence to precautionary measures
- Except as required by job function or assigned task and directed by operational SOPs and operational orders, prohibition of any casual investigation of structures, containers, or substances that may result in release of or contact with hazardous agents
- Strict adherence to site control constraints regarding both prescribed and proscribed behavior, activities and clothing, such as eating, drinking, smoking, washing, sleeping, use of sanitary facilities, use of designated footwear and other clothing (including personal protective clothing and equipment)

In no circumstance should any response personnel ignore or otherwise acquiesce to any infringement by any person of any rule or practice implemented to control exposure. This is particularly important with regard to site visits by governmental and other VIPs—a stricture that, unfortunately, is too often ignored for the sake of political grandstanding before the TV audience.

Incompatible and Synergistic Chemical Contaminants

Incompatible chemicals (see Table 9.1) are those that react with one another with the release of dangerous energy and other products, such as heat, explosion,

TABLE 9-1

EXAMPLES OF CHEMICAL INCOMPATIBILITIES (ADAPTED FROM THE DANGEROUS CHEMICAL CODE, BUREAU OF FIRE PREVENTION, CITY OF LOS ANGELES FIRE DEPARTMENT)

Chemical	Avoid Contact with...
Acetic acid	Chromic acid, nitric acid, hydroxyl compounds, ethylene glycol, perchloric acid, peroxides, permanganates
Acetylene	Chlorine, bromine, copper, fluorine, silver, mercury
Alkaline metals	Water, chlorinated hydrocarbons, carbon dioxide, halogens
Ammonia (anhydrous)	Mercury, chlorine, calcium hypochlorite, iodine, bromine, hydrofluoric acid (anhydrous)
Ammonium nitrate	Acids, metallic powders, flammable liquids, chlorates, nitrites, sulfur, finely divided organic or combustible materials
Aniline	Nitric acid, hydrogen peroxide
Bromine	Ammonia, acetylene, butadiene, butane, methane, propane, hydrogen, sodium carbide, turpentine, benzene, finely divided metals
Carbon (activated)	Calcium hypochlorite, all oxidizing agents
Chlorates	Ammonium salts, acids, metallic powders, sulfur, finely divided organic or combustible materials
Chromic acid	Acetic acid, naphthalene, camphor, glycerin, turpentine, alcohol, flammable liquids
Chlorine	Ammonia, acetylene, butadiene, butane, methane, propane, hydrogen, sodium carbide, turpentine, benzene, finely divided metals
Chlorine dioxide	Ammonia, methane, phosphine, hydrogen sulfide
Cumene hydroperoxide	Acids (organic and inorganic)
Flammable liquids	Ammonium nitrate, chromic acid, hydrogen peroxide, nitric acid, sodium peroxide, halogens
Fluorine	Isolate from all other chemicals
Hydrocarbons	Fluorine, chlorine, bromine, chromic acid, sodium peroxide
Hydrocyanic acid	Nitric acid, alkaline chemicals
Hydrofluoric acid (anhydrous)	Ammonia (aqueous or anhydrous)
Hydrogen peroxide	Copper, chromium, Iron, alcohols, acetone, aniline, nitromethane, flammable liquids, combustible materials, most other organic materials
Hydrogen sulfide	Fuming nitric acid, oxidizing gases
Iodine	Acetylene, ammonia, hydrogen
Mercury	Acetylene, fulminic acid, ammonia
Nitric acid (concentrated)	Acetic acid, aniline, chromic acid, hydrocyanic acid, hydrogen sulfide, flammable liquids, flammable gases and liquids
Oxalic acid	Silver, mercury
Perchloric acid	Acetic anhydride, bismuth and its alloys, alcohol, paper, wood
Potassium	Carbon tetrachloride, carbon dioxide, water

Note. The list of chemicals is not complete, nor are all incompatible substances for each chemical shown.

fire, and toxic gases. Synergistic chemicals are those that, when introduced into the human body simultaneously (e.g., inhalation of certain pesticide particulates as well as petrochemical vapors), result in an unpredictable enhancement of the toxic or other harmful effects of one or more the components of the mixture.

In any incident involving chemicals, there is always the possibility that response personnel will be exposed to both incompatible and synergistic chemicals. It is this possibility that informs the on-site implementation of risk management practices and the selection of appropriate PPC and PPE. However, the use of any PPC, PPE, and other incident-related equipment itself defines the need for subsequent decontamination so as to (a) avoid hazards other than those immediately presented by the incident, and/or (b) avoid additional hazards derived from response operations, and/or (c) manage chemical risks to personnel regardless of the source of risk.

The importance of decontamination as the means of avoiding incompatible and synergistic reactions beyond those immediately presented by the incident is two-fold:

1. Decontamination minimizes the probability that incompatible chemicals will be mixed as a direct result of response operations and, therefore, the likelihood that the risks associated with the incident will not be compounded by additional risks (either in kind or degree).
2. Decontamination minimizes the probability that response personnel will be unknowingly exposed to chemicals that could result in health and safety risks beyond those routinely expected by virtue of task assignment.

Decontamination Stations and Facilities

The location and layout of on-site decontamination stations must be based on site-specific conditions of the actual incident, including:

1. Precise nature of physical, chemical, and/or biological contaminants.
2. Specific resources that must undergo decontamination, including victims, response personnel, equipment and supplies, response vehicles, and any other site-related or incident-related materials (e.g., construction debris).
3. Number and type of activities and associated equipment necessary to implement proper decontamination procedures (see Figure 9.2).
4. Isolation from clean areas.
5. Required protection of decontamination processes from the ongoing incident, response-operations, and weather conditions.
6. Ease of containing and otherwise managing decontamination wastes (e.g., runoff from washing contaminated vehicles) prior to their final treatment and disposal.

Depending upon constraints of time, personnel, and/or the availability of proper equipment, decontamination may not be completed on the site of the incident. In fact, it is generally most advisable that on-site field decontamination be restricted (wherever possible) to:

FIGURE 9-2

EQUIPMENT USED FOR DECONTAMINATION OF PERSONNEL AND PERSONAL PROTECTIVE CLOTHING (ADAPTED FROM NIOSH, USCG, AND EPA, 1985: OCCUPATIONAL SAFETY AND HEALTH GUIDANCE MANUAL FOR HAZARDOUS WASTE ACTIVITIES)

- Drop cloths of plastic or other suitable materials on which heavily contaminated equipment and outer protective clothing may be deposited
- Collection containers, such as drums or suitably lined trash cans, for storing disposable clothing and heavily contaminated personal protective clothing or equipment that must be discarded
- Line box with absorbents for wiping or rinsing off gross contaminants and liquid contaminants
- Large galvanized tubs, stock tanks, or children's wading pools to hold wash and rinse solutions. These should be at least large enough for a worker to place a booted foot in and should have either no drain or a drain connected to a collection tank or appropriate treatment system
- Wash solutions selected to wash off and reduce the hazards associated with the contaminants
- Rinse solutions selected to remove contaminants and contaminated wash solutions
- Long-handled, softbristled brushes to help wash and rinse off contaminants
- Paper or cloth towels for drying protective clothing and equipment
- Lockers and cabinets for storage of decontaminated clothing and equipment
  Metal or plastic cans or drums for contaminated wash and rinse solutions
- Plastic sheeting, sealed pads with drains, or other appropriate methods for containing and collecting contaminated wash and rinse solutions spilled during decontamination
- Shower facilities for full body wash or, at a minimum, personal wash sinks (with drains connected to a collection tank or appropriate treatment system)
- Soap or wash solution, wash cloths, and towels for personnel
- Lockers or closets for clean clothing and personal item storage

- Persons (including victims and response personnel) who will not realize increased risk to life or health due to any delay of access to off-site professional medical services
- Personal protective clothing and equipment that must be readily available for subsequent use in response operations
- Any other materials or items that cannot be moved off-site without endangerment of the general public or environmental resources

Where materials are removed to off-site locations and facilities for subsequent decontamination, it is necessary, of course, that suitable precautionary measures be taken at those locations and facilities to assure the effective containment and management of contaminants prior to ultimate disposal (see Figure 9.3).

Decontamination Methods and Equipment

Basic decontamination methods involve physical and/or chemical removal, detoxification, and disinfection/sterilization of contaminants.

FIGURE 9-3

PRECAUTIONARY MEASURES FOR DISINFECTING FACILITIES IN FIRE AND EMS FACILITIES ADAPTED FROM U.S. FIRE ADMINISTRATION, 1997: SAFETY AND HEALTH CONSIDERATIONS FOR THE DESIGN OF FIRE AND EMERGENCY MEDICAL SERVICES STATIONS [FA 168])

- Design disinfecting facilities in stations with proper lighting, separate ventilation to the outside environment, fitted with floor drains connected to a sanitary sewer system, and to prevent contamination of other station areas.

- Within disinfecting facilities, install a minimum of 2 sinks with hot and cold water faucets and a sprayer attachment, and with drains connected to a sanitary sewer system. Sink faucets should not require the user to grasp, with hands, to turn on or off. All surfaces should be nonporous material with continuous molded counter top and splash panel surfaces

- Equip disinfecting facilities with rack shelving of nonporous materials. Shelving should be provided above sinks for drip-drying of cleaned equipment. All drainage from shelving should either go into a sink or drain directly into a sanitary sewer system.

- If possible, select front loading industrial laundry machines designed for the type of cleaning required for protective clothing.

- When exposure occurs, clean the equipment and store the waste water from this process in a double wall tank where it can then be pumped to waste transfer vehicles for appropriate disposal

- Provide a designated cleaning area in each station for the cleaning and disinfection of protective clothing, protective equipment, portable equipment, and other clothing. This cleaning area should have proper ventilation, lighting, and drainage connected to a sanitary sewer system.

- Physically separate the designated cleaning area from areas used for food preparation, cleaning of food and cooking utensils, personal hygiene, sleeping, and living areas; also physically separate the designated cleaning area from the emergency medical disinfecting facility.

- Store station emergency medical supplies/equipment, other than that stored on vehicles, in a dedicated, enclosed room protected from the outside environment.

- Store protective clothing and protective equipment in a dedicated, well-ventilated area or room

- Do not store reusable emergency medical supplies and equipment, protective clothing, and protective equipment in a kitchen, living, sleeping, or personal hygiene areas, nor shall it be stored in personal clothing lockers

Physical/Chemical Removal

Gross dirt and caked mud (e.g., lodged in tire treads) containing nonadhesive contaminants can be physically dislodged by brushing, scraping, and pounding, followed by either simple flushing or pressurized air or rinse water (see Image 9.3). Adhesive contaminants may require more stringent physical treatment prior to physical dislodgment, including freezing (e.g., by dry ice or ice water), melting, adsorption onto other surfaces (e.g., sand), and absorption into inert materials such as kitty litter or powdered lime. Various ultra-sound devices may be useful for dislodging small amounts of either adhesive

IMAGE 9-3

FRANKLIN, VA, SEPTEMBER 21, 1999: HURRICANE FLOYD LEFT DOWNTOWN FRANKLIN UNDER WATER. EMERGENCY WORKERS HAD TO WASH THEMSELVES OFF TO AVOID ILLNESS FROM OIL, GAS, PROPANE, AND CHEMICAL CONTAMINANTS IN THE WATER

Source: Liz Roll/FEMA News Photo

and nonadhesive materials. Steam jets are useful for cleaning many types of adhesive contaminants.

Dusts and vapors of contaminants that collect in small openings in clothing and equipment may simply be washed free with water or blown free with an air jet. Contaminants that cling more tenaciously to materials by virtue of electrostatic forces may be more easily removed by water or air from materials that have been treated with commercially available antistatic solutions and sprays.

Liquid contaminants may be wiped off, absorbed into inert substances, or, in the case of volatile liquid, air dried, followed by water rinse. Warm air or steam jets may be used to facilitate evaporation. The physical removal of liquefied contaminants may also be facilitated by solidification, which may involve the use of absorbents (e.g., clay, powdered lime) to remove moisture, dry ice, or water to cause freezing, or the addition of chemical catalysts that effect polymerization (i.e., chemical joining together of similar molecules into long chains).

Regardless of the type of process employed, the physical removal of gross contaminants should be followed by appropriate wash and rinse of contaminated materials.

Primary wash solutions typically contain cleaners that depend upon the action of a variety of solvents (which dissolve contaminants) and surfactants (which, like household detergents, reduce the forces of adhesion between the contaminant and the material contaminated). Cleaners therefore should be selected on the basis of the compatibility of solvent types (see Figure 9.4) with the structural materials of items to be decontaminated. Manufacturers of PPC,

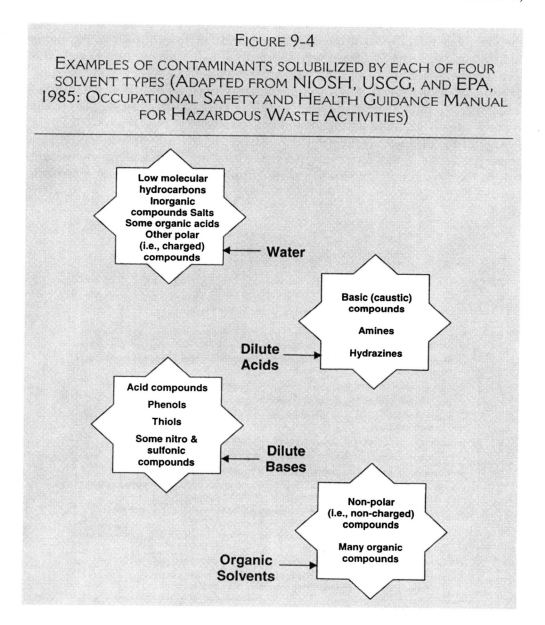

FIGURE 9-4

EXAMPLES OF CONTAMINANTS SOLUBILIZED BY EACH OF FOUR SOLVENT TYPES (ADAPTED FROM NIOSH, USCG, AND EPA, 1985: OCCUPATIONAL SAFETY AND HEALTH GUIDANCE MANUAL FOR HAZARDOUS WASTE ACTIVITIES)

Low molecular hydrocarbons Inorganic compounds Salts Some organic acids Other polar (i.e., charged) compounds ← Water

Basic (caustic) compounds

Amines

Hydrazines

Dilute Acids →

Acid compounds

Phenols

Thiols

Some nitro & sulfonic compounds ← Dilute Bases

Non-polar (i.e., non-charged) compounds

Many organic compounds

Organic Solvents →

PPE, and other types of response equipment typically supply information on solvent compatibility with the engineering specifications of manufactured items.

Types of equipment needed for each type of physical/chemical removal (see Figure 9.5) are essentially containers (including storage tanks for wash and rinse solutions), sprayers, brooms, and brushes, as well as items used for containment of contaminants (e.g., curtains, booths) during the decontamination process. All equipment used for decontamination must be dedicated equipment (i.e., not to be used for any other purpose) that is also compatible with cleaning and solvent solutions.

Physical decontamination also includes removal of contaminated surfaces as opposed to the removal of contaminants from surfaces. This process involves the disposal of materials and items (clothing, protective coverings, floor mats). However, before their final disposal, such items typically must undergo either chemical deactivation or disinfection to ensure that they do not contaminate environmental resources. No materials can be finally disposed except in conformance with applicable hazardous waste regulations. It is therefore necessary to coordinate directly with hazardous waste authorities and RCRA permitted transporters and treatment, storage, and disposal (TSD) facilities to determine appropriate disposal requirements and constraints.

Any means of physical/chemical removal (especially those involving heat or pressurized steam) may result in vapors, particles, and liquids that may present risk to the person employing them through inhalation or through eye or skin

FIGURE 9-5

EQUIPMENT USED FOR DECONTAMINATION OF HEAVY EQUIPMENT AND VEHICLES (ADAPTED FROM NIOSH, USCG, AND EPA, 1985: OCCUPATIONAL SAFETY AND HEALTH GUIDANCE MANUAL FOR HAZARDOUS WASTE ACTIVITIES)

- Storage tanks of appropriate treatment systems for temporary storage and/or treatment of contaminated wash and rinse solutions
- Drain or pumps for collection of contaminated wash and rinse solutions
- Long-handled brushes for general exterior cleaning
- Wash solutions selected to remove and reduce the hazards associated with the contamination
- Rinse solutions selected to remove contaminants and contaminated wash solutions
- Pressurized sprayers for washing and rinsing, particularly hard-to-reach areas
- Curtains, enclosures, or spray booths to contain splashes from pressurized sprays
- Long-handled brushes, rods, and shovels for dislodging contaminants and contaminated soil caught in tires and the undersides of vehicles and equipment
- Containers to hold contaminants and contaminated soil removed from tires and the undersides of vehicles and equipment
- Wash and rinse buckets for use in the decontamination of operation areas inside vehicles and equipment
- Brooms and brushes for cleaning operator areas inside vehicles and equipment
- Containers for storage and disposal of contaminated wash and rinse solutions, damaged or heavily contaminated parts, and equipment to be discarded

contact. Physical removal methods typically do not alter the chemical attributes of contaminants, nor affect the viability of pathogenic organisms. Caution therefore must be used with all methods, with appropriate attention paid to (a) personal protective clothing and equipment (e.g., gloves, goggles, respiratory protection) to be used by the person performing the decontamination procedure, (b) containment of all dislodged contaminants to prevent entry into environmental resources or contact with other persons, and (c) additional treatment (as may be required) to deactivate or disinfect or dislodge materials prior to final disposal.

Detoxification

Detoxification is essentially a chemically, physically, or biologically mediated change in the molecular structure of a contaminant molecule, or in the chemical dynamics of a contaminant mixture or solution to achieve a less hazardous substance or material.

Commonly used chemical detoxification procedures involve the removal of halogen atoms (e.g., chlorine, bromine, fluorine, iodine) from a contaminant molecule (e.g., carbon tetrachloride, trichlorethane), which is known as *halogen stripping*; the addition or removal of electrons (or hydrogen atoms) from the contaminant molecule (e.g., transformation of an alcohol to organic acid), which is known as *oxidation* (when electrons are added) or *reduction* (when electrons or hydrogen atoms are removed); and the addition of acids or bases to a contaminant solution to adjust either the acidic or the alkaline nature of the solution toward a less corrosive state, which is called *neutralization*. The most commonly used physical detoxification procedure is *thermal degradation*, in which heat is used to transform a contaminant molecule into a less hazardous molecule.

Other chemical and physical processes may be used (or are under ongoing investigation and development) to effect changes in the structure and/or chemical dynamics of contaminant molecules, such as *chelation*, which involves the addition of chemicals that tightly bind to contaminant molecules, thereby facilitating their removal or otherwise reducing their chemical reactivity. Biological agents also may be used to degrade contaminants to less hazardous substances, as in the use of selected species of bacteria and yeasts to degrade certain types of petrochemicals and pesticides—a rapidly expanding technology known as *bioremediation*.

Emergency response services are well advised to be extremely cautious with regard to the use of any detoxification method. There are several reasons for this caution:

- Even standard detoxification methods (e.g., neutralization, halogen stripping) typically are employed only when there is detailed chemical knowledge and understanding of the chemical dynamics involved
- The effectiveness and safety of detoxification methods is dependent on a large number of factors, including the concentration of target contaminants, the concentrations of potentially interfering chemical species in a contaminant mixture, and the ambient conditions under which the procedure is performed

- The treatment of any hazardous or potentially hazardous waste can be legally undertaken only by U.S. EPA permitted facilities that conform to strict procedural and technical requirements

In light of these considerations, response services should rely on the advice of regulatory authority and, as necessary, seek out the professional services of properly licensed contractors who are legally, scientifically, and technically competent to undertake detoxification.

Disinfection/Sterilization

Both disinfection and sterilization involve the killing of microorganisms, but differ with respect to the range of microbes actually killed. As shown in Table 9.2, disinfection may be carried out at various levels of efficacy as defined by the types of microbes (e.g., bacteria, viruses, fungi) affected and whether or not the disinfecting agent (e.g., heat, germicide) destroys bacterial

TABLE 9-2

DECONTAMINATION METHODS FOR EQUIPMENT USED IN THE PRE-HOSPITAL HEALTH-CARE SETTING (ADAPTED FROM THE U.S. FIRE ADMINISTRATION, 1992: GUIDE TO DEVELOPING AND MANAGING AN EMERGENCY SERVICE INFECTION CONTROL PROGRAM [FA-112])

Process	Application	Details
Sterilization	Target	All forms of microbial life, including high numbers of bacterial spores
	Methods(s)	Steam under pressure (autoclave), gas (ethylene oxide), dry heat, or immersion in EPA-approved chemical sterilant for prolonged period of time (e.g., 6-10 hours or according to manufacturers' instructions). Liquid chemical sterilants should be used only on those instruments that are impossible to sterilize or disinfect with heat.
	Use	For those instruments or devices that penetrate skin or contact normally sterile areas of the body (e.g., scalpels, needles). Disposable invasive equipment eliminates the need to reprocess these types of items. When indicated, however, arrangements should be made with a healthcare facility for reprocessing of reusable invasive instruments
High-Level Disinfection	Target	All forms of microbial life except high numbers of bacterial spores.
	Method(s)	Hot water pasteurization (80-100°C for 30 minutes) or exposure to an EPA-registered sterilant chemical (as above), except for a short exposure time (10-45 minutes, or as directed by the manufacturer).

Continued

TABLE 9-2—*Continued*

DECONTAMINATION METHODS FOR EQUIPMENT USED
IN THE PRE-HOSPITAL HEALTH-CARE SETTING
(ADAPTED FROM THE U.S. FIRE ADMINISTRATION, 1992:
GUIDE TO DEVELOPING AND MANAGING AN EMERGENCY
SERVICE INFECTION CONTROL PROGRAM [FA-112])

Process	Application	Details
	Use	For reusable instruments or devices that come into contact with mucous membranes (e.g., laryngoscope blades, endotracheal tubes).
Intermediate-Level Disinfection	Target	*Mycobacterium tuberculosis*, vegetative bacteria, most viruses, and most fungi; does not kill bacterial spores.
	Method(s)	EPA-registered "hospital disinfectant" chemical germicides that have a label claim for tuberculocidal activity; commercially available hard-surface germicides or solutions containing at least 500 ppm free available chlorine (a 1:100 dilution of common household bleach; approximately ¼ cup bleach per gallon of tap water).
	Use	For those surfaces that come into contact only with intact skin (e.g., stethoscopes, blood pressure cuffs; splints), and which have been visibly contaminated with blood or bloody body fluids. Surfaces must be pre-cleaned of visible material before the germicidal chemical is applied for disinfection.
Low-Level Disinfection	Target	Most bacteria, some viruses, some fungi, but not *Mycobacterium tuberculosis* or bacterial spores.
	Method(s)	EPA-registered "hospital disinfectant" (no label claim for tuberculocidal activity).
	Use	These agents are excellent cleaners and can be used for routine housekeeping or removal of soiling in the absence of visible blood contamination.
Environmental Disinfection		Environmental surfaces that have become soiled should be cleaned and disinfected using any cleaner or disinfectant agent which is intended for environmental use. Such surfaces include floors, woodwork, ambulance seats, countertops, etc.

Important: To assure the effectiveness of any sterilization or disinfection process, equipment and instruments must first be thoroughly cleaned of all visible soil.

spores (i.e., reproductive structures of bacteria). Only sterilization, which involves the use of an autoclave, or a sterilant gas, liquid sterilant, or dry heat, can destroy all forms of microbial life as well as a large proportion of bacterial spores. It should be noted, however, that even sterilization techniques do not necessarily kill all bacterial spores.

With the exception of certain emergency medical service organizations, very few emergency response services possess the equipment to carry out effective sterilization. Even where appropriate sterilization equipment is available, it is impractical (if not impossible) to sterilize large equipment of PPC and PPE. Wherever possible, therefore, disposable clothing and equipment should be used, with appropriate planning for safe disposal through incineration. No infected clothing or equipment should be deposited in a landfill unless it has been sterilized.

Disinfection by means of commercially available and U.S. EPA registered germicides, household bleach, and cleaning agents (see Figure 9.6) is effective only when proper attention is given to the prior removal of gross dirt and grime. This is because a disinfectant must come into direct contact with a microbe in order to kill it, and dirt, grime, and other materials prevent direct contact between the disinfectant and the target microbe. The duration of contact between the disinfectant and the target organism is also important—the

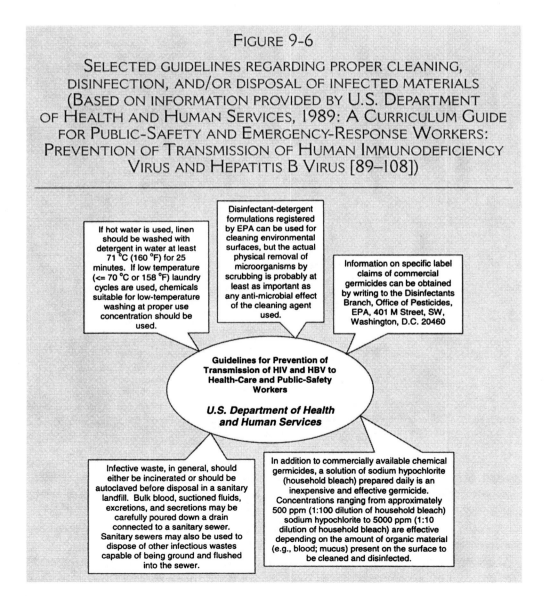

FIGURE 9-6

SELECTED GUIDELINES REGARDING PROPER CLEANING, DISINFECTION, AND/OR DISPOSAL OF INFECTED MATERIALS (BASED ON INFORMATION PROVIDED BY U.S. DEPARTMENT OF HEALTH AND HUMAN SERVICES, 1989: A CURRICULUM GUIDE FOR PUBLIC-SAFETY AND EMERGENCY-RESPONSE WORKERS: PREVENTION OF TRANSMISSION OF HUMAN IMMUNODEFICIENCY VIRUS AND HEPATITIS B VIRUS [89–108])

If hot water is used, linen should be washed with detergent in water at least 71 °C (160 °F) for 25 minutes. If low temperature (<= 70 °C or 158 °F) laundry cycles are used, chemicals suitable for low-temperature washing at proper use concentration should be used.

Disinfectant-detergent formulations registered by EPA can be used for cleaning environmental surfaces, but the actual physical removal of microorganisms by scrubbing is probably at least as important as any anti-microbial effect of the cleaning agent used.

Information on specific label claims of commercial germicides can be obtained by writing to the Disinfectants Branch, Office of Pesticides, EPA, 401 M Street, SW, Washington, D.C. 20460

Guidelines for Prevention of Transmission of HIV and HBV to Health-Care and Public-Safety Workers

U.S. Department of Health and Human Services

Infective waste, in general, should either be incinerated or should be autoclaved before disposal in a sanitary landfill. Bulk blood, suctioned fluids, excretions, and secretions may be carefully poured down a drain connected to a sanitary sewer. Sanitary sewers may also be used to dispose of other infectious wastes capable of being ground and flushed into the sewer.

In addition to commercially available chemical germicides, a solution of sodium hypochlorite (household bleach) prepared daily is an inexpensive and effective germicide. Concentrations ranging from approximately 500 ppm (1:100 dilution of household bleach) sodium hypochlorite to 5000 ppm (1:10 dilution of household bleach) are effective depending on the amount of organic material (e.g., blood; mucus) present on the surface to be cleaned and disinfected.

longer the contact, the greater the probability the microbe will be killed. Detailed information concerning the precleaning of infected items required for the effective use of germicides, as well as recommended contact times, is printed on germicide labels. Additional information concerning germicide labels can also be obtained directly from the U.S. EPA.

Prevention of Contamination of Clean Areas

During an actual incident, the most important means of preventing the contamination of clean areas is the establishment of *work zones*. Depending upon the nature of the incident, any number of clearly demarcated work zones may be designated. In most incidents involving hazardous waste sites, three zones are typically used: exclusion, contaminant reduction, and support zones.

Exclusion Zone

This is the contaminated area or an area that is likely to become contaminated in the development of the incident. The outer boundary of this zone is the *hotline*, which must be clearly marked. The precise location of the hotline is determined on the basis of various considerations, including (a) known or possible routes of dispersion of contaminants (including surface runoff and wind dispersion), (b) the amount of contaminants, (c) the relationship of site topography to the actual and potential area of contamination, (d) distances necessary to prevent any possible explosion or fire within the exclusion zone from affecting personnel outside of the exclusion area, (e) area necessary to conduct response operations, including the use of response vehicles and heavy equipment, (f) on-site meteorological conditions, including both current and projected conditions, and (as time may permit) (g) field monitoring data (including data derived from air, water, soil, and contaminant sampling).

As the incident develops and more information becomes available about the nature of contaminants and other potential risks (e.g., underground tanks, subsurface transformer stations, underground electrical cables), it may become necessary or advisable to adjust the delimitation of the hotline to better protect personnel.

Access control points must be established on the hotline to manage the movement of response personnel and equipment into and out of the exclusion zone.

The exclusion zone itself may be subdivided on the basis of different types of hazards and/or degree of risk encountered or expected. The type of personal protective clothing and equipment required within the exclusion zone must be clearly indicated at access points and, as appropriate, within each subdivision.

Contaminant Reduction Zone

This is the zone where decontamination occurs; it is essentially a buffer area between the highly contaminated exclusion zone and the contaminant-free area of the site (see Figure 9.7).

All access to and through the contaminant reduction zone is through the *contamination reduction corridor*, which extends from the contaminant-free area of

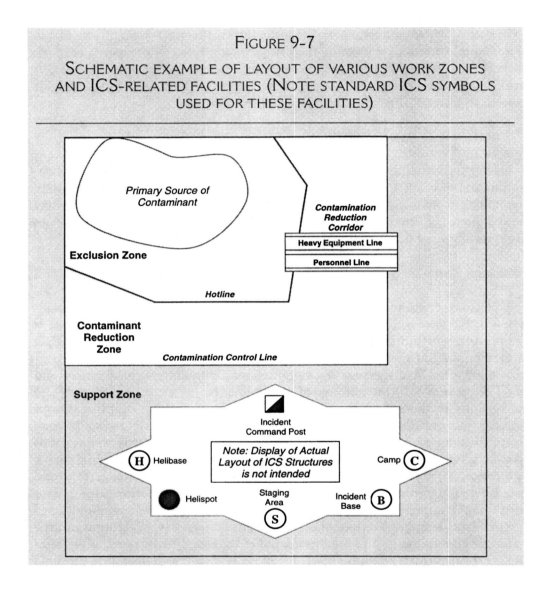

FIGURE 9-7

SCHEMATIC EXAMPLE OF LAYOUT OF VARIOUS WORK ZONES AND ICS-RELATED FACILITIES (NOTE STANDARD ICS SYMBOLS USED FOR THESE FACILITIES)

the site through the hotline. Access points must be established on the "clean side" of this corridor to control entry and exit. Within the contamination reduction corridor, at least two lines of decontamination should be established—one for personnel, and one for equipment. As personnel and equipment move from work assignments in the exclusion line toward the clean area, decontamination takes place and, therefore, the risk of contaminating the clear area diminishes.

The essential layout of the contamination reduction zone is based on the need to decontaminate personnel and equipment (in the contamination reduction corridor), but it also must accommodate other important functions, including:

- Emergency response, including the transport of injured response personnel and emergency first aid
- Emergency containment equipment and operations
- Resupply of operational equipment, including PPC and PPE, sampling equipment, and tools
- Packaging and preparation of samples (e.g., soil, debris, hazardous wastes) for subsequent analysis
- Rest and recovery areas for response personnel, including toilet facilities, potable water, and washing facilities
- Safe containment of all liquids and other materials used in the decontamination process

Support Zone

This is the clean area in which all response-related administrative duties and all other operations that need not be performed in either the exclusion or contaminant reduction zones are performed. The support zone is operationally protected by the contamination control line, but it should be noted that, as the emergency develops, necessary adjustments to the location of the contamination control line could result in adjustments to the location of support zone facilities, with consequent interference in overall incident management. The location of support zone facilities therefore must be based as much as possible on a worst-case analysis of the developing incident, while still maintaining practical administrative control of all on-site activities. Other factors in locating specific support zone facilities (e.g., incident command post, staging area, incident base, camps, helibase, and helispot) include:

- Accessibility for emergency vehicles and equipment
- Availability of electrical power, telephones, shelter, potable water, and roads
- Line-of-site visibility of incident-site operations (while still locating facilities as far away from the exclusion zone as possible)
- Wind direction and topography (which could influence intrusion of wind-blown or runoff-entrained contaminants into the support zone

Removal of PPC and PPE

Depending on the nature of an actual incident and types and degrees of risk, the incident safety officer establishes requirements for protective ensembles and equipment to be used by response personnel in the exposure zone, and specific steps for removing and disinfecting each article in the process of moving from work areas within the exposure zone to the support zone.

As depicted in Figure 9.8, decontamination procedures extend (via the contamination reduction corridor) into both exclusion and support zones, even though they are most vigorous in the contamination reduction zone. Decontamination procedures designated for the exclusion zone are intended to prevent gross contamination of the contamination reduction zone; those designated for the contamination reduction corridor, to remove all remaining contamination prior to entry to the support zone.

FIGURE 9-8

MAXIMUM DECONTAMINATION LAYOUT: LEVEL A PROTECTION (ADAPTED FROM NIOSH, USCG, AND EPA, 1985: OCCUPATIONAL SAFETY AND HEALTH GUIDANCE MANUAL FOR HAZARDOUS WASTE ACTIVITIES)

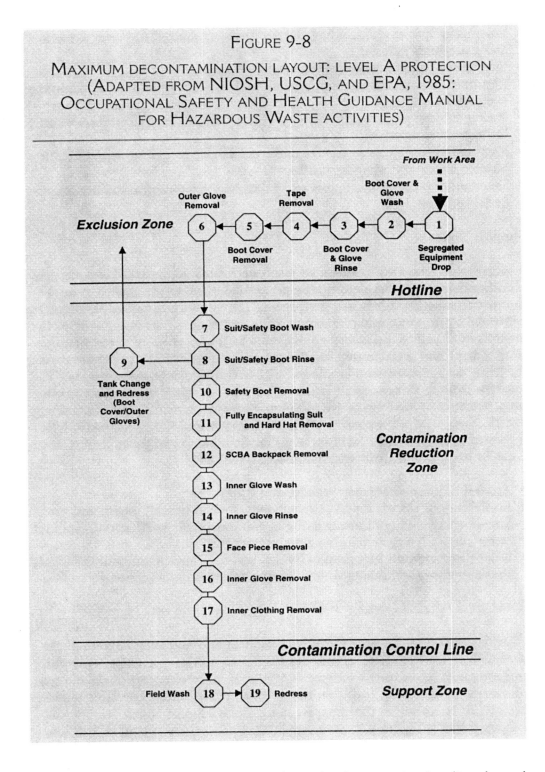

The number and sequence of steps along the decontamination line depend upon actual site conditions, becoming more numerous and stringent with increasing hazards and risks associated with the exclusion zone. In some instances, it may be necessary to establish resting/cooling stations along the decontamination line, for example, when response operations must be undertaken under ambient conditions of high heat and humidity.

Whereas typical layouts of decontamination stations have long been established for incidents that involve hazardous waste, similar configurations (adaptations thereof) may be employed for incidents involving nonhazardous waste sites but which nonetheless present the hazard of contamination with physical, chemical, or biological agents. Of course, depending upon the nature of the contaminant hazard, different types of decontamination activities at the various stations of the decontamination line may have to be implemented.

Disposal of Contaminated Materials

A key consideration with regard to any decontamination line is the waste that is generated as a consequence of performing the various decontamination procedures, including:

- Neutralization spray solution
- Detergent wash solution
- Disinfectant rinse solution
- Rags and paper wipes
- Clumps and scrapings of contaminated mud
- Disposable PPC and PPE

The decontamination program must specify how these various wastes are to be packaged and disposed in compliance with applicable federal, state, and local regulations. It is important to distinguish between contaminated waste generated by the incident and that generated by on-site decontamination procedures deigned for response personnel and equipment.

In many instances, for example, contaminated construction debris (e.g., from a collapsed building or other structure) presents a major obstacle to the primary response effort (e.g., search and rescue) and must be moved out of the way as quickly as possible. In such a situation, it is unlikely that there will be time or resources to decontaminate debris; on-site storage, with provision for containment, must then be implemented. Once the primary response effort has been concluded by the lead responding agency, the management of bulk contaminated debris (including decontamination and disposal) typically becomes the responsibility of another agency having appropriate jurisdictional authority.

In some instances, contaminated PPC, PPE, and other response equipment are wrapped or placed in secure bulk containers for subsequent decontamination at the home-base of the response service.

EMERGENCY DECONTAMINATION

Emergency decontamination may have to be undertaken as a result of any on-site accident involving response personnel, whether that accident results in relatively minor or major injury to personnel. Figure 9.9 includes a standard

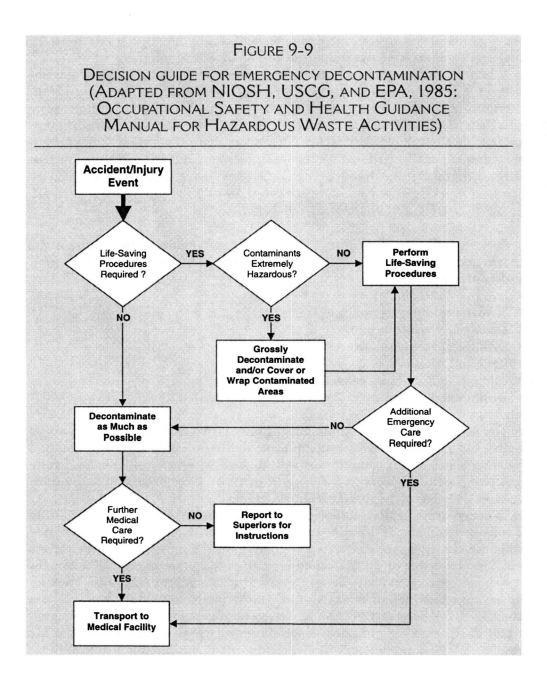

FIGURE 9-9

DECISION GUIDE FOR EMERGENCY DECONTAMINATION
(ADAPTED FROM NIOSH, USCG, AND EPA, 1985:
OCCUPATIONAL SAFETY AND HEALTH GUIDANCE
MANUAL FOR HAZARDOUS WASTE ACTIVITIES)

decision guide for determining the steps to take in such a situation. The various steps included in this guide are designed to maximize the administration of appropriate on-and/or off-site medical response to such an emergency, but it is also important that the medical response not result in the contamination of on-site clean areas or of off-site medical personnel, facilities, or ambulance services.

Of course, in a life-threatening situation, it is the health of the injured worker that must be given priority; this sometimes requires a highly abbreviated on-site decontamination effort, consisting primarily of wrapping the victim in a protective covering (e.g., blanket, plastic sheet) that can contain gross contamination and that later can be decontaminated and disposed. However, before wrapping the contaminated victim, it is vitally important to remove or cut off as much grossly contaminated clothing as possible in order to minimize contact of contaminants with the victim's body.

Any clothing gives some degree of protection against contamination and, of course, the protective clothing worn by response personnel is specifically selected to maximize protection. However, all clothing is subject to permeation by contaminants, the degree of permeation being subject to various factors (see Figure 9.10), including (a) the time interval in which contaminants are in contact with the clothing, (b) the physical state and chemical nature of the contaminant, (c) the concentration of the contaminant, and (d) the ambient temperature. In an emergency situation, it is therefore most advisable to strip the victim of any obviously contaminated clothing or equipment as quickly as possible. No attempt should be made to wash or rinse gross contamination off of the victim's clothing because this most likely would increase the rate of permeation of the contaminant through that clothing. Depending on the nature of the contaminant, a wash or rinse solution also could result in enhanced risk due to unsuspected chemical or physical reactions of the contaminant with rinse solutions—a situation that usually cannot be reliably assessed in the press of an emergency.

In all situations involving the spillage of blood or the release of body fluids, appropriate decontamination (see Figure 9.11) should be implemented to

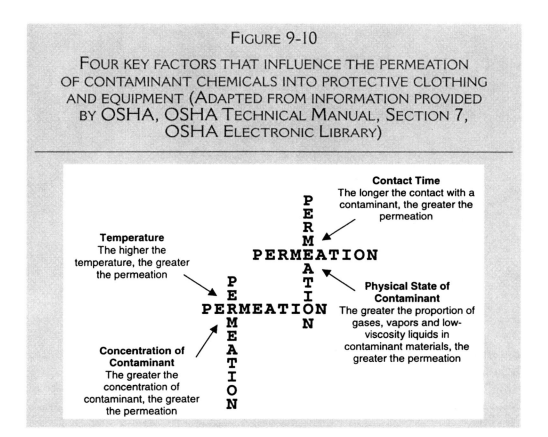

FIGURE 9-10

FOUR KEY FACTORS THAT INFLUENCE THE PERMEATION OF CONTAMINANT CHEMICALS INTO PROTECTIVE CLOTHING AND EQUIPMENT (ADAPTED FROM INFORMATION PROVIDED BY OSHA, OSHA TECHNICAL MANUAL, SECTION 7, OSHA ELECTRONIC LIBRARY)

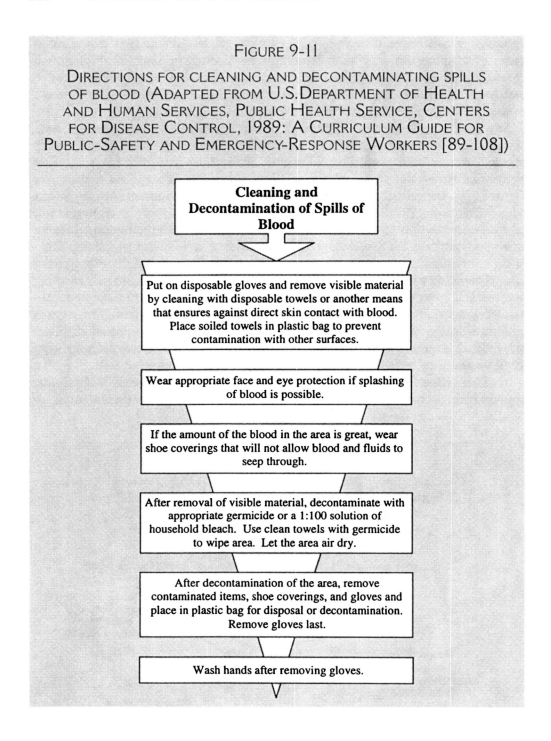

FIGURE 9-11

DIRECTIONS FOR CLEANING AND DECONTAMINATING SPILLS OF BLOOD (ADAPTED FROM U.S. DEPARTMENT OF HEALTH AND HUMAN SERVICES, PUBLIC HEALTH SERVICE, CENTERS FOR DISEASE CONTROL, 1989: A CURRICULUM GUIDE FOR PUBLIC-SAFETY AND EMERGENCY-RESPONSE WORKERS [89-108])

Cleaning and Decontamination of Spills of Blood

Put on disposable gloves and remove visible material by cleaning with disposable towels or another means that ensures against direct skin contact with blood. Place soiled towels in plastic bag to prevent contamination with other surfaces.

Wear appropriate face and eye protection if splashing of blood is possible.

If the amount of the blood in the area is great, wear shoe coverings that will not allow blood and fluids to seep through.

After removal of visible material, decontaminate with appropriate germicide or a 1:100 solution of household bleach. Use clean towels with germicide to wipe area. Let the area air dry.

After decontamination of the area, remove contaminated items, shoe coverings, and gloves and place in plastic bag for disposal or decontamination. Remove gloves last.

Wash hands after removing gloves.

minimize the potential for the spread of blood-borne diseases. This procedure, including the universal precautions always to be associated with handling blood and body fluids, is as applicable (and necessary) in an emergency situation as in any nonemergency situation, and need not in any way interfere with the timely emergency treatment of the victim.

Where it is impossible to decontaminate the victim properly prior to off-site transport, it is important that the victim be accompanied to the off-site medical facility by a person who is fully knowledgeable of (a) the type of contaminant to which the victim has been exposed, (b) the nature of emergency decontamination that was performed on the victim while on-site, and (c) the appropriate types of decontamination and other precautions that medical, rescue, and other emergency personnel should employ during and subsequent to transporting, handling, and treating the victim.

Because the potential for injury of response personnel is always high, emergency decontamination and all associated activities (e.g., rescue), procedures (e.g., emergency first aid), and equipment (e.g., ventilation equipment in ambulances) must be viewed as integral components of the site safety plan. The site safety officer therefore must ensure that all possible needs for decontamination as a result of response operations and contingencies be thoroughly examined and appropriate proactive policies and SOPs be established. Moreover, because the need for emergency decontamination may (depending upon the nature of the incident) instantaneously escalate due to uncontrollable exposure of the general public, extensive planning is required to deal effectively with the decontamination of not only individual response personnel but also large community populations.

Given the potential scope of emergency decontamination in any incident (and, especially, in terrorist incidents), effective proactive planning depends heavily upon effective coordination with local, state, and federal rescue, response, and support resources (see Table 9.3) that are available to provide a wide range of services, including on-site emergency treatment of victims, the control and stabilization of hazardous conditions that could result in increased instances of exposure to contaminants, and the provision of technical assistance and equipment.

It is typically the responsibility of the safety officer of the lead response organization to provide for the basic on-site equipment needed to control potential contaminants and to implement emergency decontamination (see Table 9.4), however no local response agency or organization can be expected to maintain sufficient equipment to meet the potential decontamination needs of large off-site populations. This situation underscores the importance of interagency and interorganizational liaison to ensure the regional stockpiling or other timely availability (e.g., Federal Emergency Management Agency, regional response team stockpiles) of equipment, materials, and supplies that can be used both to control the spread of contaminants into the general population and to effect, as needed, decontamination of large numbers of persons and vital community resources.

With regard to the local availability of equipment and supplies required proactively and reactively for the control of contaminants and incident-related decontamination, municipal managers should consider the use of municipal-wide inventories of potential industrial contaminants to identify potential equipment and supply needs, including not only the types and amounts of equipment and supplies that should be stockpiled or made available via requisition, but also needs regarding (a) the maintenance, upkeep, and replacement of identified equipment and supplies, (b) the timely transport of supplies

TABLE 9-3

TYPES OF SUPPORT SERVICES TYPICALLY AVAILABLE THROUGH FEDERAL, STATE, AND LOCAL SUPPORT SERVICES

Agency or Organization	Rescue	Response	Support
Federal			
Army Corps of Engineers			☐
Coast Guard		☐	☐
Department of Defense (DOD)		☐	☐
Department of Transportation			☐
Environmental Protection Agency (EPA)		☐	☐
Federal Aviation Administration (FAA)			☐
Federal Emergency Management Agency (FEMA)			☐
National Institute for Occupational Safety and Health (NIOSH)			☐
Occupational Safety and Health Administration (OSHA)			☐
State			
Civil defense			☐
Department of Health			☐
Department of Labor			☐
Environmental Agency		☐	☐
Office of the Attorney General			☐
State Police	☐		☐
Local			
Ambulance and rescue services	☐	☐	☐
Cleanup contractor	☐	☐	☐
Disposal companies	☐	☐	
Fire department	☐	☐	☐
Hospital			☐
Police	☐		☐
Red Cross			☐
Salvation Army			☐
Transporters			☐
Utility companies			☐

under worst-case conditions, and (c) back-up supplies that may be needed in the event of simultaneous incidents.

Toward the same objective, industrial managers should explore the potential for mutual assistance programs among industrial partners, with particular emphasis given to mock incidents as a means of training multifacility personnel to ensure the rapid delivery of fully operational contamination control equipment, supplies, and, as appropriate, personnel to the incident scene.

STUDY GUIDE

True or False

1. In the absence or malfunction of properly selected PPC and PPE, decontamination prevents personal exposure during donning and doffing procedures.

TABLE 9-4

BASIC ON-SITE EQUIPMENT USED TO CONTROL EXPOSURE TO CONTAMINANTS AND TO IMPLEMENT EMERGENCY DECONTAMINATION PROCEDURES

Personal Protection
- Escape SCBA or SCBA, which can be brought to the victim to replace or supplement his or her SCBA
- Personal protective equipment and clothing specialized for known site hazards

Medical
- Air splints
- Decontamination solutions appropriate to onsite hazards
- Reference books containing basic first-aid procedures and information on treatment of specific Injuries

- Antiseptics
- Emergency eye wash
- Resuscitator
- Safety Harness
- Stretchers
- Water, in potable containers

- Blankets
- Emergency showers or wash stations
- Wire basket litter (Stokes litter) which can be used to carry victim in bad weather and on difficult terrain, and is itself easy to decontaminate

Hazard Mitigation
- Fire-fighting equipment and supplies
- Spill-containment equipment, such as absorbents and oil booms
- Special hazardous-use tools, such as remote pneumatic impact wrenches, non-sparking wrenches and picks
- Containers to hold contaminated materials

2. The first responsibility of every emergency response provider is protect him- or herself.

3. Oxidation involves the removal of halogen atoms from a contaminant molecule.

4. Sterilization is the killing of all forms of microbial life, including high numbers of bacterial spores.

5. The outer boundary of the exclusion zone is the hotline.

6. The contaminant reduction zone is essentially a buffer area between the highly contaminated exclusion zone and the contaminant-free area of the site.

7. The contamination reduction corridor extends from the contaminant-free area of the site through the hotline.

8. The support zone is the clean area in which all response-related administrative duties and all other operations that need not be performed in either the exclusion or contaminant reduction zones are performed.

9. The greater the concentration of a contaminant, the lesser the permeation of the contaminant into personal protective clothing.

10. Incompatible chemicals are those that, when introduced into the human body simultaneously, result in an unpredictable enhancement of the toxic or other harmful effects of one or more of the components of a mixture.

11. The first step in decontamination is to minimize contact.

12. One important objective of decontamination is to protect the families of response personnel from carry-home contamination.

Multiple Choice

1. Decontamination procedures are designed to:
   A. protect on-site response personnel from direct bodily exposure to contaminants
   B. prevent the mixing of incompatible or synergistic chemical contaminants derived from different response activities
   C. protect off-site personnel from exposure in the process of treating victims, servicing equipment, or handling and transporting incident-related debris or other materials
   D. protect families of response personnel from carry-home contamination
   E. protect environmental resources and the general public from incident-related release of contaminants
   F. all of the above

2. On-site field decontamination should be restricted to:
   A. persons who will not realize increased risk to life or health due to any delay of access to off-site professional medical services
   B. personal protective clothing and equipment that must be readily available for subsequent use in response operations
   C. any materials or items that cannot be moved off-site without endangerment of the general public or environmental resources
   D. all of the above

3. Basic types of solvents used for decontamination include
   A. water
   B. dilute acids
   C. dilute bases
   D. organic solvents
   E. all of the above

4. Commonly used chemical detoxification procedures include halogen stripping, oxidation, reduction, and
   A. ammonification
   B. chelation
   C. neutralization

5. The most commonly used physical detoxification procedure is
   A. thermal degradation
   B. denitrification
   C. coagulation

6. Standard detoxification methods are employed
   A. only when there is detailed chemical knowledge and understanding of the chemical dynamics involved
   B. on a best-guess basis
   C. on the basis of that single chemical contaminant that is present in the highest concentration

7. No infected clothing or equipment should be deposited in a landfill unless it has been
   A. disinfected
   B. sterilized
   C. bagged and tagged

8. All clothing is subject to permeation by contaminants, the degree of permeation being subject to the time interval in which contaminants are in contact with the clothing, the physical state and chemical nature of the contaminant, the concentration of the contaminant, and
   A. relative humidity
   B. ambient temperature
   C. reflective index of the clothing

Essays

1. The text discusses a situation in which it might be necessary to severely abbreviate a planned decontamination plan. Identify other circumstances in which it might be necessary to short-circuit the most desirable decontamination plan.
2. Should such conditions as discussed in #1 be provided for in written alternative decontamination plans? Give an example.

Case Study

Use the Internet to familiarize yourselves with the essentials of the Three Mile Island Incident (1979) and the Chernobyl Incident (1986).

1. How do these incidents highlight the practical problems of implementing any plan of containment and decontamination?
2. Do you find any similarities between these incidents and the collapse of the twin towers of the WTC on September 11, 2001?

DATA AND INFORMATION MANAGEMENT

INTRODUCTION

Though often used synonymously to refer to "things that are known," the terms *data* and *information* may be variously distinguished. For example, some prefer to reserve the use of data (plural of datum) for reference to quantitative knowledge and information for qualitative knowledge. Others prefer to distinguish between them not on the basis of type of knowledge but, rather, on whether or not individual pieces of knowledge have been integrated into a larger understanding—an approach that reserves the use of data (or datum) for singular pieces of knowledge and information for data that have been connected to other knowledge bases or otherwise purposely processed for use.

By either definition, emergency response is always a data- and information-intensive effort, not only in terms of the level of detail, but also in terms of the diversity of data and information that must be collected, evaluated, and acted upon. The management of data and information is therefore absolutely critical to the success of emergency response. Moreover, because the focus of both proactive and reactive emergency response is always action—in particular, the act of making decisions—it is perhaps best to view data as either qualitative or quantitative knowledge that must be processed into precisely that information that is most useful for decision-making.

Given the widespread accessibility to global information networks and databases, and the ready availability of ever-expanding computer technology, it might appear that the management of emergency-related health and safety data and information should be a relatively simple task. However, it is well worth considering that access to global databanks and information networks enhances not only the potential for improvement in the efficiency and comprehensiveness of decision-making, but also the potential for utter confusion. Even the most sophisticated technology for retrieving and processing information is no guarantor of competence, nor can it correct the consequences of incompetence.

As happens with the application of any new technology, the application of computer technology to emergency response needs is subject to a variety of misconceptions that can actually contravene the objectives of the response service. Some of the most common misconceptions involve the following considerations:

1. Despite the continuing development of *expert programs*, computers are essentially ignorant tools. Though an extremely powerful tool in terms of flexibility, efficiency, and range of applications, a computer cannot as yet even begin to substitute for human intelligence. The practical consequence of this simple fact must be the realization that any aspect of computerized response applications must be fully conceived and developed before appropriate computerization should even be attempted, and even then, only when the specific objectives of computerization can be clearly defined in terms of the actual needs of emergency response.

2. The rapidity with which we can now access worldwide databases means that we can retrieve bad information as quickly as we can good information. In fact, we can reasonably suppose that the likelihood of retrieving bad information, pure nonsense, or at least misleading data is far greater than retrieving information that is subject to strict quality criteria and review. The practical consequence of this situation is that data and information to be used for emergency response purposes must be evaluated for its veracity and pertinence regardless of source.

3. Software marketing hoopla to the contrary, there is no single computer program that can meet all the needs of managing emergency response databases. Each program has its capacities and its limitations—and both its capacities and its limitations are inherently obstinate. The practical consequence of this must be the realization that the capacities and limitations of each program must carefully be evaluated with regard to specific objectives of the emergency response service. There is, in short, no such thing as "an excellent program" except that it meets precisely defined needs and objectives.

   Given the importance that a computer serve specific response needs and objectives and not *vice versa*, emergency response planners and safety officers seriously should consider developing, where possible, custom-made computer programs rather than simply relying upon commercially available "canned" or "off-the-shelf" programs. Although this alternative too often is given little serious attention, it should be noted that few companies or other organizations entrust their financial, inventory, or billing procedures to over-the-counter computer programs but, rather, utilize the consulting services of professional programmers. Moreover, the ready availability of powerful yet simple programming languages and tools makes it increasingly possible for in-house response personnel who are not professional programmers to develop highly useful programs.

4. In many companies and organizations having extensively computerized operations, and especially where such operations entail the use of mainframe computers, computer programs and procedures typically are

centralized in a computer operations or data processing department (or computer information services department). Even where PC networking is employed, such a centralized department usually exerts full authority over all computer hardware and software. Certainly there are very good reasons for this, including the need for data and information security and the handshaking requirements of computer networks.

However, it is reasonable to suggest that there are practical levels of flexibility required in order to ensure that the needs of corporate financial management, inventory control, and office management do not unnecessarily constrain the operational needs of health and safety programs and, in particular, those health and safety programs directly relevant to emergency response. In proposing appropriate computerization of the various elements of health, safety, and emergency response programs, particular attention must be given to defining capabilities that provide timely and practical information to emergency response personnel without conflicting the needs of other computer-assisted corporate functions and programs.

EXPERT SOFTWARE

Over the past two decades, there has been an explosive development of professionally designed software of particular importance to emergency planning and response. Some of this software has been designed by governmental agencies having broad jurisdictional responsibility and/or special expertise regarding certain types of emergencies; some has been designed by independent experts who, often in close coordination with emergency response services, focus on practical response needs that are particularly data intensive

For example, CAMEO (Computer-Aided Management of Emergency Operations) is a software suite developed originally by the U.S. Environmental Protection Agency (EPA) and the National Oceanic and Atmospheric Administration (NOAA) to assist both governmental and private sector managers in planning for and mitigation of chemical accidents, and meeting regulatory compliance objectives of the Emergency Planning and Community Right-to-Know Act of 1986 (EPCRA: SARA Title III). Detailed and continually updated information regarding CAMEO can easily be obtained via the CAMEO Web site developed by the U.S. EPA and NOAA (http://www.epa.gov/ceppo/cameo/) or via the National Safety Council (http://www.nsc.org/search2.cfm?criteria=CAMEO).

Designed for use by EPCRA-defined entities, such as Local Emergency Planning Committees (LEPCs) and State Emergency Planning Commissions (SERCs) as well as by fire departments, emergency planners, and chemical facilities, CAMEO meets four key objectives:

- Provides instant access to safety and emergency response information on thousands of chemicals (in 1998, 4,700; in 2005, over 6,000)

- Tracks chemical inventories in the community and in transit
- Provides for the electronic submission of reports submitted by regulated facilities in compliance with EPCRA
- Performs an analysis of hazards and off-site consequences of air-dispersed chemical plumes

CAMEO is composed of basically three interactive programs (see Figure 10.1):

- ALOHA, which is a sophisticated air-dispersion model
- MARPLOT, which is a mapping application for planning and managing field response operations at chemical incidents
- CAMEO chemical database, which includes chemical-specific information on fire and explosive hazards, health hazards, fire fighting techniques, cleanup procedures, and protective clothing. The U.S. National Safety Council (NSC) provides extensive professional training on the use of CAMEO.

Other commercially available expert programs are also easily accessed via the Internet, including programs that meet both general and highly specific emergency response operational and management needs, such as:

- 3-D modeling of groundwater flow and contaminant plumes
- Hydrodynamic simulation of hydraulic flow in open channels
- Simulation of contaminant flow in lakes, estuaries, and coastal waters
- Modeling of soil-vapor interactions
- Modeling of contaminant flow, transport, and environmental fate in saturated and unsaturated soils

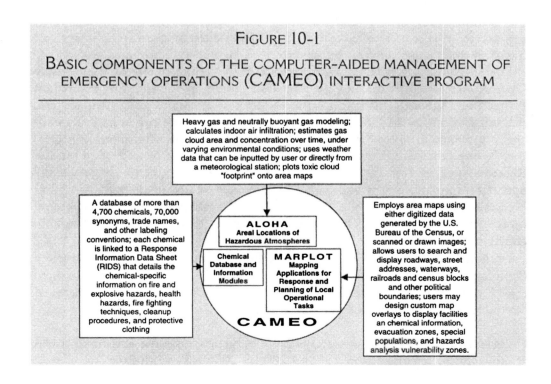

FIGURE 10-1

BASIC COMPONENTS OF THE COMPUTER-AIDED MANAGEMENT OF EMERGENCY OPERATIONS (CAMEO) INTERACTIVE PROGRAM

Heavy gas and neutrally buoyant gas modeling; calculates indoor air infiltration; estimates gas cloud area and concentration over time, under varying environmental conditions; uses weather data that can be inputted by user or directly from a meteorological station; plots toxic cloud "footprint" onto area maps

A database of more than 4,700 chemicals, 70,000 synonyms, trade names, and other labeling conventions; each chemical is linked to a Response Information Data Sheet (RIDS) that details the chemical-specific information on fire and explosive hazards, health hazards, fire fighting techniques, cleanup procedures, and protective clothing

ALOHA
Areal Locations of Hazardous Atmospheres

Chemical Database and Information Modules

MARPLOT
Mapping Applications for Response and Planning of Local Operational Tasks

CAMEO

Employs area maps using either digitized data generated by the U.S. Bureau of the Census, or scanned or drawn images; allows users to search and display roadways, street addresses, waterways, railroads and census blocks and other political boundaries; users may design custom map overlays to display facilities an chemical information, evacuation zones, special populations, and hazards analysis vulnerability zones.

- Simulation of environmental transformations of hazardous chemicals

These and many other such expert programs can be accessed most easily through the use of such search phrases as "computer models," "dispersion models," or "transport simulations," or through the home pages of individual providers (e.g., Scientific Software Group: http://www.scientificsoftware-group.com).

Of course, the most important sources of information regarding highly useful expert programs for emergency response are practicing emergency response professionals. For example, one of the most important software tools under continuous development over the past five years for firefighters and other rescue personnel utilizes virtual reality technology to provide response services an operational management tool as well as a highly effective training technique by integrating site-specific structural, locational, and hazard information with simulated optical feedback. This software was developed under the direction of Fire Master James Jameson, Strathclyde Fire Brigade, U.K.

IN-SERVICE DATA AND INFORMATION BASE

The data and information base for emergency planning and response must serve the site-specific needs of an actual incident; however, certain types of data and information have universal relevance and easily can be implemented without the aid of expert programs and techniques derived from external sources. Examples of minimal types of data and information bases and the necessary cross-referencing among individual databases may be briefly summarized as follows.

1. *Persons and Personnel at Potential Risk*
   - Response personnel by name and operational task category, with cross-reference to potential sources of risk or hazard, pertinent regulations, training needs, required protective equipment, communication needs, decontamination and waste disposal requirements, emergency medical treatment, required medical surveillance, personal susceptibilities, or other factors of special relevance to health and safety
   - Other on-site persons, including facility employees, contractors, consultants, and other support personnel, with cross-reference to specific health and safety precautions, evacuation and temporary shelter needs, task-related restrictions and constraints, and documentation requirements
   - Off-site persons who may be exposed to hazards associated with incident and response operations (see Image 10.1), including property abutters, downwind or downstream residents and communities, with cross-reference to environmental mechanisms of dispersal of hazardous materials and substances, automatic and manual alarm devices and systems, and evacuation procedures

IMAGE 10-1

NEW ORLEANS, LA, SEPTEMBER 2, 2005: A FIREFIGHTING TEAM
ASSISTS IN THE MOVEMENT OF LITTER-BORNE PATIENTS INTO A
U.S. AIR FORCE MEDEVAC PLANE THAT WILL TAKE THEM TO A
HOSPITAL IN SHREVEPORT, LA AS NEW ORLEANS IS EVACUATED AS
A RESULT OF FLOODING CAUSED BY HURRICANE KATRINA

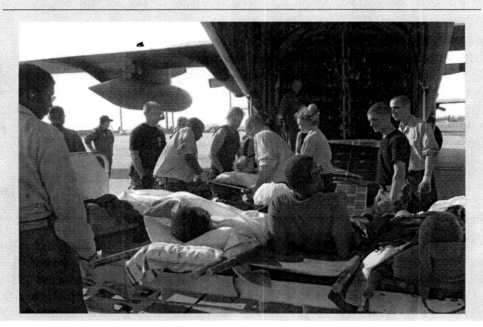

Source: Win Henderson/FEMA

2. *Inventory and Assessment of Hazards*
   - Types of physical, chemical, and biological hazards, with cross-reference to source, modes of exposure, chronic and acute effects, signs and symptoms of exposure, rehabilitation requirements, emergency and follow-up medical treatment and surveillance
   - Sources of routine and emergency hazards, with cross-reference to required engineering and managerial controls, required use of personal protective clothing and equipment, routine and emergency ambient monitoring requirements, inspection schedules, and evaluation criteria
3. *Incident Response Operations*
   - Description of individual health and safety incidents (including routine and emergency incidents), with detailed assessment of cause, and cross-reference to pertinent regulatory requirements, organizational policies, and specific requirements of the health and safety plan
   - Assessment of frequency and magnitude of incidents, with cross-reference to review and modification of health and safety program, notification of regulatory authorities, and personnel training requirements

4. *Support Resources*

- Consultant, contractor, regulatory, and other available personnel having special knowledge and experience relevant to health and safety and response operations (see Image 10.2), with cross-reference to specific data, information, material, equipment, and other needs of support resources
- Hard copy and electronic sources of regulatory, technical, and scientific data and information, with cross-reference to routine and emergency need for information, including up-to-date information on health and safety standards, chemical toxicity and compatibility, personal protective equipment, monitoring devices, and medical treatment and surveillance
- In-place maps, schematics, and diagrams for all facility structures and properties that locate all primary sources of hazards, routes of ingress and egress, potential pathways and collection systems for spills or releases of hazardous materials, with cross-reference to specific regulatory requirements (e.g., underground storage tanks; hazardous waste storage areas; electrical transformers)

IMAGE 10.2

NEW ORLEANS, LA, SEPTEMBER 3, 2005: SURVIVORS OF HURRICANE KATRINA ARRIVE AT NEW ORLEANS AIRPORT FROM WHERE THEY WILL BE FLOWN TO SHELTERS IN OTHER STATES

Source: Michael Rieger/FEMA

It must be stressed that these types of data and information should be immediately available to the incident commander or other responsible persons whether or not the data and information are computerized. However, the cross-referencing required to meet the pressing needs of an actual incident, a facility inspection by regulatory or response personnel, or even routine operational decision-making clearly emphasizes the importance of well-designed and highly integrated computerized files.

What is meant by "highly integrated" is that the data contained in one file allow the user to identify (i.e., through appropriate cross-reference) other data that may be contained in other files. For example, a file that contains information on the general technologies available to technical rescue personnel (see Figure 10.2) can be cross-referenced to files that contain data (e.g., design spec-

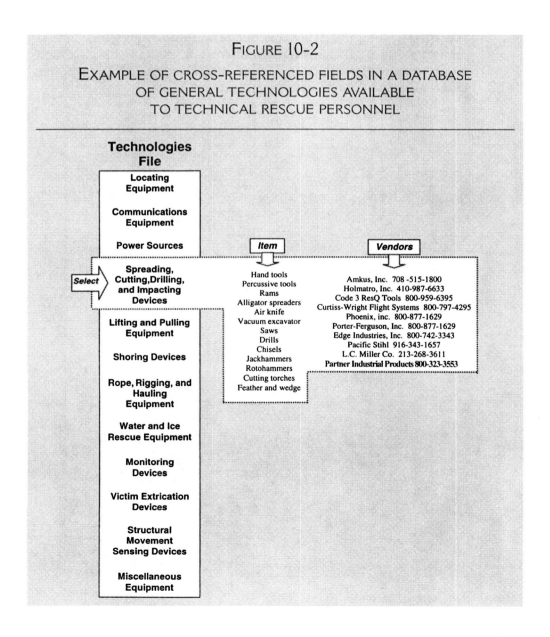

FIGURE 10-2

EXAMPLE OF CROSS-REFERENCED FIELDS IN A DATABASE
OF GENERAL TECHNOLOGIES AVAILABLE
TO TECHNICAL RESCUE PERSONNEL

ifications) on specific equipment that is available within a type of technology as well as vendors who can provide that equipment. Such cross-referencing of files (or integration) can be done, of course, without the use of computers, although computers do provide for a much more efficient management (e.g., updating, correction, correlation) of the relevant databases.

External Databases

Electronic publishing is a rapidly expanding phenomenon that commercial companies, professional organizations, and governmental agencies increasingly use to make technical and scientific information and data more easily available at little or no cost. Powerful search-and-retrieve programs, CD ROMs, and worldwide networking provide essentially instantaneous access to data and information pertinent to all aspects of emergency planning and response, including state, national, and international regulations, health and safety standards, epidemiological and laboratory studies, personal protective clothing and equipment, ambient and personal monitoring systems, and medical surveillance protocols.

Although it is important to explore the full range of available databases, it is equally important to consider the following:

1. Even a brief perusal of health and safety standards is sufficient to determine that standards are highly variable from one legal jurisdiction to another. It goes without saying that the safety officer must ensure compliance with the specific legal authority having jurisdictional precedence; however, it may very well be appropriate to adopt a more stringent standard proposed by some other authority. Such an approach is consistent with not only the principle of minimizing health and safety risk, but also the recognition that there is often a significant lag between scientific findings and regulatory reform. Of course, there are instances in which standards become less stringent precisely because of advances in scientific understanding of hazards and risks—a consideration that nonetheless should be weighed against state-of-the-art practice.

2. There are many CD-ROM databases on chemical hazards and risk, some of which are available through chemical manufacturers and some through commercial sources, including companies that specialize in the production of *material safety data sheets* (MSDSs). A comparison of MSDSs prepared by different companies for the same chemical substance or product often will reveal differences regarding not only specific hazards, but also routes of entry, target organs, and recommended protective clothing and equipment.

   The adoption of the findings, determinations, and recommendations made by any purveyor of information does not absolve the buyer or user of such information from potential liabilities that might accrue to errors of fact or judgment on which that information is based. It is therefore necessary that comparisons of alternative databases be examined and, where differences do occur, discrepancies be resolved.

In many instances, discrepancies in hazard determinations and the toxicology of chemicals are not due to oversight or error but to differences in the interpretation of highly technical data. Where this is the case, guidance must be sought from regulatory and competent scientific authority.

3. Although many commercial electronic databases are offered as part of a subscription service, which ensures periodic updating of information, updating does not ensure veracity. Confidence in a database is warranted only when efforts are made to review that database in light of recognized legal, professional, and scientific standards as promulgated through a wide range of governmental agencies and professional organizations, including (but not limited to) the National Fire Protection Association (NFPA), Federal Emergency Management Agency (FEMA), National Institute of Occupational Safety and Health (NIOSH), Occupational Safety and Health Administration (OSHA), Agency for Toxic Substances and Disease Registry (ATSDR), and the U.S. Environmental Protection Agency (EPA).

Internal Databases

As important as external databases are, they cannot substitute for those databases that must be compiled on a facility- or incident-specific basis and that represent the operational details of emergency response operations. Encompassing information on all service-related health and safety policies and procedures, personal protective clothing and equipment, engineering and managerial controls, personal risk factors, ambient monitoring, medical surveillance, operations auditing and inspection, and personnel training, internal databases must not only accommodate the documentation needs of the response service but also meet the day-to-day operational and planning needs of that service, including:

- Scheduling of key activities (e.g., personnel assignment, ambient monitoring, personnel training, medical surveillance, internal audits of operations)
- Assessment of incident response
- Revision of pertinent programs and policies in light of incidents, changes in health and safety standards, changes in operational capacity, changes in regulatory requirements, development of new materials and technologies
- Personnel actions required to enforce health and safety policies and procedures
- Assessment of effectiveness of response strategies and tactics
- Evaluation of in-place policies, procedures, and equipment with regard to the current and developing state-of-the-art

In constructing internal databases, which include selected data and information obtained from external sources, it is vital that emphasis be given to the specific information that is most likely to be needed immediately during response operations (e.g., nature of biological hazards in facility; likelihood of release of a toxic vapor or mist into the atmosphere). Only by defining the particular informational needs engendered by actual situations can specific cross-reference and retrieval capabilities be incorporated into diverse databases—capabilities that ultimately determine the actual usefulness of those databases and that, unfortunately, are too often overlooked by personnel so mesmerized by the sheer volume of information at their command that they forget that no data or information are of any value whatsoever unless they can readily be integrated with specific planning and response objectives.

Integrating Data with Objectives

Data processing is the means by which individual pieces of information become integrated with planning and response objectives. It consists of both the analysis and presentation of data so as to facilitate purposeful decision-making. Data processing long precedes the advent of computer technology; however, computers not only greatly reduce the time and effort required to produce a competent analysis and presentation of data, but also significantly increase the sophistication of data processing.

For example, the U.S. Fire Administration developed a comprehensive text (Fire Data Analysis Handbook) on statistical and graphic methods of data processing for use by fire services based on the following premise:

> Turning data into information is neither simple nor easy. It requires some knowledge of the tools and techniques used for this purpose. Historically, the fire service has had few of these tools at its disposal and none of them has been designed with the fire service in mind.
> —from the Forward in *Fire Data Analysis Handbook*, U.S. Fire Administration

Today, all the methods and techniques discussed in this excellent text are readily available through low- to moderate-priced commercial software. In fact, a good

TABLE 10-1

NUMBER OF FIRES (BOSTON, 1988) ON BASIS OF TIME OF DAY (ADAPTED FROM U.S. FIRE ADMINISTRATION, FIRE DATA ANALYSIS HANDBOOK)

Time Period	Number	Time Period	Number
Midnight -1 AM	478	Noon-1 PM	307
1 AM-2 AM	420	1 PM-2 PM	316
2 AM-3 AM	360	2 PM-3 PM	363
3 AM-4 AM	273	3 PM-4 PM	381
4 AM-5 AM	192	4 PM-5 PM	417
5 AM-6 AM	127	5 PM-6 PM	433
6 AM-7 AM	122	6 PM-7 PM	492
7 AM-8 AM	139	7 PM-8 PM	514
8 AM-9 AM	156	8 PM-9 PM	540
9 AM-10 AM	168	9 PM-10 PM	622
10 AM-11 AM	206	10 PM-11 PM	510
11 AM-Noon	242	11 PM-Midnight	547

number of these techniques are standard components of popular word processing programs that can automatically convert tabular data (see Table 10.1) typically compiled by fire service personnel into alternative graphic presentations (see Figure 10.3) having direct relevance to specific operational objectives (e.g., most efficient scheduling of fire response personnel over a 24-hour period).

Essentially one step up in technical sophistication from the simple data processing afforded by word processing programs are those spreadsheet and database management programs that typically are included in home and office program suites. Yet another step up from these are simple programming languages (e.g., Visual Basic) that can nonetheless perform rigorous data analysis and informational packaging with relatively little investment in time, training, or money. More complex and sophisticated programming tools are, of course, available.

Figure 10.4 is a representation of a chemical database that may be established by any readily available and simple data management program. Such a base may include from several hundred (e.g., for a small manufacturing facil-

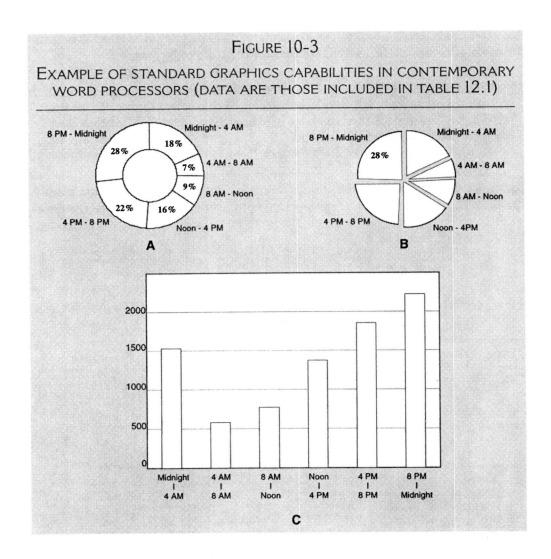

FIGURE 10-3

EXAMPLE OF STANDARD GRAPHICS CAPABILITIES IN CONTEMPORARY WORD PROCESSORS (DATA ARE THOSE INCLUDED IN TABLE 12.1)

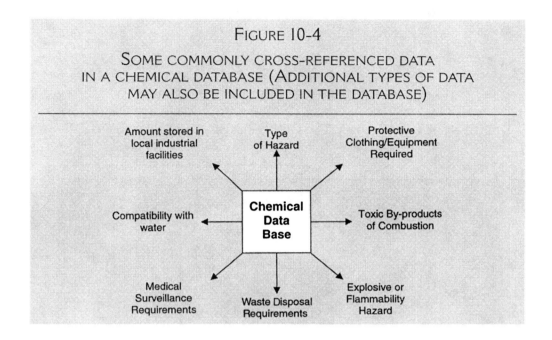

FIGURE 10-4

SOME COMMONLY CROSS-REFERENCED DATA
IN A CHEMICAL DATABASE (ADDITIONAL TYPES OF DATA
MAY ALSO BE INCLUDED IN THE DATABASE)

ity) to several thousand or more individual chemicals (e.g., for a community response facility and/or large manufacturer), each being correlated with appropriate information, such as waste disposal requirements, toxic byproducts, type of hazard, and other types of information important for planning for or responding to a chemical incident. Thus, relevant information about a particular chemical may be retrieved on the basis of any of the types of information associated with that chemical. Of course, in an actual incident, the precise chemical involved in a spill or release may not be known, or mixtures of chemicals may be involved. Because of this, it is important to include not only the names of pure chemicals in the database, but also categorical descriptors (e.g., flammable solvents, toxic gases) that correspond to chemical mixtures likely to be encountered in a site-specific incident.

Whatever the tool used (i.e., whether provided by spreadsheets, database programs, or programming languages), the most important element in any data processing is the clear understanding of how individual pieces of data can be linked to one another, and how those linked data can be formatted and actually used to provide essential input into decision-making.

In short, the difficulty today is not the analytical techniques of data processing—these are already prepackaged for almost instantaneous use in user-friendly format; rather, the difficulty is matching informational needs with planning and response objectives. Thus, the data to be included in the database depicted in Figure 10.4 must be selected from among voluminous available data—selected precisely because only certain types of information can be acted upon, whether for planning, operational, or training purposes.

During an actual incident, informational needs will vary with a wide range of factors, including type of incident, weather conditions, and timing of the incident with respect to work schedules and community activities (see Image 10.3). Informational needs may also vary during any particular incident as addi-

IMAGE 10.3

DES MOINES, IOWA, JULY 9, 1993: DURING A DISASTER, FEMA
PROVIDES MUCH NEEDED FINANCIAL ASSISTANCE AND ENSURES
THAT FRESH WATER, FOOD, SHELTER, AND COMMUNICATIONS ARE
AVAILABLE IN THE FLOOD STRICKEN AREA. AFTER THE 1993
MIDWEST FLOODS, A TOTAL OF 534 COUNTIES IN NINE STATES
WERE DECLARED FOR FEDERAL DISASTER AID

Source: Andrea Booher/FEMA News Photo

tional details become evident. The substantive design of a database is therefore essentially a function of managerial strategy rather than of technical or
scientific necessity.

MODULAR APPROACH TO DATABASE DESIGN

The practical approach to building databases having high relevance to the
practical needs of emergency planning and response is to focus on the construction of individual modules—databases that correlate information of a particular type, such as the chemical database shown in Figure 10.4, or a database
for protective clothing (see Figure 10.5), or for decontamination (see Figure
10.6), or for any other major category of operational concern. This compartmentalization of databases, especially if undertaken by in-service personnel,

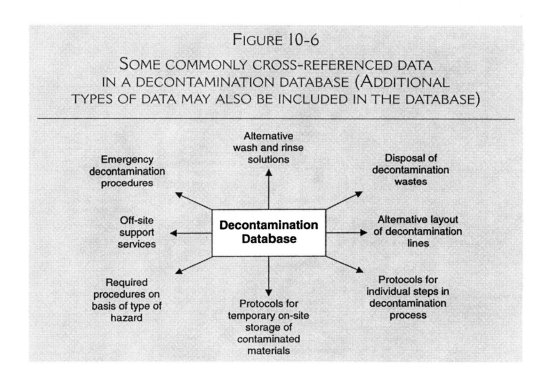

FIGURE 10-5

SOME COMMONLY CROSS-REFERENCED DATA
IN A PROTECTIVE CLOTHING DATABASE (ADDITIONAL TYPES
OF DATA MAY ALSO BE INCLUDED IN THE DATABASE)

Pre- and post-use
inspection
requirements

Alternative
Vendors

Procedures for
decontamination

Maintenance and
Replacement
schedule

**Protective
Clothing
Database**

Required personal
protective
equipment

Personnel
rehabiliation
requirements

Specifications on
basis of type of
hazard

Donning and
Doffing
Protocols

FIGURE 10-6

SOME COMMONLY CROSS-REFERENCED DATA
IN A DECONTAMINATION DATABASE (ADDITIONAL
TYPES OF DATA MAY ALSO BE INCLUDED IN THE DATABASE)

Emergency
decontamination
procedures

Alternative
wash and rinse
solutions

Disposal of
decontamination
wastes

Off-site
support
services

**Decontamination
Database**

Alternative layout
of decontamination
lines

Required
procedures on
basis of type of
hazard

Protocols for
temporary on-site
storage of
contaminated
materials

Protocols for
individual steps in
decontamination
process

typically results in less time required for not only the design of databases, but also for revision required by the development of new material, equipment, standards, and regulations. It also allows for the use of the overall format used in one module as a design template for other modules. Finally, it allows individual personnel (either singly or in small groups) to focus on topical areas in which they have particular expertise, experience, and interest.

Another key attribute of the modular approach is that it allows for more concentrated and efficiently performed debugging and testing to ensure that modules contain precisely the information required and correctly performs retrieval, updating, and corrective functions. An important disadvantage of this approach is that the final product could be nothing more than a series of disconnected databases that must be individually accessed to retrieve information sets that are operationally interdependent. It is therefore vital that the design of individual modules allow for multidimensional access into related modules.

For example, in a given incident, the type of chemical may be known and can be used (either by name or hazard category) to access response-relevant information (e.g., reactivity with water, type of health or safety hazard) from a chemical database. However, hazard type can also be used as a means of entry into the database module designed for protective clothing. If this second module (i.e., protective clothing module) contains information on procedures for decontamination, then such procedures may also serve as a means of access to a

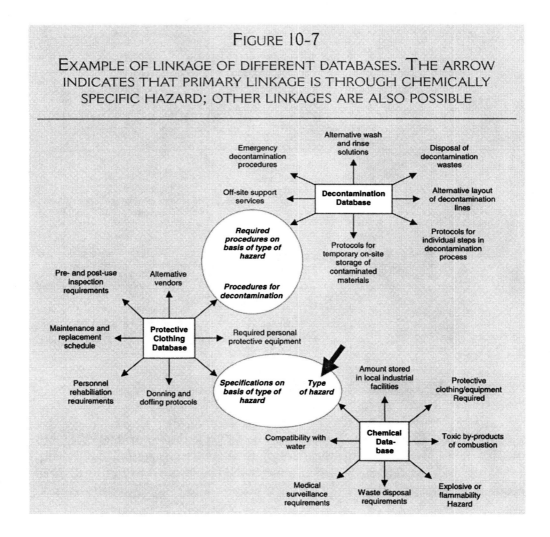

FIGURE 10-7

EXAMPLE OF LINKAGE OF DIFFERENT DATABASES. THE ARROW INDICATES THAT PRIMARY LINKAGE IS THROUGH CHEMICALLY SPECIFIC HAZARD; OTHER LINKAGES ARE ALSO POSSIBLE

third module, which may be a database for decontamination. In short, the design of individual modules can be accomplished easily so that one piece of information (e.g., hazard category) triggers the retrieval of relevant hazard-specific information from a range of different types of databases (see Figure 10.7)—information that singularly and collectively provides essential direction to decision-making.

Of course, in the example presented in Figure 10.7, this approach might appear to result in an unnecessary duplication of files (e.g., decontamination procedures in both protective clothing and decontamination databases). However, this is not the case. After all, any database ultimately consists of interrelated files; the arrows shown in Figure 10.7 do not point to actual duplicate files but, rather, to the same file. In this context, it is important to emphasize that the construction of interconnected modules depicted in Figure 10.7 is not at all dependent upon computer technology—such a system can be constructed just as solidly out of traditional file folders and paper labels as out of electrons. What is depicted is simply a logic, not a technology—a way of organizing data to meet decision-making needs for information, not an exercise in arcane engineering. The computer is simply a highly efficient means for consistently exercising a defined logic on organized data to produce information in a selected format. Yes, the technology is new. But the manner of thinking is not.

TEAM APPROACH TO DATABASE DESIGN

Ideally, the construction of any database and associated retrieval systems should be undertaken by a team composed of experienced response personnel who, because of the depth and diversity of their experience, are best able to (a) identify the types of information that are operationally critical to both emergency and normal operations, (b) identify and collect the types of data that must be organized to provide that information, and (c) define and implement design criteria for data and information management that are fully consistent with the practical constraints of time and resources imposed by emergency incidents.

The importance of using experienced response personnel to design data processing systems cannot be overstressed. Systems designed primarily by computer and other information-processing specialists are very likely to meet technical criteria of excellence and coding elegance, but they are also prone to be impractical in terms of the actual needs and constraints of response services. In the early development of PC technology, there was an obvious need to rely upon computer specialists—after all, such specialists were the only people who had the necessary knowledge of the software and hardware intricacies of electronic data management. However, this is no longer true. Today, not only is that knowledge more widely dispersed, but it also has become encapsulated in user-friendly technology as readily available to grade-school students as it is to practicing professionals.

In addition to the use of an experienced and diverse response team, it is critical that the design of a data and information management system proceed in close coordination with other key members of the community partnership for emergency planning and response. Whether it is called coordination, communication, liaison, or networking, the ongoing functional interaction and interdependence of all team members must be structured into the vital decision-making processes of each member—and nothing is more critical to coordinated decision-making than data and information processing, a fact that is most obvious in the midst of an actual incident where immediate, coordinated, and

FIGURE 10-8

DATA AND INFORMATION PROCESSING SYSTEMS UNDERLIE THE DYNAMIC LINKAGE AMONG DIFFERENT RESPONSE SERVICES, THEIR SUPPORT RESOURCES, AND ON-SITE INCIDENT RESPONSE. SUPPORT RESOURCES INCLUDE A WIDE RANGE OF BOTH PUBLIC AND PRIVATE ORGANIZATIONS, INCLUDING INDUSTRIAL AND OTHER COMMUNITY RESOURCES

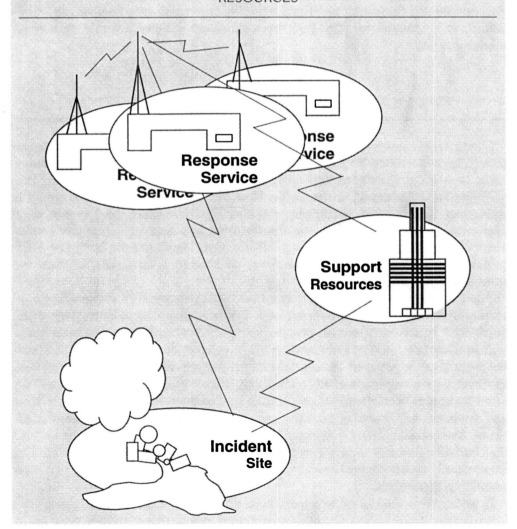

complementary response action must be taken by a wide range of response services and support resources. Precisely the same teamwork required to manage an actual incident should be manifest in the day-to-day management of critical data and information processing systems that influence the decision-making of separate response services and support resources (see Figure 10.8). Toward the achievement of this objective, it is necessary that concerted effort be made to ensure:

- That local industries invite community response services (e.g., fire, medical, municipal) to review, comment upon, and suggest modification of in-house response-related data and information processing systems
- That community response services as well as municipal and other jurisdictional authorities similarly involve local industry as well as private support contractors and vendors in the design and implementation of relevant service and municipal databases and retrieval systems
- That all members of the community response partnership actively maintain information sharing programs with their global colleagues, with particular emphasis on the sharing of ideas and approaches to the design of practical and effective data processing strategies and techniques

TESTING DATABASE DESIGN

Both modular and integrated databases should be tested using evaluation criteria germane to their operational uses under normal and emergency conditions. Clearly, criteria for assessing a database designed to facilitate decision-making regarding the nonemergency purchase of clothing, equipment, and supplies would not be appropriate criteria for assessing a database designed to facilitate decision-making regarding on-site deployment of PPC decontamination stations. Some criteria, however, may be consistently applied to all databases and processing software, including (but not limited to):

- Consistency of output with different users and different makes/models of auxiliary equipment (e.g., printers, monitors, fax modems, e-mail programs)
- Flexibility of output formats (e.g., monitor and printed page formats, text, graphics)
- Memory requirements
- Search time
- Ease of correcting, upgrading, deleting, and archiving
- Susceptibility to crashing and user misuse
- Security
- Automatic documentation of use

Once appropriate evaluation criteria are established, serious consideration should be given to the use of table-top or other simulation exercises (see Chapter 13) as a means of assessing the data and information management system.

A very practical approach to such an assessment is to define a variety of scenarios (e.g., accidental spill of bulk hazardous liquid, emergency exposure of response personnel to a biological agent, on-site entrapment of technical rescue personnel, release of toxic vapor in a residential area, fire in a pharmaceutical R&D laboratory) that would require immediate access to particular types of information. In such an exercise, the objective is not simply to retrieve the appropriate information, but also to test the range of factors that may influence the successful retrieval and subsequent processing or use of the retrieved information. Of particular concern should be such considerations as:

- How does the on-site person faced with a particular problem determine which information in the database is required or, at least, most relevant?
- Can the information be retrieved by persons most likely to be available at the time, or must it be retrieved by a limited number of personnel who may not be available in a timely fashion?
- How can the data required in a critical situation be retrieved or processed in the case of a power failure? For example, are printed records containing the needed information readily available under any emergency condition?
- Does the retrieved information direct the person who retrieves it how to act upon it, or is it simply assumed that available personnel will know what to do with the information?
- What information do I need? How do I get it? Is the information I retrieve good information? What do I do with it?

These are the necessary questions that must be addressed to any data and information management system and, as yet, they cannot be obviated by even our most sophisticated electronic tools. Unasked or unanswered, or posed imprecisely or unclearly, they transform even the most extensive database and most elegantly conceived information processing system into simply so many gigabits of pure nonsense.

STUDY GUIDE

True or False

1. CAMEO is an expert program that contains a database of over 6,000 chemicals, including chemically specific information on hazards, cleanup procedures, and appropriate protective clothing.
2. ALOHA is a component of CAMEO that provides a footprint of toxic atmospheric plumes on a local map.
3. Many of the commercial expert programs deal with transport simulations (i.e., movement of plumes in water, air and soil).
4. In the event of an emergency, databases regarding personnel at potential risk, inventory and assessment of hazards, incident response operations, and support services (e.g., consultant, contractor) should be immediately available to the incident commander.

5. Integrated databases are those constructed with cross-referenced information related to a general topic.
6. The purveyor of any database is solely responsible for the accuracy of information; the user of the database incurs no liability.
7. All databases should be constructed so as to provide specific information that can be acted upon in an emergency situation; information that does not meet the specific needs of emergency response decision-making should not be included.
8. The major difficulty in data management for emergency response is to match informational needs with planning and response objectives.
9. A key attribute of the modular approach to database design is that it allows for more concentrated and efficiently performed debugging and testing to ensure that it contains precisely the information required and correctly performs retrieval, updating, and corrective functions.
10. Ideally, the construction of any database and associated retrieval systems should be undertaken by an experienced response team.

Multiple Choice

1. Data are either qualitative or quantitative knowledge that must be processed into precisely that information
   A. required by regulatory authority
   B. which is most useful for decision-making
   C. both A and B
2. CAMEO provides
   A. an extensive database of chemicals
   B. an atmospheric model of the movement of hazardous plumes
   C. local maps
   D. all of the above
3. In-service databases most useful for inventorying and assessing hazards would focus on types of physical, chemical, and biological hazards, with cross-reference to
   A. specific sources of the hazards
   B. modes of exposure
   C. chronic and acute effects
   D. signs and symptoms of exposure
   E. emergency and follow-up medical treatment and surveillance
   F. all of the above
4. It is possible to create very useful emergency-related databases by utilizing
   A. spreadsheet programs typically included in home and office program suites
   B. simple programming languages
   C. both A and B
5. In addition to the use of an experienced and diverse response team, it is critical that the design of a data and information management system proceed in close coordination with
   A. corporate executive-level personnel
   B. members of the community partnership for emergency planning and response
   C. external expert consultants

Essays

1. What do you think are the important first steps in constructing a database to be used for emergency planning and response?
2. Comment on the problems incidental to the loss of power that may be associated with an emergency incident.
3. In addition to the use of selected databases for purposes of actual emergency response, what other uses might such databases serve?

Case Study

The chemical inventory of your plant is on the order of 300 chemicals. What kind of information do you think it is most important for a fire chief to have concerning these chemicals?

Note: There is a vast amount of information that can be collated with respect to even a single chemical, and the fire chief does not have the time, in the midst of an emergency, to study the entire inventory in detail.

MONITORING STRATEGIES AND DEVICES

INTRODUCTION

Chemical surveillance of the American workplace is standard practice, with specific requirements defined not only by regulatory authority (e.g., emergency response, laboratory standard, confined space entry, respiratory protection, hazard communication, chemical process safety) but also by corporate insurance carriers, corporate legal counsel, health and safety professional organizations, and employee safety committees. Chemical surveillance is, in fact, as intrinsic to modern business practice as loss control, total quality management, and human resource development.

Given the broad legal political, economic, and ethical ramifications of exposure to workplace chemicals, it is useful to distinguish between *chemical monitoring* and *chemical surveillance*.

Chemical monitoring connotes the technical and methodological aspects of any qualitative or quantitative analysis of process or fugitive chemicals. It may be undertaken for a variety of reasons, including not only the management of potential human exposure, but also to control production processes or the quality of intermediate and finished products. Chemical surveillance is a much broader, programmatic approach to the management of human exposure to chemicals. Surveillance includes monitoring, as well as a variety of other efforts, such as the control of chemical inventories, waste minimization, chemical substitution, and process management.

In the United States, chemical monitoring for the purpose of managing human risk is historically linked to industrial compliance with chemical-specific health and safety standards (e.g., 29 CFR 1900 Subpart Z), with generic hazard and risk management regulations (e.g., laboratory standard, hazard communication), and with standards established for hazardous waste sites (RCRA; CERCLA) and operations involving especially hazardous chemicals and chemical processing (e.g., EPCRA: SARA Title III).

The rapid development of numerous workplace standards in the 1980s and 1990s coincided with two other significant developments: the explosive progress in materials sciences and electronic engineering (which provide the physical means for much of monitoring technology), and the growing recognition of the potential chemical risks associated with modern industry, including not only normal design and operational risks, but also those engendered by natural catastrophe (see Image 11.1), human error, and acts of terrorism.

All three factors (i.e., regulatory standards, monitoring technology, and catastrophic chemical release) continue to underscore the vital importance of monitoring capability in modern emergency planning and response—a capability that, although necessarily responsive to potential chemical risk, is not limited to chemical monitoring but includes a range of diverse sensor systems necessary to achieve proactive and reactive response objectives.

CHEMICAL MONITORING TECHNOLOGIES

Common techniques for monitoring hazardous chemicals may be conveniently divided into three basic types:

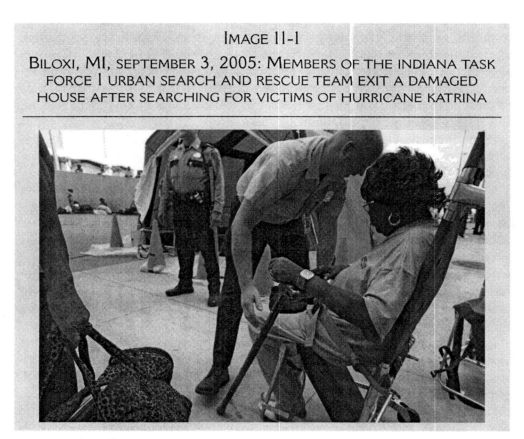

IMAGE 11-1

BILOXI, MI, SEPTEMBER 3, 2005: MEMBERS OF THE INDIANA TASK FORCE 1 URBAN SEARCH AND RESCUE TEAM EXIT A DAMAGED HOUSE AFTER SEARCHING FOR VICTIMS OF HURRICANE KATRINA

Source: Mark Wolfe/FEMA

1. *Ambient air monitoring*: techniques that provide rapid on-site detection or measurement of chemicals that are present in the air as dusts, vapors, gases, or mists.
2. *Ambient materials testing*: techniques that typically require off-site laboratory analysis of samples, including solids (e.g., soil samples) and liquids (e.g., groundwater, hazardous waste mixtures).
3. *Personal monitoring*: techniques that involve the detection or measurement of chemicals within body tissues, such as blood or urine, or in the immediate vicinity of a worker equipped with a personal monitor to measure cumulative exposure over specific period of time.

Ambient Air Monitoring Devices

By far the most commonly used chemical monitoring devices, ambient air monitoring devices provide a rapid, direct reading of chemical concentrations in air (see Image 11.2). However, there are usually significant limitations associated with any particular device, including:

IMAGE 11-2

OSHA TEAM FROM THROUGHOUT THE UNITED STATES DISCUSSES AIR SAMPLING PROCEDURES AT THE WORLD TRADE CENTER SITE. SAMPLES WERE ANALYZED FOR ASBESTOS, SILICA, LEAD AND OTHER HEAVY METALS, CARBON MONOXIDE, AND NUMEROUS ORGANIC AND INORGANIC COMPOUNDS. NOISE WAS ALSO MONITORED

Source: Shawn Moore/OSHA News Photo

- Most detect or measure only specific chemicals or chemical classes; none detect all possible chemicals
- Though the sensitivity of such devices is always subject to the development of new technology, they generally are incapable of detecting airborne concentrations of chemicals below 1 mg/m^3
- Many can give false readings (false positives) because, although designed to detect one particular substance, they are subject to interference (i.e., "poisoned") by the presence of other chemicals

Colorimetric Indicator Tube

This relatively low-priced and easily used device consists of (a) a tubular glass ampoule containing an *indicator chemical* that reacts with a specific ambient contaminant of interest, and (b) a manual or motorized pump to draw a calibrated amount of air through the ampoule. The reaction of the indicator chemical and the air contaminant changes the color of the indicator chemical. The linear length of the color change in the ampoule is proportional to the concentration of the air contaminant. Calculating the air concentration of the contaminant requires a simple mathematical operation involving the calibrated length of the color reaction in the ampoule and the volume of air pumped through the device. Depending upon the specific chemical being measured, the calculation also may require correction for barometric pressure. The measurement of certain air contaminants also may require the simultaneous use of a second ampoule, which is affixed to the indicator tube.

Manufacturers of colorimetric indicator tubes provide detailed information on the limits of each indicator tube that is specific to the ambient gas or vapor of interest (see Table 11.1). In addition to such chemical-specific limitations, all colorimetric tubes share certain general limitations:

- Although each indicator tube is specific to a particular ambient chemical, other ambient contaminants can interfere with the indicator chemical
- Most tubes can be affected by high humidity, thereby giving false readings
- Tubes available from different manufacturers may have different sensitivities, thereby providing different measurements of ambient concentrations
- Because of the variability of color perception among persons, different personnel may make different judgments as to the length of the color stain within the indicator ampoule

Another common problem associated with colorimetric indicator devices is the problem of false negatives. If, after use, an ampoule shows no color reaction, the negative result may be due to the concentration of the ambient contaminant being lower than the sensitivity of the tube or the indicator tube being defective. In such a case, the user is well advised to test the negative ampoule with a known high concentration of the contaminant vapor.

Finally, it must be stressed that the volume of air perfused through the indicator tube is typically very small and therefore represents a tiny portion of the ambient atmosphere. The location of the air intake to the detection device is therefore critical with regard to estimating air quality in the total volume of

TABLE 11-1

DATA AND INFORMATION TYPICALLY PROVIDED BY MANUFACTURERS OF COLORIMETRIC DEVICES USED FOR THE MONITORING OF COMMON INDUSTRIAL CHEMICALS

Gas or Vapor To Be Monitored	Catalog Number	Range (ppm/hours)	Detection Limit (ppm; 8-hour)	Color Change	Storage Temp.	Shelf-Life (years)
Ammonia	3D	25–500 ppm	1.0	Purple to Yellow	Room Temp.	3
Carbon Dioxide	2D	0.2–8.0% hr.	0.015%	Blue to White	Room Temp.	2
Carbon Monoxide	1D	50–1000ppm	2.5	Yellow to Dark Brown	Room Temp.	2
Chlorine	8D	2–50 ppm	0.13	White to Yellow	Room Temp.	2
Formaldehyde	91D	1–20 ppm	0.06	Yellow to Red Brown	Refrigerate	1
Hydrogen Cyanide	12D	10–200 ppm	0.5	Orange to Red	Room Temp.	2
Hydrogen Sulfide	4D	10–200 ppm	0.25	White to Dark Brown	Room Temp.	3
Nitrogen Dioxide	9D	1–30 ppm	0.06	White to Yellow	Refrigerate	1
Sulfur Dioxide	5D	5–100 ppm	0.13	Green to Yellow	Room Temp.	2

breathable air. It is, therefore, always advisable to conduct colorimetric monitoring within the immediate breathing space of personnel at risk.

High-flow personal samplers are increasingly available at reasonable cost and should be considered as an important adjunct to any monitoring program. Single pumps typically are housed in a lightweight plastic case that clips to the user's belt. Multiple pumps are also available and can be used for simultaneous monitoring of atmospheric samples taken at different locations within the same general area. Battery packs for both single and multiple samplers allow continuous sampling over the typical work day.

Electronic Devices

A large variety of electronic devices are available for the detection of specific chemicals and broad categories of chemicals, and typically include such additional capabilities as data retrieval and storage, database searching, automatic calculations, and the formatting and printing (or modem transmission) of written reports and display graphics. Some, such as the combustible gas indicator, flame ionization detector, portable infrared spectrophotometer, and ultraviolet photoionization detector (see Table 11.2), have broad application for compliance with numerous health and safety standards. Others, such as an oxygen meter or sound meter, obviously have a much more narrow application.

Electronic instruments typically combine monitoring capabilities for different chemicals. Examples of such combined capabilities include:

- A combustible gas meter that also measures oxygen, hydrogen sulfide, and carbon monoxide
- A toxic gas meter than can detect oxygen and combustible gases and that can also be equipped to monitor hydrogen cyanide, hydrogen chloride, nitrogen dioxide, nitrous oxide, and sulfur dioxide

TABLE 11-2

BASIC TYPES OF ELECTRONIC MONITORING DEVICES (ADAPTED FROM NIOSH, USCG, AND EPA, 1985: OCCUPATIONAL SAFETY AND HEALTH GUIDANCE MANUAL FOR HAZARDOUS WASTE ACTIVITIES)

Combustible Gas Indicator (CGI)

- Measures the concentration of a combustible gas or vapor
- A filament, usually made of platinum, is heated by burning the combustible gas or vapor; the increase in heat is measured
- Accuracy depends, in part, on the difference between the calibration and sampling temperatures
- Sensitivity is a function of the differences in the chemical and physical properties between the calibration gas and the gas being sampled
- The filament can be damaged by certain compounds, such as silicones, halides, tetraethyl lead and oxygen-enriched atmospheres

Continued

TABLE 11-2—*Continued*

BASIC TYPES OF ELECTRONIC MONITORING DEVICES
(ADAPTED FROM NIOSH, USCG, AND EPA, 1985:
OCCUPATIONAL SAFETY AND HEALTH GUIDANCE MANUAL
FOR HAZARDOUS WASTE ACTIVITIES)

Flame Ionization Detector (FID) with Gas Chromatography Option

- In *survey mode*, detects the total concentrations of many organic gases and vapors; all organic compounds are ionized and detected at the same time
- In *GC mode*, identifies and measures specific compounds; volatile species are separated
- Gases and vapors are ionized in a flame; a current is produced in proportion to the number of carbon atoms present
- Does not detect inorganic gases and vapors, or some synthetics; sensitivity depends on the compound
- Should not be used at temperatures < 40 deg. F (4 deg. C)
- Difficult to identify compounds absolutely; specific identification requires calibration with the specific compound of interest
- High concentrations of contaminants or oxygen-deficient atmospheres require system modification
- In *survey mode*, readings can be only reported relative to the calibration standard used

Portable Infrared (IR) Spectrophotometer

- Measures concentration of many gases and vapors in air
- Passes different frequencies of IR through the sample; the frequencies absorbed are specific for each compound
- In the field, must make repeated passes to achieve reliable results
- Not approved for use in a potentially flammable or explosive atmosphere
- Water vapor and carbon dioxide interfere with detection
- Certain vapors and high moisture may attach to the instrument's optics, which must then be replaced

Ultraviolet (UV) Photoionization Detector (PID)

- Detects total concentrations of many organic and some inorganic gases and vapors; some identification of compounds is possible if more than one probe is used
- Ionizes molecules using UV radiation; produces a current that is proportional to the number of ions
- Does not detect a compound if the probe used has a lower energy level than the compound's ionization potential
- Response may change when gases are mixed
- Other voltage sources may interfere with measurements; response is affected by high humidity

- A hazardous gas detector that provides continuous measurements of many gases or vapors, including acetone, ammonia, arsine, benzene, carbon monoxide, ethylene oxide, and formaldehyde

Of critical importance in emergency response is the combustible gas indictor. As shown in Figure 11.1, the combustion of a substance depends upon an adequate supply of both burnable fuel and oxygen. The relative amounts of oxygen and fuel that will support combustion are described in terms of both the *lower explosive limit* (LEL) and the *upper explosive limit* (UEL). Below the LEL, there is insufficient fuel to support combustion; above the UEL, there is insufficient oxygen.

FIGURE 11-1

LOWER AND UPPER EXPLOSIVE LIMITS. AS THE AIR IN A CONTAINER IS PROGRESSIVELY DISPLACED BY A FLAMMABLE VAPOR (UPPER PORTION OF FIGURE), THE CONCENTRATION OF OXYGEN IN THE CONTAINER DECREASES WHILE THE CONCENTRATION OF POTENTIAL FUEL INCREASES. AT CONCENTRATIONS OF FUEL (EXPRESSED AS PERCENTAGE OF ATMOSPHERE) BELOW THE LEL (LOWER EXPLOSIVE LIMIT), THERE IS TOO LITTLE FUEL TO SUPPORT BURNING; AT CONCENTRATIONS OF FUEL ABOVE THE UEL (UPPER EXPLOSIVE LIMIT), THERE IS TOO LITTLE OXYGEN TO SUPPORT BURNING. EXPLOSION MAY OCCUR ONLY WHEN THE RELATIVE CONCENTRATIONS OF FUEL AND OXYGEN ARE AT OR BETWEEN THE LEL AND UEL. ON A SCALE WHERE 100% REPRESENTS THE LEL, 10% IS TYPICALLY USED AS THE TRIGGER FOR IMPLEMENTING PERSONNEL EVACUATION

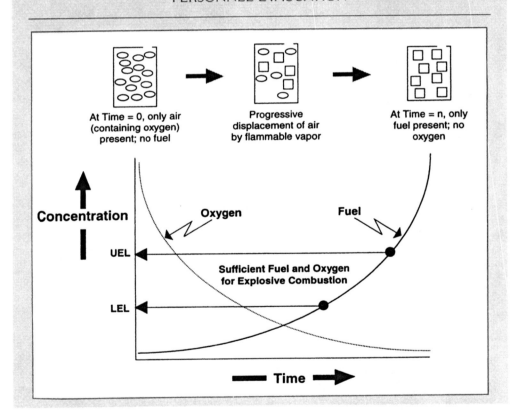

The readout of the combustion meter usually is given in terms of percentage of LEL, with 100% indicating that the mixture of fuel and oxygen meets the minimal requirements for explosion. On such a scale, a reading of 10%, which is very often used to trigger the evacuation of an area, means that the concentration of flammable vapor is one-tenth of that required for a state of imminent explosion.

Portable detectors (see Table 11.2) can, with proper calibration, detect hundreds of individual toxic vapors and gases. FID (flame ionization detector) and PID (photoionization detector) units can also be operated to measure total concentrations of ambient chemicals without regard to chemical species. This mode of operation (i.e., *survey mode*) is useful because its lack of chemical specificity provides an inherent safety factor.

For example, in the survey mode, a reading of, say, 250 ppm, which could represent a total concentration of potentially hundreds of organic compounds, could also be interpreted to represent the concentration of a particular toxic compound of concern. Such a worst-case interpretation of the reading might or might not be realistic in a particular circumstance, but such an interpretation always has very real value as a criterion for further investigatory (if not corrective) action or even the initiation of area evacuation.

An oxygen meter (and, in some circumstances, a toxic gas and/or combustible gas meter), is a basic requirement in any situation in which confined space entry is required. There are many different designs of oxygen meters; however, it is imperative that the meter be provided with a long probe that can be lowered or otherwise extended into a confined space without the operator becoming exposed to an atmosphere that is potentially oxygen deficient and/or toxic. It is also important that the operator understand that an oxygen meter is typically sensitive to environmental factors, including barometric pressure and ambient concentrations of carbon dioxide and other oxidizing agents (e.g., ozone).

Various types of radiation survey meters are readily available for the detection of alpha, beta, and gamma radiation and X rays. It should be noted that, although alpha particles (energetic helium nuclei) are relatively large and slow-moving particles that are easily stopped by the outer layer of skin, they are also very hazardous if inhaled (e.g., via contaminated dust). Gamma and X rays (electromagnetic radiation) and beta particles (high-speed electrons) may have high penetration capabilities.

It must be stressed that, despite their apparent simplicity of design and operation, all electronic detectors are sophisticated instruments and require a precise understanding of their inherent limitations and requirements regarding calibration, care, and maintenance. Users are well advised to ensure that the manufacturer of any electronic detector provide proper instruction in all aspects of operation and maintenance, with particular emphasis given to the routine documentation of calibration, precision, and accuracy.

Ambient Materials Testing

In some situations a safety officer may require analyses of site materials and resources that typically cannot be performed on site, such as the analysis of:

- Potable water supplies, including wells, public water supply mains and lines, and other resources that might be contaminated with biological or chemical agents
- Surface or groundwater supplies that might become contaminated as a result of an incident and/or incident response activities

- On-site soils and dusts possibly contaminated with heavy metals and hazardous contaminants
- Structural and other on-site materials containing toxic substances (e.g., asbestos, pesticide residue, lead-based paint)

Such analyses typically require the use of specialized laboratories, including commercial water testing laboratories and materials testing laboratories. Where such professional analytical services are required, only those vendors who are certified by legal authority should be selected, and, more precisely, only those who are specifically certified for the particular analysis to be performed.

For example, in the United States, a water testing laboratory may be certified through a state agency under the aegis of the U.S. Environmental Protection Agency; however, the certification typically is highly specific on the basis of the different types of analyses required by the Federal Safe Drinking Water Act, with certification of the analysis of heavy metals, for example, being separate and distinct from certification for the analysis of microorganisms.

Having procured the professional services of an appropriately certified laboratory, the safety officer must ensure that all samples are collected, stored, and delivered in full compliance with relevant regulations. Although it is generally desirable that the contracted certified laboratory itself collect and handle samples, this is typically impossible during an emergency. The site safety officer therefore should obtain written directions from the certified laboratory for the proper procedures for collecting, handling, packaging, preserving, transporting, and documenting samples.

PERSONAL MONITORING TECHNOLOGIES

Personal monitoring devices include such devices as badges, monitors, dosimeters, and open diffusion detector tubes, all of which can easily be clipped or otherwise attached to personal clothing. Monitors may be dedicated to a particular chemical species (e.g., mercury vapor, trichloroethylene) or provide detection of a broad class of chemicals (e.g., organic vapors). In some designs, monitors that detect classes of chemicals may be processed to yield specific exposure data regarding a limited number of specific chemicals out of several dozen possibilities. Some devices give direct readouts of timed exposure (see Table 11.3) or require simple comparison of color changes with standard color charts or data sheets; others require off-site laboratory processing, which introduces delays of several or more days in obtaining results.

Because of the necessary delays in obtaining data from badges that require off-site processing, it is imperative that such devices not be used in emergency response where exposure can exceed health or safety standards. The basic rule is that personal monitoring devices be used only after a comprehensive survey of an area has established, by means of ambient monitoring devices, potential worst-case exposures—and even then, only when there is assurance that worst-case exposures cannot exceed health and safety standards within the estimated time frame of on-site personnel work schedules.

TABLE 11-3

EXAMPLES OF GASES AND VAPORS THAT
CAN BE MONITORED BY MEANS OF OPEN
DIFFUSION DETECTOR TUBES

Gas or Vapor	8 Hr. Measuring Range (ppm)
Acetic acid	1.25–25
Ammonia	2.5–187.5
Butadiene	1.25–37.5
Carbon dioxide	62.5–2500
Carbon monoxide	6.25–75
Ethanol	125–3000
Ethyl acetate	62.5–1250
Hydrochloric acid	1.25–25
Hydrocyanic acid	2.5–25
Hydrogen sulfide	1.25–37.5
Nitrogen dioxide	1.25–25
Olefin	12.5–250
Perchloroethylene	25–187.5
Sulfur dioxide	.63–18.8
Trichloroethylene	25–125

OTHER MONITORING TECHNOLOGIES

In addition to the technologies already discussed, which are standard technologies employed by industrial hygienists, regulatory compliance officers, and hazardous waste emergency response personnel, other monitoring capabilities should be assessed for their relevance to the health and safety of response personnel.

Some of the more common monitoring devices that should be considered have long been used by industrial hygienists in the routine surveillance of workplace conditions and, depending upon incident site conditions, may be important tools for managing site-specific risks, including:

- Temperature and relative humidity monitors (especially important when incident conditions and operations enhance the risk of heat stress and dehydration; also available are heat stress monitoring systems that, though essentially hand-held units, can be connected to fixed or remote sensors)
- Sound meters (for continuous and intermittent noise that may not only result in significant damage to hearing, but also interfere with communication as well as concentration among site personnel)
- Particle analyzers (concentration and size distribution; an important adjunct to respirator use)
- Hand-held gas tracers and leak (gas and liquid) detectors (including colorimetric developers that can detect specific liquids, such as oil and chlorine)

Depending upon the nature of the incident and site operations, other more specialized but long developed technologies may be appropriate, such as:

- Vibration and dynamic strain sensor, as well as a variety of structural motion and/or level sensors (e.g., during technical rescue operations inside unstable structures)
- Piezoelectric films, cables, and other devices (which, by converting mechanical stress or strain into proportionate electrical energy, may also be adapted to monitor the stability of on-site structural features)
- Personal alert safety system (PASS; to monitor motion of isolated personnel working under hazardous conditions and, as necessary, sound rescue alarm)
- Hand-held thermal sensor (to detect hot spots in overhaul and other response-related operations)
- Fixed and portable traffic monitors (which, in situations of public panic or situations involving multiple incidents in dense population centers, can prove to be of inestimable value for achieving both strategic and tactical response objectives)
- Meteorological monitors (that can supplement regionally available data with site-specific details or, in the case of communication failure or multiple incidents, provide basic data—for example, wind speed and direction or barometric pressure—and forecast capability needed for operational management)

In addition to existing technologies, critical assessment of developing monitoring and alert technologies must be maintained as an ongoing emergency management planning effort. This is especially important with regard to the monitoring of chemical and biological warfare agents that may purposely be used by terrorists or inadvertently released during a nonterrorist incident.

Both manual and electronic techniques are employed (and are under constant development) for detecting warfare agents, including:

- Detection paper, which is impregnated with dyes and pH indicators and yields characteristic colors when in contact with specific chemical warfare agents
- Detection tubes, which (as discussed earlier) involve chemically specific, colorimetrically defined reactions between known substrate and target agents
- Detection tickets, which consist of enzyme- and substrate-impregnated papers that mediate detectable reactions with various nerve gases
- Ion mobility spectroscopy (IMS) system, such as the Chemical Agent Monitor (CAM), which is currently in worldwide use for the detection of nerve gas, blister and choking agents
- Flame photometric detector (FPD), which can detect certain chemical warfare agents through the photometric analysis of air-hydrogen combustion residues
- Biosensors, which utilize enzymes and/or bioreceptors to detect chemical agents

DESIGN AND IMPLEMENTATION OF MONITORING PROGRAM

Different response services have different monitoring needs that may or may not overlap in the progress of a particular incident. The design and implementation of an effective monitoring program is therefore essentially a two-tiered effort:

the first, focusing on specific in-service (or, as in the case of industry, in-plant) informational needs; the second, on mutual assistance among community services and support resources to provide a comprehensive monitoring capability.

In-Service Monitoring Capability

The design of an in-service monitoring capability should be viewed as basically an ongoing process of matching operational decision-making needs with available technologies, and, of course, an assessment of potential applications within the constraints of costs and available personnel. Costs must include not only capital costs but also direct and indirect costs associated with maintenance and replacement, quality control testing and calibration, and personnel training requirements. Basic steps in the design and assessment process should, at a minimum, include the following.

Conduct a Needs Analysis

This is a methodical examination of all decision-making needs during the conduct of response operations. In catastrophic situations (e.g., flood, earthquake), it is likely that the needs analysis will include consideration of both ambient and personal monitoring of both responding personnel and large numbers of civilian victims (see Image 11.3). Decision-making needs must be clearly

IMAGE 11-3
HURRICANE KATRINA SURVIVORS SHELTERED IN THE RED CROSS SHELTER AT THE ASTRODOME

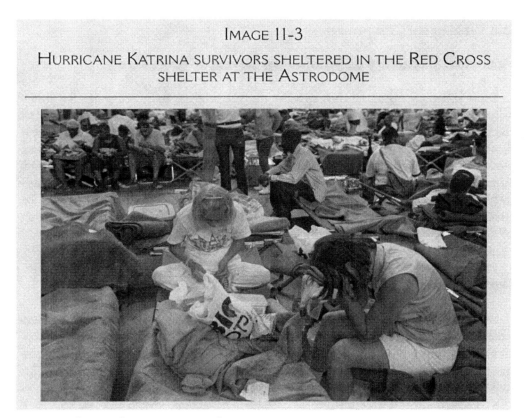

Source: Andrea Booher/FEMA News Photo

and specifically defined with respect to (a) which monitoring data actually are used to choose from among which alternative courses of action, and (b) alternative sources and/or types of data and information that may serve the same decision-making objective. A proper needs analysis includes not only a detailed examination of the linkage of certain types of data to certain types of operational decisions, but also an evaluation of the relative importance or criticality of both the data and the decisions under emergency conditions.

For example, it may not at first appear to be necessary to use a field detector to monitor for toxic organic fumes if it is definitely known that toxic organic fumes are present and that search and rescue teams will wear SCBA—and, perhaps most importantly, that there are victims that must be recovered as quickly as possible. However, a properly conducted needs analysis will force additional considerations that can significantly increase the criticality rating of monitoring data, such as, in the following:

- If a SCBA systems fails or there is a physical tearing or ripping in protective clothing, what is the possible health consequence to the wearer, and what is the time frame in which these consequences may become irreversible?
- Are the permeability ratings of fully functional protective clothing adequate under actual response conditions?

The answers to these questions depend, of course, on knowing just what the organic fume is, its concentration and physical and chemical properties, its combustion products, and other ambient conditions that affect its permeation into clothing (e.g., temperature).

Document Reliability of Monitoring Device or Procedure

Once it is demonstrated that specific monitoring data are necessary for operational decision-making, it is necessary to demonstrate the field reliability of the monitoring device or procedure selected to provide that data. This is best done by documenting its actual field use by other agencies and organizations, testing laboratories, and regulatory agencies, with particular attention given to observed failures and limitations. Design specifications should also be submitted to prospective vendors to ensure that devices and procedures can be successfully used under typical field conditions. Finally, actual field testing of the device or procedure should be conducted under simulated emergency conditions.

Develop Standard Operating Procedures

Once there is demonstrable assurance that the device or procedure meets practical decision-making needs under actual field conditions, comprehensive SOPs should be developed with regard to proper use, storage, maintenance, cleaning and decontamination, calibration, and any other protocol that may be necessary to ensure effective readiness and reliability.

SOPs also should give particular attention to action levels; that is, specified monitoring data (e.g., concentration, % LEL, radiation level) that require the implementation of predetermined actions by specifically identified personnel. SOPs should also specify personnel training requirements, with appropriate attention given to ensuring the ready availability of properly trained personnel under likely field conditions.

Finally, it is vital that SOPs specify backup or fail-safe requirements for each device or procedure. For example, it may be deemed necessary to require the on-site availability of two or more fully functional devices so that the failure of any one device will not result in the loss of key monitoring capability. In some instances, it may be possible to provide manual monitoring capability (e.g., indicator tube technology) as a backup to an electronic monitoring device (e.g., photoionizing device).

Mutual Assistance Monitoring Programs

While there is much monitoring technology that is moderately priced, there is also much that is expensive not only terms of capital costs, but also in servicing and maintenance costs. No response service therefore is likely to have the full primary and backup monitoring capability that may be called for in a particular incident. It is necessary therefore that careful attention be given to developing mutual assistance programs among both governmental and private sector organizations at local, regional, and even national levels. Just who should initiate this action and take responsibility for developing such programs is arguable. However, because any incident presents potential risks to the community as a whole, it is reasonable to suggest that municipal authority assume lead responsibility. In the United States, this approach is consistent with the establishment of local emergency response commissions (LERCs) under the Emergency Planning and Community Right-to-Know Act of 1986. However, more regionally or even nationally centralized authorities are most appropriate to the political structures of many other countries. As is evident in the United States, specifically with respect to September 11, 2001 and the ongoing threat of terrorism, it is likely that federal authority, whether under the rubric of homeland security or some other appropriate response to terrorist aggression, will continue to exert significant impact at both regional and municipal levels. Although this is certainly true with respect to terrorist incidents, the fact that it is impossible, at the onset of any major incident, to determine whether or not terrorism is involved, it is likely that federal authority will continue to influence response efforts even at the local municipal level.

Whatever its position in the sociopolitical hierarchy, the lead authority must ensure the on-site ready availability of an appropriate, fully functioning monitoring capability that is likely to be composed of different resources supplied by a variety of response and support organizations.

In many instances, monitoring devices are only part of the total resources (which may include heavy equipment and/or transport vehicles as well as radiation meters available through a civil defense or state emergency response center) to be mobilized to meet incident response needs. However, there are also circumstances in which vital monitoring equipment may be most readily available only through a particular source that has no other resource of potential response value (e.g., local pharmaceutical company, industrial chemical manufacturer, research and development facility).

The questions to be addressed by the lead organization for mutual assistance are essentially simple, including: Who has what device? Where is it? How will it be delivered to where it is needed? Who takes responsibility for delivery? How long will it take to obtain it? Will it be functional when it arrives? Who knows how to use it?

Obtaining clear, reliable answers, of course, is very difficult and requires significant, long-term commitment to preincident planning, coordination, and (especially) training.

PROACTIVE INDUSTRIAL MONITORING

Because the industrial facility must always be considered a potential source of risk to the surrounding community, in-plant monitoring of potentially hazardous substances, which is subject to extensive regulatory scrutiny in the United States, should be considered not simply a means of achieving regulatory compliance but also a key step in proactive emergency planning at the community level.

Whereas monitoring technologies are equally available to community response services and industrial facilities, the industrial safety officer and incident safety officer have essentially different objectives, even if both must be concerned primarily with the health and safety of their respective personnel. In particular, the industrial safety officer must focus on the use of monitoring as a primary means of preventing or correcting a situation that can otherwise evolve into a full-blown, community emergency.

Identification of Monitor Parameters

The essential first step to implementing an effective in-plant monitoring program is to identify potential parameters to be monitored on the basis of (a) regulatory requirements (e.g., 29 CFR 1910 Subpart Z, hazard communication regulations, confined space regulations, OSHA laboratory standard, hazardous waste regulations, chemical process safety regulations), (b) requirements imposed by corporate insurance carriers, (c) state-of-the-art practices within the industry, (d) concerns of personnel, and (e) recommendations of a facility safety committee. This typically is accomplished by means of a comprehensive hazard assessment of the total facility, which usually (and preferably) includes a prioritization of hazards. The prioritization of hazards is important for two basic reasons:

1. Prioritization facilitates the implementation of appropriate mitigation techniques, including administrative and engineering control measures and the required use of personal protective clothing and equipment.
2. Because even a moderate size manufacturing facility may have a chemical inventory of several thousand chemicals, prioritization of hazards

requires not only an assessment of the degree of hazard, but also the detailed assessment of the technological feasibility of monitoring individual chemical species.

In identifying potential chemical agents and available technologies for monitoring those agents, the safety officer must ensure that chemical byproducts and processing intermediaries are considered along with feedstock product chemicals.

Finally, it must be noted that the availability of a monitoring technology does not necessarily imply the existence of a health or safety standard that can be used to interpret generated data. The safety officer must ensure that the final selection of parameters and monitoring technology is based on clear criteria for acting upon monitoring results.

Establishing Baseline Conditions

The concentration of ambient substances (e.g., chemical fumes, dusts) in the workplace typically varies greatly over the workday. Before any schedule of ambient monitoring or sampling is established, it is necessary to conduct a baseline study to identify the range of variation that may be correlated with routine workplace production schedules, seasonal patterns of temperature and humidity (which may directly influence in-plant ventilation) and plant production levels. A baseline study must also establish variations in ambient concentrations with regard to nonroutine situations, as in the case of power outages and staged shutdowns of ventilation for equipment repair and replacement.

The importance of establishing a comprehensive baseline cannot be overemphasized simply because it is the deviation of monitored concentrations away from baseline conditions that may indicate an impending or actual emergency situation. In some situations, deviations away from in-plant baselines may even indicate off-site emergencies that have resulted in atmospheric plumes that subsequently have become entrained in the make-up air of downwind facilities.

Action Levels

Some regulations (e.g., 29 CFR 1910 Subpart Z) require specific actions to be taken if monitoring data regarding certain chemicals (e.g., formaldehyde) meet or exceed certain limits (i.e., action levels). Even though the number of chemicals having action levels defined by safety and health regulations is small, the key requirement for any monitoring program is that corporate action levels be clearly defined for all monitored parameters. In other words, if you can't take action on the basis of a monitored parameter, don't measure that parameter!

In the typical situation, routine monitoring data are collected, recorded, and, over some period of time involving days, weeks, or months, processed and eventually filed. Oftentimes, the employee who conducts the monitoring is not the employee responsible for reviewing the data. This situation is in direct opposition to the objective of using monitoring data to protect human health and safety. Persons who conduct the monitoring and have first access to the resultant data must be equipped with clear criteria for immediately initiating any protective or correction action. Data that meet or exceed established action levels are not to be used to call a meeting to discuss the ramifications of the data; they must trigger immediate response to protect personnel and the general public, and to correct a hazardous situation—actions that already have been assessed, fully formulated, and coordinated with external authorities.

Carry-Home Contamination

Most often overlooked in corporate monitoring programs is the contamination that may be carried home or elsewhere from the workplace by hair, clothes, shoes, and other personal clothing. Even where the company attempts to control such carry-home contamination by the use of site-restricted shop uniforms and specially required workplace practices (e.g., workplace showers, hair nets), it is advisable to consider including personal clothing and other items in a comprehensive monitoring program. Even where the potential for such carry-home contamination is negligible, periodic monitoring data can provide documentation that might prove important in a legal proceeding involving any claim of corporate negligence. Such data are also important in setting contractual responsibilities with company out-service contractors, including corporate laundry services.

Quality Control

In addition to routinely scheduled monitoring activities, the safety officer is well advised to consider implementing unscheduled or even randomized monitoring efforts. Such unscheduled monitoring, whether conducted by in-plant personnel or external consultants, is a very useful means of ensuring the adequacy and quality of routine monitoring. Where external consultants are used, data sets generated by those consultants should be carefully compared to data sets generated internally. Consideration should also be given to using external consultants (including representatives of local response services) to conduct or to participate in a comprehensive annual review and assessment of the entire monitoring program.

ALARP Principle

Regulatory standards must always inform and guide any program of in-plant monitoring and surveillance; however, the objective of keeping workplace ambient concentrations of hazardous substances *as low as reasonably possible* (ALARP) is internationally recognized as a universally relevant objective. Elevated to a principle within a globally competitive business community, ALARP properly emphasizes that regulatory standards should be considered maximum allowable limits. However, within those regulatory limits, the company should endeavor to set action levels for monitored data that minimize all workplace and environmental exposure to hazardous substances within the constraints of available technology and economic reasonableness.

But, of course, what is "economic reasonableness"? There are many who would argue that the objective should be to keep exposures *as low as possible* (ALAP)—that economic reasonableness should not enter into any formula that relates human health and safety to employment.

In establishing a program of workplace monitoring that is consistent with either ALARP or ALAP, companies are well advised to consider that, notwithstanding the necessity of employing legally enforceable health and safety standards, specific technological (and, possibly, economic criteria) for setting action levels are also to be assessed (by the public at large and/or legal authority) in terms of the state-of-the-art—that is, what the "best" companies actually do.

Because what happens in the workplace can so easily happen to the community that surrounds that workplace, it can be expected that state-of-art practice will increasingly become the measure of competent industrial planning and management rather than, as is too often now the case, economic practicality.

STUDY GUIDE

True or False

1. Chemical surveillance includes chemical monitoring, but also the control of chemical inventories, waste minimization, chemical substitution, and process management.
2. Ambient air monitoring techniques typically measure only specific chemicals or chemical classes; none detect all possible chemicals.
3. Ambient air monitoring using colorimetric indicator tubes can give false results due to the presence of ambient chemicals that can interfere with the indicator chemical.
4. It is always advisable to conduct colorimetric monitoring within the immediate breathing space of personnel at risk.
5. A Flame Ionization Detector (FID) detects inorganic gases and vapors.
6. A reading of 10% on a combustion meter means that the concentration of flammable vapor is one-tenth of that required for a state of imminent explosion.
7. Flame ionization detectors and photoionization detectors can be operated to measure total concentrations of ambient chemicals without regard to individual chemical species.
8. An oxygen meter is typically insensitive to environmental factors such as barometric pressure or ambient concentrations of carbon dioxide and other oxidizing agents.
9. Some devices used for personal monitoring require off-site laboratory processing.
10. Piezoelectric cables can be adapted to monitor the physical stability of on-site structural features.
11. A personal alert safety system (PASS) monitors the motion of personnel working under hazardous conditions and, as necessary, sounds a rescue alarm.

12. Detection tickets consist of enzyme- and substrate-impregnated papers that respond to various atmospheric irritants.
13. At concentrations of fuel above the UEL (upper explosive limit), there is too little oxygen to support burning.
14. Operating a flame ionization detector or a photoionization detector in the "survey mode" is useful because its lack of chemical specificity provides an inherent safety factor.
15. On a combustion meter scale where 100% represents the LEL, 25% typically is used as the trigger for implementing personnel evacuation.

Multiple Choice

1. Common techniques of monitoring hazardous chemicals include
   A. ambient air monitoring
   B. ambient materials testing
   C. personal monitoring
   D. all of the above
2. False negatives when using colorimetric methods may be due to
   A. the concentration of the ambient contaminant being lower than the sensitivity of the tube
   B. a defect in the indicator tube
   C. both A and B
3. Depending upon the specific chemical being measured by a colorimetric indicator tube, it may be necessary to correct for
   A. ambient temperature
   B. barometric pressure
   C. both A and B
4. Most colorimetric indicator tubes can be affected by
   A. ambient temperature
   B. high barometric pressure
   C. ambient humidity
5. For a flammable vapor, explosion can occur only
   A. below the LEL
   B. above the UEL
   C. between the UEL and LEL
6. The chemical agent monitor (CAM) is used for the detection of
   A. nerve gas
   B. blister agents
   C. choking agents
   D. all of the above
7. A baseline study of ambient substances in the workplace must identify the range of variation that may be correlated with
   A. routine workplace production schedules
   B. seasonal patterns of temperature and humidity
   C. both A and B

Essays

1. Where would you begin to collect information on appropriate monitoring devices for your in-plant facility?
2. Comment on how you would incorporate quality control mechanisms and action level procedures into your in-plant monitoring system.

Case Study

You are responsible for designing and implementing a surveillance program (including ambient and personal monitoring) in your furniture manufacturing plant.

1. How do you begin to determine what you should be monitoring?
2. What criteria might you use in selecting from among alternative means of measuring particular parameters?
3. What particular difficulties might you expect as a result of a personal monitoring program?

TERRORISM

INTRODUCTION

The U.S. Federal Bureau of Investigation (FBI) defines *terrorism* as "the use of force or violence against persons or property in violation of the criminal laws of the United States for purposes of intimidation, coercion or ransom," and distinguishes between *international terrorism* (involving "groups or individuals whose terrorist activities are foreign-based and/or directed by countries or groups outside the United States or whose activities transcend national boundaries") and *domestic terrorism* (involving "groups or individuals whose terrorist activities are directed at elements of our government or population without foreign direction").

Emergency response services may become involved in terrorist acts in two ways: first (and most obvious), simply because they respond to any community emergency, whatever its cause; second (and too often overlooked), because they themselves may become the object of terrorist acts or otherwise unwittingly become pawns in terrorist strategy.

To emphasize the difference in these two types of involvement, it is instructive to consider that, whereas "intimidation, coercion or ransom" may in fact be key motivational dimensions of documented terrorist acts, other emotions, volitions, and psychological states as well can serve to unleash wanton disregard for human life, including revenge, anger, and frustration, and even (albeit perversely and pathologically misguided) a sense of excitement or challenge.

From this perspective, response services can become the objects of terrorist violence for several and diverse reasons:

- Because they are high-profile targets of opportunity
- Because their dependable response to an emergency situation makes them predictable and therefore susceptible to the advanced planning of those intent on murder

- Because they are so central to community safety and health (and yet finite in number), they become that first shield of defense to break apart before releasing a more concerted onslaught on the community itself
- Because their field operations necessarily entail physical disruptions of normal traffic flow and so occupy the attention of other community services and the public that they can be used both as tactical bottlenecks and feints in a many-layered stratagem for community-wide destruction

As these examples demonstrate, to focus on "intimidation, coercion or ransom" as the sole or even primary (or even necessary) motivations of terrorism is, perhaps, to define the risk presented by terrorism to emergency services (and, thence, to the public at-large) too narrowly, essentially ignoring not only the range of human motivation for inventing horror, but also a basic corollary of any humanly contrived and directed violence—that *if you play a tactical role against the interests of an adversary, then you yourself become of special interest in that adversary's strategy.*

It is important to be very clear here—the intent is not to derogate the FBI's definition of terrorism but, rather, to argue as strongly as possible against automatically and uncritically extending that definition into the total province of emergency response. Definitions, after all, are precisely whatever we choose them to be, and the FBI and other organizations define things as they do to meet the organizational and legal constraints of their activities. But the objectives, constraints, and activities of the FBI are not precisely identical to those of community response services, which should quickly come to understand that they are subject to the hazards imposed not only by their response to terrorist-caused incidents, but also by themselves becoming targets—that they are at risk not only of the physical, chemical, and biological agents wielded by terrorists, but also of strategies contrived by very clever social psychopaths.

POTENTIAL WEAPONS

Regardless of underlying motivation, a terrorist act (see Image 12.1) is most brutally characterized by the willful use of weapons of mass and indiscriminate murder and destruction. Historically, the typical terrorist weapon has been an explosive device. However, there is a wide range of weaponry that is increasingly available to terrorists. The U.S. Department of Justice and the Federal Emergency Management Agency (FEMA) use the acronym B-NICE to denote this range, the letters standing for Biological, Nuclear, Incendiary, Chemical, and Explosive incidents.

Biological Incidents

Biological agents include microorganisms (i.e., bacteria, rickettsia, and viruses) and byproducts of microbiota (e.g., *Botulinum* toxin) that can be disseminated within human populations by such means as atmospherically released aerosols and particulates, the contamination of food, water, and other

IMAGE 12-1

ARLINGTON, VA, SEPTEMBER 12, 2002: EXTERIOR OF THE CRASH SITE FOLLOWING THE ATTACK ON THE PENTAGON

Source: Jocelyn Augustino/FEMA News Photo

common-use items (e.g., cosmetics, clothing), and the use of living carriers (humans and animals). Biological weaponry is not, as many think, a new development or the product of the Cold War—rather, such weaponry has been used for centuries.

Documented instances of the use of biological weapons in warfare include the use of plague-infested corpses as a means of disseminating the Bubonic Plague throughout a besieged city (1346 A.D.), the use of virus-contaminated blankets to spread smallpox among American Indian loyalists to the French (1759 A.D.), and the long established use (from Roman times through the American Civil War) of both human and animal corpses to contaminate the water supplies of opposing armies (see Lt. Col. Terry Mayer, USAF, *The Biological Weapon: A Poor Nation's Weapon of Mass Destruction*).

Beginning with World War I and extending through all subsequent wars to the present, the potential for biological warfare resulted in concerted research and development regarding both the use of and the response to biological warfare agents. Despite the unilateral action of the United States in the early 1970s to destroy its offensive biological warfare capability, and despite subsequent actions by the world community of nations to remove the threat of biological warfare, the relevant technologies for the production, storage, and delivery of such agents are well developed and, because of their essential simplicity as well as low cost, are readily available.

The World Health Organization(WHO) has estimated that a single incident involving a deadly biological agent could result in approximately 100,000 deaths and an even greater number of incapacitated victims; however, it is a mistake to assume that a terrorist incident involving biological agents would necessarily be directed against large populations. They can as easily be used against a small population, targeted, perhaps, as part of a more community-wide strategy or, perhaps, simply out of personal revenge.

For example, in the mid 1980s, 751 people in Oregon became severely ill after dining at several restaurants. Two years later, a member of a religious cult confessed that these illnesses were the direct result of a plot to contaminate the salad bars in these restaurants with the pathogenic bacterium *Salmonella*. The objective of the plot was to incapacitate enough of the local population to influence the results of an election in which one of the key issues was a land-use dispute between the sect and local officials. Other similar incidents involving the purposeful use of biological agents (e.g., the bacterium *Shigella*) against relatively small numbers in a selected population have been documented more recently, including hospital and laboratory staff members (Texas) who evidently became targets of personal revenge rather than, as in the Oregon case, of political strategy.

Examples of microbial diseases that could be employed by terrorism include (but are not limited to):

- Anthrax: contagious disease of warm-blooded animals caused by *Bacillus anthracis* bacterium; characterized by fever, prostration, malignant pustules on exposed skin, and internal hemorrhage
- Cholera: infectious disease of the small intestine caused by *Vibrio cholerae* bacterium; characterized by profuse watery diarrhea, vomiting, muscle cramps, severe dehydration, and depletion of electrolytes
- Bubonic Plague: contagious, often fatal epidemic diseased caused by *Yersinia pestis* bacterium; transmitted from person to person or by the bite of fleas from an infected host, especially a rat; characterized by chills, fever, vomiting, diarrhea, and the formation of buboes (i.e., inflamed, tender swellings of lymph nodes, especially in the area of the armpit or groin)
- Tularemia: infectious disease caused by *Francisella tularensis* bacterium; chiefly affects rodents but can also be transmitted to human beings by bite of various insects or contact with infected animals; characterized by intermittent fever and swelling of the lymph nodes
- Salmonellosis: infection caused by intestinal bacteria of the genus *Salmonella*; characterized by nausea, abdominal pains, diarrhea, and fever; can lead to death, especially in people with impaired immune systems
- Staphylococcus infections: any of a number of infections caused by bacteria of the genus *Staphylococcus*; characterized by abscesses, boils, and other inflammations of the skin; can also produce infection in any organ of the body (e.g., staphylococcal pneumonia in the lungs)
- Q Fever: infectious diseased caused by *Coxiella burnetii* rickettsia; characterized by fever, general malaise, and muscular pains
- Epidemic Typhus: any of several infectious diseases caused by rickettsia (e.g., *Rickettsia prowasecki*); typically transmitted by fleas, lice, or mites;

characterized by severe headache, sustained high fever, depression, delirium, and the eruption of red skin rashes

- Smallpox: highly contagious, sometimes fatal viral disease; characterized by a high fever and successive stages of severe skin eruptions
- Lassa Fever: often fatal viral disease endemic to West Africa; characterized by high fever, headache, ulcers of the mucous membranes, and disturbances of the gastrointestinal tract
- Hemorrhagic Fever: types of fever characterized by profuse bleeding from internal organs and rapid wasting and death; caused by a variety of viruses (e.g., *Ebola* virus, *Marburg* virus)
- Venezuelan Equine Encephalitis: viral infection of the central nervous system, with potential fatal swelling of the brain
- Hanta Disease: viral infection due to any member of the genus *Hantavirus*; transmitted by rodents; characterized by flu-like symptoms and, in more severe cases, shock, kidney failure, internal bleeding, fluid accumulation in the lungs, and death

Among the metabolic byproducts of microbial growth are potent chemicals collectively known as toxins, which can also be used as biological weapons. Such toxins include bacterial toxins, plant toxins, and other molecules (e.g., bio-regulators), which, though not (in a strict sense) toxins, can exert toxic-like effects on the body. Examples of such substances include:

- Diphtheric toxin: potent toxin produced by the bacterium *Corynebacterium diphtheriae*, which causes tissue destruction and the formation of a gray membrane in the upper respiratory tract that can detach to cause asphyxiation; toxin may also enter into blood and subsequently damage tissues elsewhere in the body
- Botulin toxin: extremely potent toxin typically associated with food poisoning; produced by *Clostridum botulinum* bacterium; characterized by disturbances in vision, speech, and swallowing and, within a few days, paralysis of respiratory muscles and death by suffocation
- Clostridian toxin: toxin produced by the bacterium *Clostridium perfringens*, the causative agent of gas gangrene; characterized by slow asphyxiation and subsequent necrosis (cellular death) of living tissue
- Staphylococcus Enterotoxin Type B (SEB): toxin produced by bacterium *Staphylococcus aureus*; most commonly associated with food poisoning; characterized by stomach cramps, diarrhea, and vomiting
- Saxitoxin; produced by marine blue-green alga (i.e., cyanobacterium), which serves as food supply for various shellfish that are immune to effects of the toxin, but pass it on to higher order consumers (e.g., humans); in humans, toxin acts on central nervous system to produce paralysis; at high doses, death can occur in less than 15 minutes
- Ricin: mixture of poisonous proteins produced by the castor oil plant; plant gene controlling the production of ricin has been successfully transferred to the bacterium *Escherichia coli*; ricin interferes with the body's normal synthesis of proteins; symptoms include decreased blood pressure, with death occurring most often through heart failure

- Substance P: a protein closely related to normally produced proteins in the body; may cause pain, or act as anesthetic, or affect blood pressure; rapid loss of blood pressure in victim may cause unconsciousness

With regard to biological weapons, it is necessary to emphasize several facts that underscore the growing concern about their potential use by terrorists:

1. Many of the diseases of primary concern have long been recognized for their potency as human pathogens. Anthrax, for example, reached epidemic proportions in the Roman Empire, resulting in sharp reductions in human populations of a period of five years. Similar epidemic outbreaks have occurred throughout history and, even with the invention of effective vaccines for animals (Pasteur; 1861) and humans (Koch; 1883), modern outbreaks have persisted—most notably in the USSR in 1979, when there was an accidental release of dry anthrax spores at the Microbiology and Virology Institute, a Soviet biological warfare facility. This incident resulted in the contamination of an area with a radius of at least 3 kilometers. Despite strict Soviet censorship concerning this incident, it is estimated that from several hundred to several thousand people died after inhaling the spores and contracting pulmonary anthrax.

   The decision of the U.S. government in 1997 to inoculate troops that possibly could be assigned to duty in the Persian Gulf in opposition to Iraq, which was considered to have developed anthrax as a biological weapon, underscores the fear of this ancient disease. However, as potent as the causative bacterial agent (*Bacillus anthracis*) of anthrax is, there can be no doubt that modern biotechnologies can be used to develop and mass-produce even more potent bacterial strains.

2. Whether the intent is to mass-produce known lethal microbial byproducts (e.g., ricin), existing pathogens, or newly engineered pathogens, the relevant biotechnologies (e.g., fermentation techniques, DNA amplification, genetic engineering) are universally available and relatively cheap, require little space, can be implemented without the use of highly sophisticated equipment, and are essentially impossible to detect or to distinguish from legitimate use, except through their lethal consequences.

3. As with any means of mass destruction, the delivery of a weapon to an intended target is a primary constraint on its actual use. At one extreme is a thermonuclear weapon, which (at least at the national level) requires highly engineered, sophisticated rockets. Far less sophisticated engineering is required for the delivery of explosive devices, which can be efficiently delivered by means of homemade mortars, cars, and trucks as well as by a single person harnessed to a bomb. With regard to biological weapons, the primary requisite sophistication for effective delivery is in planning and execution, not in engineering.

   Whether the target is a large or small population, many biological agents can be effectively dispersed with minimum dependence on mechanical contrivance, in all probability, dependent more on simple access to a vulnerable population than anything else. Of course, in the case of viable pathogens (as opposed to toxins), access can be quite indirect because, once infected, even a single victim becomes a disease vector within the larger community.

4. Easily engineered to maximize potency, easily manufactured and hidden from detection, and easily dispersed into large and small target populations, biological agents are highly cost-effective, with killing rates equivalent to those of conventional means at comparatively tiny fractional cost and with a manifold increase in certitude.

5. Whereas modern terrorist incidents have been characterized essentially by a clearly defined instant of horror (e.g., as in the attack and collapse of the WTC twin towers, biological incidents (especially those involving the use of viable microbes) are much more likely to be discernible only after a period of several days or weeks—a period referred to as the *incubation period*, which is the time required for a disease to result in clinically defined symptoms. Depending on the length of the incubation period, as well as on the virulence and clinical severity of the disease, a terrorist incident involving biologicals would therefore tend to evolve through various distinct (though also overlapping) phases, most of which could impose profound restraints on normal social interactions.

The first phase would extend from the time of the release of the microbial agent to the time at which clinical symptoms become recognized and defined. During this period, the disease would spread surreptitiously and essentially unencumbered through the targeted population. Once clinical symptoms become obvious, a second phase would likely be characterized by a (possibly overwhelming) press on limited and most likely unprepared community hospitals and public health services, accompanied by growing public fear if not panic. Subsequent phases would be characterized by concerted efforts to determine preventive and treatment alternatives, to contain the spread of disease, and to manage secondary exposure; to secure or, as necessary, develop sufficient stockpiles of antidotes/vaccines; and finally, to implement populationwide preventive/treatment methods.

Nuclear Incidents

Potential nuclear agents are of two basic types: actual thermonuclear devices and conventional explosive devices that structurally incorporate nuclear materials (i.e., radiological dispersal device, or RDD) and therefore, which could be used to disperse hazardous nuclear materials over an extended area. It is also possible to achieve the same effects of an RDD by detonating conventional explosives in the immediate vicinity of normal sources of nuclear materials, such as nuclear power plants or transport vehicles carrying nuclear cargo. Although the access of terrorists to thermonuclear devices cannot be ruled out as a possibility, by far the more likely possibility is the terrorist use of RDDs or, as described earlier, RDD-equivalent incidents involving normal sources of nuclear materials.

The health impact of nuclear materials is due to three types of radiation characteristically emitted as a result of the natural radioactive decay of nuclear materials:

- *Alpha particles*, which are indistinguishable from the nuclei of helium atoms (2 protons and 2 neutrons). Being very heavy and relatively slow moving, alpha particles travel only a small distance (e.g., a few inches) before they become absorbed. They cannot penetrate human skin. They are dangerous, therefore, only when they enter the body through the ingestion of contaminated food or water, or the inhalation of dusts or other contaminated materials—either of which results in the direct exposure of internal body organs to alpha particles.
- *Beta particles*, which are fast moving electrons. Because they are much smaller than alpha particles and travel at very high velocity, beta particles can penetrate human skin tissue. At high levels, they can cause skin burns. As with alpha particles, the danger of beta particles is also through direct exposure of internal organs after ingestion and/or inhalation of contaminated materials.
- *Gamma radiation (or rays)*, which is a form of high-energy (i.e., high-frequency) electromagnetic radiation that is indistinguishable from energetic X-rays. Traveling at the speed of light, gamma radiation can penetrate the human body as well as most materials, causing severe and even fatal injury to tissues and organs. Early symptoms of high exposure include skin burns, nausea, vomiting, high fever, and hair loss; later symptoms include the development of various types of cancers, and diminished immunological capacity.

In an RDD-incident, first responders are at immediate risk from all three types of radiation. Where the incident involves the generation of large amounts of dust and smoke that can be wind-driven over great distances, large populations become subject to the risks attendant to the inhalation and ingestion of contaminated materials, resulting in significant, long-term interruption of normal daily life and consequent severe (and possibly overwhelming) strain on community resources and services.

Incendiary Incidents

An incendiary device is any mechanical, chemical, and/or electrical device specifically designed to start a fire. Whatever the arrangement of mechanical, chemical, and/or electrical components, any incendiary device consists of (a) an igniter or fuse, (b) a container, and (c) a flammable or combustible accelerator that, once ignited, serves as a source of fire for surrounding combustible materials. The igniter or fuse may be as simple as a lighted cigarette or a chemically impregnated fuse, or as complex as a sophisticated electrical circuit that incorporates pressure-, light-, or sound-detectors as well as radio-frequency components. Containers (for the accelerator) may be of any shape and any material and, therefore are easily camouflaged to appear as an ordinary item in any type of surrounding. Accelerators are typically liquid or solid, but may be gaseous.

The range of possible designs of incendiary devices is essentially infinite, limited only by the inventiveness of the designer. Regardless of the sophistication of actual design, essential mechanical, chemical, and electrical components are readily available at low cost, being indistinguishable from legitimate items in daily commerce. In fact, highly effective and reliable incendiary devices can easily be composed using materials and items typically found in any American home. Moreover, the knowledge needed to contrive such devices is commonplace in the printed and electronic libraries of both general and specialized references that are available to anyone.

Chemical Incidents

Chemical agents may be variously classified. For example, they may be classified on the basis of their volatility, with volatile agents being those that can be used to contaminate the atmosphere and nonvolatile agents being those that can coat surfaces. They may also be classified on the basis of whether they are intended to result in death (i.e., lethal agents) or in incapacitation of victims (e.g., nausea, disorientation, visual problems). Most often, chemical agents are classified on the basis of the types of effects they cause in victims. This is the scheme commonly used by the U.S. Department of Justice and the Federal Emergency Management Agency—a scheme that recognizes five basic classes of chemical agents:

- Nerve agents (disruption of transmission of nerve impulses)
- Blister agents (cause severe burns to eyes, skin, and respiratory tract)
- Blood agents (interfere with capacity of blood to transport oxygen)
- Choking agents (cause severe stress on respiration)
- Irritating agents (causing sufficient respiratory distress, tearing, and/or skin pain to temporarily incapacitate victim)

The development of modern biotechnologies obscures historical distinctions between biological and chemical agents that have been perpetuated by international law and convention as well as by military practice. For example, even though certain toxins and bioregulators are several thousand times more potent than even the most lethal nerve gases, they are not classed as chemical warfare agents. Because the terminology related to warfare armatoria is typically applied to the weaponry of terrorism, biologically derived toxins and bioregulators that may be used by terrorists are classified as biological agents despite their being chemicals in precisely the same physical sense as mustard gas. Also, similarly excluded from this list of agents are pesticides that, whether they are directed against plants (i.e., herbicides) or animals (e.g., piscicides, molluscicides), are also potential terrorist weapons.

On the basis of these considerations, it is imperative that emergency response personnel clearly understand that the five classes of chemical agents listed earlier represent selected categories from among a much larger number of types of chemical hazards potentially imposed by terrorist activity, and that they have been selected because they represent, for the emergency responder as well as for military personnel, immediate personal risk.

Nerve Agents

In their pure form, nerve agents are colorless, odorless, and tasteless chemicals having a wide range of volatility. They enter the body primarily through inhalation and/or absorption through the skin; however, nerve agents may also be consumed via contaminated food and water.

Poisoning is usually most rapid as a result of inhalation, which facilitates the blood's distribution of the nerve agent to target organs throughout the body. Death can occur in a matter of minutes, although (depending on the specific agent and its concentration) distinct symptoms may become evident prior to death, including the following.

Symptoms of Initial Poisoning

- Increased salivation
- Contraction of pupils, dim and blurred vision, pain in the eyes
- Runny nose and pressure in chest
- Headache and nausea
- Slurred speech and hallucinations
- Unexplained tiredness

Symptoms of Progressive Poisoning

- Uncontrollable salivation, lachrymation, urination, and defecation
- Involuntary contraction of muscles
- Excessive sweating
- Coughing and difficulty in breathing
- Abdominal pain, nausea, and vomiting
- Giddiness, anxiety, and difficulty in thinking

If the primary mechanism of entry into the body is absorption through the skin, the symptoms of poisoning may not become evident for 15 to 30 minutes after initial exposure. However, at high ambient concentration, death typically occurs within a few moments after the first symptoms appear. It is therefore crucial that the earliest symptoms be immediately recognized so that appropriate antidotes (e.g., atropine, oximes) can be administered.

Because of the extreme rapidity of the action of nerve agents, it is sometimes necessary to administer *preventive antidotes*. Preventive antidotes (e.g., pyridostigmine, diazepam) are given in the form of tablets, which require up to 30 minutes to begin having a protective effect, with maximum effect realized about two hours following ingestion. Preventive antidotes are therefore most effectively used when it is judged that there is a high likelihood of exposure to nerve agents and there is sufficient time (i.e., 2–3 hours) to take appropriate preventive action. One indication of a situation in which the use of preventive antidotes may be considered would be the on-site presence of many dead insects, birds, and other animals. Another would be the discovery of supplies of nerve gas ingredients during response operations.

Nerve agents are known by American alpha denomination (e.g., GA), common name (e.g., Tabun), and scientific name (e.g., o-ethyl dimethylamidophosphorylcyanide). Except for the letter V, the first letter in alpha denominations designates the country that first developed the agent (e.g., G: Germany), and the second letter indicates the relative order in which the agent was developed (e.g., A: first). In V-designated agents, the V stands for *venom*, and X stands for one of the chemical components in the chemical compound.

The most important nerve agents are:

- GA or Tabun [o-ethyl dimethylamidophosphorylcyanide]
- GB or Sarin [isopropyl methylphosphonofluoridate]
- GD or Soman [pinacolyl methylphosphonofluoridate]
- GF [cyclohexyl methylphosphonofluoridate]
- VS [o-ethyl s-diisopropylaminomethyl methylphosphonothiolate]

It should be noted that Sarin was used by Iraq during the Iraq-Iran war (1984–1988), and it was also used by Japanese Aum Shinrikyo cult members against fellow civilians in a Tokyo subway (1995). In this incident, cult members punctured plastic bags containing Sarin in several different subway cars. Despite this primitive mode of release, the incident resulted in a dozen deaths and the serious injury of almost 6000 additional commuters.

Nerve agent devices may be manufactured in so-called ready-to-use (unitary) form, in which state the agent is fully active and need only be released, or in binary form, in which two or more ingredients must be mixed together to produce the active agent.

Because binary devices essentially contain inactive ingredients up to the moment of mixing, they are safer to manufacture, store, and transport than are fully activated, unitary devices. They are also likely to be less reliable, because the mixing of ingredients must be held within certain constraints of temperature and concentration in order to maximize the production of the active agent—a fact that is of primary concern regarding their military use, which requires a high level of dependability and efficiency for each of a large numbers of devices. However, terrorists need not worry about quality control of large stockpiles of identical weapons, whereas the relative ease of storing and transporting could be of primary concern to terrorists in the planning and execution of a specific incident.

For example, if the objective is to kill and incapacitate a large number of people in an urban setting, it would make little sense to use an explosive device to release a nerve agent—the noise of the explosion and resultant panic and confusion would serve to disperse the target population away from lethal concentrations. The more insidious and efficient approach would be to activate strategically located devices so as to release the agent silently and otherwise as unobtrusively as possible. In such a scenario, the act of placing devices becomes of critical importance, requiring perhaps an extended period of time (e.g., months), during which time it becomes necessary to ensure that there is no premature release that would serve as warning. Radio-controlled, nonexplosive binary devices installed and camouflaged into basic infrastructure (e.g., HVC components, sewer conduit, electrical conduit, building raceways, street culverts) could serve this purpose.

Blister Agents

Also known as mustard agents, blister agents are colorless and, having a characteristic garlic or onion odor, they quickly dull the sense of smell. Rapidly penetrating clothing and skin, blister agents not only produce burn- and blister-like wounds, but also interfere with a large number of essential cellular processes in living tissue.

Examples of blister agents include:

- Mustard gas (bis-(2chloroethyl) sulfide)
- O-Mustard (bis (2-chloroethylthioethyl)ether)
- Nitrogen mustard (bis(2-chloroethyl) ethylamine)
- Lewisite 1 (2-chlorovinyldichloroarsine)
- Lewisite 2 (bis (2-chlorovinyl) chloroarsine)
- Lewisite 3 (tris (2-chlorovinyl) arsine)

The timing of appearance of symptoms after exposure depends upon the specific agent, varying from immediate to delayed appearance (e.g., 2 to 24 hours after exposure). Depending upon exposure levels of specific agents, symptoms may include:

- Aching eyes and lachrymation
- Inflammation of skin
- Skin blister and necrosis
- Irritation of mucous membranes
- Hoarseness, coughing, and sneezing
- Loss of sight
- Abdominal pain, nausea, blood-stained vomiting, and diarrhea
- Severe respiratory distress due to lung lesions; pulmonary edema
- Significant injury to bone marrow, spleen, and lymphatic tissue, with resultant diminution of immune response

Death of exposed persons typically is due to complications from agent-induced injury to lung tissue and, to a lesser extent, to secondary infections as a result of agent-mediated reduction in immunological capacity.

Nonlethal effects of low-dose exposures to blister agents are not known; however, at high, long-term doses, mustard gases and Lewisites are known to increase the risk of cancer (e.g., skin, respiratory tract) as well as other dysfunctions including chronic respiratory diseases, chronic psychological disorder, and suppression of the immunological system. Although mustard gases are classified as mutagens on the basis of animal studies, it is unknown if they present significant mutagenic risk to humans.

There is no comprehensive antidote for blister agents. Although dimercaptopropanol yields good protection against minor injuries to skin and mucous membranes, the primary form of treatment consists of removal of the victim from sources of additional exposure, decontamination of the body, and the treatment of symptoms, including the use of antibiotics against secondary infections.

Blood Agents

By interfering with the transfer of oxygen between red blood cells and body tissue, blood agents (e.g., hydrogen cyanide, cyanogen chloride) cause asphyxiation of living tissue (especially heart and brain tissue), resulting in rapid death. The primary route of entry is primarily through inhalation, but hydrogen cyanide as well as cyanide salts in solution can be absorbed through the skin.

At high concentrations of hydrogen cyanide (e.g., 300 mg/m³), death occurs within a matter of seconds. At low concentrations, distinct symptoms may progressively develop over a period of several hours and, depending upon exposure time, may include:

- Restlessness
- Increased rate of respiration
- Lachrymation
- Giddiness and headache
- Heart palpitation
- Irritation of lungs, respiratory difficulty
- Vomiting
- Convulsions
- Respiratory failure

Blood agents typically are generated by the mixture of cyanide salts (e.g., sodium cyanide, potassium cyanide) and acids (e.g., hydrochloric acid) that are readily available as common industrial chemicals. As with nerve gases, terrorist devices may easily be constructed in binary mode, with the mixing of precursors achieved by simple timing mechanisms or by radio-control devices. Because blood agents are liquids while under pressure but become gaseous at normal atmospheric pressure, unitary devices can easily be activated through simple mechanical or electrical means

Low-level cyanide poisoning can be treated medically (e.g., sodium thiosulfate, sodium nitrite, demethlaminophenol); however, it is necessary—given the rapidity of toxic effects—that such treatment be rapid. Antidotes that can be used for pretreatment are under active development.

Choking Agents

These agents (e.g., chlorine, phosgene) cause severe irritation of the lungs, with the consequence that lung tissue secretes large volumes of fluids. Because of the presence of these fluids in lung cavities (the condition known as pulmonary edema), the lung cannot function to exchange oxygen and carbon dioxide, and the victim asphyxiates—literally drowning in his or her own body fluid. In addition to severe respiratory stress, symptoms include extreme irritation of the eyes.

Choking agents are readily available as common industrial gases, and are easily stored and transported in variously sized gas bottles and cylinders.

Irritating Agents

Also called riot control agents or tear gas, irritating agents such as chloropicrin, MACE, tear gas, pepper spray, and dibenzoxazepine are used to cause respiratory distress and uncontrollable tearing of the eyes. They may also cause severe skin pain, nausea, and vomiting. Although irritating agents are designed to incapacitate rather than to kill, lethality is possible in certain circumstances, such as extremely high ambient concentration, or victim hypersensitivity.

Most irritating agents are readily available from retail markets. Many may also be easily manufactured from common industrial chemicals or standard laboratory supplies.

INCIDENT SITE AS CRIME SCENE

In 1996, the U.S. Fire Administration (USFA) undertook the development of a series of training courses on emergency response to terrorism. The objective of these courses, which are offered through FEMA (http://training.fema.gov/), is to introduce first responders to the consequences of emergency response to terrorist incidents. A self-study training program (Emergency Response to Terrorism: Self-Study) is available through the National Emergency Training Center's (NETC) Virtual Campus (http://training.fema.gov/VCNew/firstVC.htm?Submit=OK). One of the points given special emphasis in these excellent courses is the fact that *any response to an incident other than a natural disaster may be a response to a crime scene* (see Image 12.2). There are several necessary precautions that follow from this dictum, as described next.

Recognition of Warning Signs

It must be understood that a terrorist objective may be not simply to cause a particular incident (e.g., an explosion or fire), which will of course result in an emergency response, but rather, to lure responding community services into an ambush. Whether the target is the immediately involved population at the incident site or the responders to that incident, emergency personnel must be constantly alert to any warning signs of terrorist involvement.

The USFA identifies various signs or signals that may warn of the presence of lethal agents included in the five categories of incidents discussed previously, including:

Biological Incidents

- Unusual numbers of sick or dying people or animals
- Dissemination of unscheduled and unusual community spraying, especially outdoors and/or at night
- Abandoned spray devices with no distinct odors

IMAGE 12-2

NEW YORK, NY, OCTOBER 5, 2001: RESCUE WORKERS CONTINUE THEIR EFFORTS AT THE WORLD TRADE CENTER

Source: Andrea Booher/FEMA News Photo

Nuclear Incidents

- Ambient radiological monitoring data
- Presence of U.S. DOT placards/labels (e.g., in rubble or containers)

Incendiary Incidents

- Hazardous materials or lab equipment that is not relevant to the occupancy
- Exposed individuals reporting unusual odors or tastes
- Explosions that disperse liquids, mists, or gases

- Explosions that seem only to destroy a package or bomb device
- Unscheduled dissemination of an unusual spray
- Abandoned spray devices
- Numerous dead animals, fish, and birds
- Absence of insect life in a warm climate
- Mass casualties without obvious trauma
- Distinct pattern of casualties and common symptoms
- Civilian panic in potential target areas

Explosive Incidents

- Obvious large-scale structural damage (see Image 12.3)
- Blown-out windows and widely scattered debris
- Shrapnel-induced trauma
- Shock-like symptoms and/or damage to victims' eardrums

Of course, it must be emphasized that any combination of lethal agents may be employed by terrorists, as well as devices primed for release in sequenced fashion. Thus, for example, the absence of any indicator of a chemical device at the site of a fire does not mean that such a device is not present, or that one could not be activated by an internal or external trigger.

IMAGE 12-3

OKLAHOMA CITY, OK, APRIL 26, 1995: A SCENE
OF THE DEVASTATED MURRAH BUILDING FOLLOWING
THE OKLAHOMA CITY BOMBING

Source: FEMA News Photo

Entry Precautions

Given the potency of biological, chemical, and nuclear agents, and given the limited resources typically available to first responders, any suspicion of possible terrorist involvement in an incident is sufficient cause to delay entry into the incident area until additional and specialized resources are available.

The USFA has emphasized that, in the face of a determined terrorist effort, perhaps the single most important task is the decontamination of equipment, personnel, survivors, and casualties—a task that, depending on the geographical extent of the incident and the nature of the hazardous agent, can easily overwhelm community and even state resources.

The on-site evaluation of needed resources (e.g., radiological monitoring, biological agent monitoring, decontamination equipment and supplies) cannot be conducted in the absence of extensive, preincident planning between local, state, regional, and federal services and authorities, as well as critical industrial and private sector facilities and organizations.

Crime Scene Precautions

Just as the possibility of arson requires fire fighters to conduct their operations in a manner consistent with the needs of a criminal investigation, so does terrorism require similar caution for all emergency response operations. The rescue of victims and community safety are always critical objectives of on-site operations; however, the preservation of physical evidence of terrorism must be given equal priority. After all, until terrorists are apprehended, they are free to target additional victims, and they cannot be apprehended and successfully prosecuted if crucial evidence is lost or destroyed as a result of emergency response operations. No better example of the importance of the preservation of physical evidence can be had than the eminently successful investigations by U.K authorities of the London bombings (actual and attempted) of the London transport system in July 2005.

In order to ensure the preservation of critical criminal evidence, all response operations must be tightly coupled with law enforcement needs, which requires (a) extensive preincident liaison between response service and criminal investigative authorities, (b) the development of relevant SOPs by response services, and (c) intensive training of response personnel.

THREAT AND RISK TARGET ASSESSMENT

Although much emphasis has been given to the need for extensive preincident operational planning by response services in close coordination with municipal, state, regional, and federal authorities (as shown earlier), equal emphasis must be given to two types of assessment that can provide crucial input not only to the formulation of operational response plans but also of preventive strategies:

- *Threat Assessment*, which is the attempt to identify groups and organizations that may pose a terrorist threat to the community
- *Risk Target Assessment*, which attempts to identify specific facilities, activities, organizations, and groups that might become targets of terrorists.

Examples of potential terrorist groups might include (but are not limited to) ethnic separatist and émigre groups; left-wing radical organizations; right-wing racist, anti-authority, survivalist groups; foreign terrorist organizations; and single-issue oriented groups, such as animal-rights groups, extremist environmental groups, extremist religious groups, and anti-abortionists. Potential community targets might include (but are not limited to) military or governmental installations, industries that are part of the military-industrial complex, industries that manufacture environmentally sensitive products or operate in politically sensitive counties, major financial institutions, major components of social infrastructure (e.g., transportation, communication, utilities), sports arenas, shopping centers, special events (e.g., parades, public concerts), and, of course, emergency response services.

Although both Threat and Risk Target Assessments should be coordinated through local, state, regional, and federal law enforcement authorities, the primary responsibility for conducting these assessments should be assumed (ideally) by the local emergency planning and response authority, which is best situated to have detailed knowldedge of potentially senstive local targets and can take meaningful steps toward developing practical parternships among both public and private resources, services, and organizations. This approach is, of course, entirely consistent with the National Response Plan.

NATIONAL RESPONSE PLAN

The National Response Plan (NRP) is essentially the response of the federal government to the terrorist attacks of September 11, 2001 as well as to the ongoing threats of human-caused and natural hazards. This plan (March 1, 2004) grew out of the National Incident Management System (NIMS), which assures an integration of the capabilities and resources of diverse governmental agencies, incident response disciplines, nongovernmental organizations, and the private sector. The plan, which can be downloaded in its entirety from the Department of Homeland Security (DHS) (http://www.dhs.gov/interweb/assetlibrary/NRPbaseplan.pdf):

- Establishes a comprehensive, national, all-hazards approach to domestic incident management
- Presumes a nationwide means for enabling governmental and nongovernmental responders to respond to all domestic incidents
- Provides the structure and mechanisms for national-level policy and operational coordination for domestic incident management
- Does not alter or impede the ability of fedeal, state, local, or tribal departments and agencies to carry out their specific authorities

- Assumes that incidents typically are managed at the lowest possible geographical organizational and jurisdictional level

The NRP distinguishes between incidents requiring DHS coordination (Incidents of National Significance, or INS) and all other incidents that typically are handled by usual jurisdictional authorities. INS are high-impact events that require a coordinated response by some combination of federal, state, local, tribal, private-sector, and nongovernmental organizations.

Specific coordinated entities that manage INS under the NRP include:

- Incident Command Post
- Area Command (Unified Area Command)
- Local Emergency Operations Center (LEOC)
- State Emergency Operations Center(SEOC)
- Homeland Security Operations Center (HSOC)
- Interagency Incident Management Group (IIMG)
- National Response Coordination Center (NRCC)
- Regional Response Coordination Center (RRCC)
- Strategic Information and Operations Center (SIOC)
- Joint Field office (JFO)
- Join Operations Center (JOC)

The two Homeland Security Presidential Directives (HSPD) that provide key policies for implementation of the nationally coordinated, all-hazard approach for responding to emergency incidents are HSPD-5 (Management of Domestic Incidents) and HSPD-8 (National Preparedness).

EMERGENCY OPERATIONS PLAN

Whenever a response service determines that the magnitude of an incident (whether a terrorist incident or a natural diaster) is beyhond its routine response responsbility and capability, that response service must implement the community Emergency Operations Plan (EOP). The EOP is a written plan that:

- Assigns responsibility to organizations and individuals for carrying out specific actions at projected times and places
- Sets forth lines of authority and organization relationships, and shows how all actions will be coordinated
- Describes how people and property will be protected in emergencies and disasters
- Identifies personnel, equipment, facilities, supplies, and other resources available—within the jurisdiction or by agreement with other jurisdictions—for use during response and recovery operations
- Identifies steps to address mitigation concerns during response and recovery activities

As a local plan, the EOP is intended to address specific needs to be provided to local authorities by state and fedeal authorities in order to ensure the protection of the public. Although emergency response in the United States is the primary responsbility of local government, the EOP details precise procedures wherein the state provides essential assisitance:

1. Providing direct response assistance to local jurisdictions whose capabilities are overwhelmed by an emergency.
2. Providing state response services as primary response authority in certain types of emergencies.
3. Coordinating with federal authorities to secure additional assistance.

Federal emergency response assistance to state and local governments is authorized by the Federal Robert T. Stafford Diaster Relief and Emergency Assistance Act (Public Law 93-288, as amended). This assistance is provided according to the provisions of the U.S. Federal Response Plan (FRP; as amended or superceded by the NRP), which is activated when the state governor, having determined that emergency response needs exceed state resources, requests federal asssistance. Once activated, the FRP assigns federal lead agencies to coordinate federal assistance in each of 12 functional areas, as follows:

- Transportation (U.S. Department of Tranportation (DOT))
- Communications (National Communication System)
- Public Works and Engineering (U.S. Department of Defense, Army Corps of Engineers)
- Firefighting (U.S. Department of Agriculture, Forest Service)
- Information and Planning (Federal Emergency Management Agency (FEMA))
- Mass Care (American Red Cross)
- Resource Support (General Services Administration (GSA))
- Health and Medical Services (U.S. Department of Health and Human Services, Public Health Service)
- Urban Search and Rescue (FEMA)
- Hazardous Materials (Environmental Protection Agency (EPA))
- Food (U.S. Department of Agriculture, Food and Nutrition Service)
- Energy (U.S. Department of Energy (DOE))

The FRP is amended by Presidential Decision Directive 39 (PDD-39; 1995), United States Policy on Counterterrorism. If the FRP is activated in response to a state governor's request (through FEMA) for assistance in a terrorist incident, federal assistance will be provided in conformance with the provision of the Terrorism Incident Annex to the FRP. This annex to the FRP, reflecting the directives of PDD-39, assigns specific responsibility for two key aspects of operations response to any terrorist incident:

- Crisis management (law-enforcement efforts that focus on the criminal aspects of the incident)
- Consequence management (response efforts that focus on alleviating damage, loss, hardship, or suffering related to the incident)

Crisis management activities are the responsibility of the Federal Bureau of Investigation (FBI), which coordinates all relevant local, state, and federal legal authorities. Consequence management activies are the responsibility of FEMA, which coordinates federal, state, and local volunteer and private agencies

Upon a state governor's request for asssitance and a Presidential Declaration of Disaster, the sequence of eventous would be as follows:

1. FEMA would use its emergency authority to notify appropriate fedeal agencies, activate the FRP, begin coordinating the delivery of federal assistance, and establish liaison operations with the FBI.
2. The FEMA director would consult with the governor of the affected state to determine the scope and extent of the incident.
3. An emergency response team, made up of representatives from each of the primary federal agencies, would be assembled and deployed to the field to establish a disaster field office and initiate operations.

In 2002, FEMA issued additional guidance to states and local governments for updating their EOP in light of lessons learned from the 9/11/01 terrorst attacks. This guidance was issued to ensure that each jurisdiction receiving federal funds to update its EOP would adequately address all-hazards operations (see *Guide for All-Hazard Emergency Operations Planning* [SLG-101]; FEMA) with special emphasis on *weapons of mass destruction* (WMD) terrorist incidents. Also, states and local governments are advised as to how to implement the following:

- Identification and protection of critical infrastructure
- Inventory of critical respponse equipment and teams
- Interstate and intrastate mutual aid agreements
- Resource typing
- Resource standards to include interoperability protocols and a common incident command system
- State and local continuity of operations (COOP) and continuity of government (COG)
- Citizen and family preparedness, including citizen corps and other volunteer initiatives for responding to an incident

STUDY GUIDE

True or False

1. A radiological dispersal device (RDD) is a conventional explosive device that structurally incorporates nuclear materials.
2. The death of persons exposed to blister agents typically is due to complications from injury to lung tissue and, to a lesser extent, to secondary infections as a result of the reduction in immunological capacity.
3. Blood agents interfere with the transfer of oxygen between red blood cells and body tissue.

4. Risk target assessment is the attempt to identify specific facilities, activities, organizations, and groups that might become targets of terrorists.
5. There is a wide variety of antidotes for blister agents.
6. Choking agents cause severe irritation of the lungs, with the consequence that lung tissue secretes large volumes of fluids, literally resulting in the exposed person drowning in his or her own body fluid.
7. Crisis management focuses on alleviating damage, loss, hardship, or suffering related to a terrorist incident.
8. Threat assessment is the attempt to identify groups and organizations that may pose a terrorist threat to the community.
9. If a nerve agent is inhaled, death typically occurs within minutes.
10. Operational response to an act of terrorism is comprised of crisis management and consequence management.

Multiple Choice

1. Potential terrorist weapons include
   A. biological agents
   B. nuclear agents
   C. incendiary agents
   D. chemical agents
   E. explosive agents
   F. all of the above
2. Signs or signals that may warn of an impending terrorist attack using biological agents include
   A. unusual numbers of sick or dying people or animals
   B. dissemination of unscheduled and unusual sprays, especially outdoors and/or at night
   C. abandoned spray devices with no distinct odors
   D. all of the above
3. Beta particles are a form of radiation that
   A. is very heavy and relatively slow moving
   B. consists of fast-moving electrons
   C. is a form of high-energy electromagnetic radiation
4. In an RDD event, first responders are at immediate risk from
   A. alpha particles
   B. beta particles
   C. gamma radiation
   D. all of the above
5. Under the Federal Response Plan (FRP), crisis management activities are the responsibility of
   A. the FBI
   B. FEMA
   C. the EPA

Essays

1. Comment on why it is often said that biological weapons are a poor nation's weapon of mass destruction.
2. Comment on why it is important to consider that community response teams may well be the targets of terrorist attack.

Case Study

Use the Internet to review terrorist-related incidents in the United States over the past 20 years. On the basis of this review, would you say that perhaps the greatest terrorist threat in the U.S. is due to internal terrorists? Regardless of your answer, be sure that you carefully define the criteria you are using in comparing one incident to another.

PERSONNEL TRAINING

INTRODUCTION

Over the past two decades, personnel training requirements have become the key requirements in regulations related to workplace health and safety and environmental quality. Their importance derives from several distinct although interrelated factors that influence not only the American but also the global workplace:

- The need for employees to develop the skills and behavioral patterns required to achieve and maintain safe work conditions
- The right of employees to participate in decision-making that affects their well being
- The cognition by the public at large that whole communities are increasingly at risk of their continually expanding and increasingly complex industrial base

Given the diversity of health and safety regulations and the growing awareness of health and safety risks as well as of alternative methods for controlling those risks, personnel training has become a complex undertaking for even small business and, in large corporations, typically demands a significant investment of time and money. In both large and small corporations, effective personnel training is directly relevant not only to broad health and safety objectives, but also to the converging economic and marketing interests that underlie any modern business.

Whatever the objectives of any specific type of corporate training in health and safety, all such personnel training is today best viewed in the context of *corporate risk management*, which is inclusive of all corporate effort to control losses in productivity, capital resources, human resources, and market performance. However today, an effective corporate risk management program is the necessary first step not only toward achieving business objectives, *but also to achieving proactive management control of hazards that could result in communitywide emergencies.* In short, good business management practices today are based upon good hazard management practices.

PROACTIVE MANAGEMENT OF HAZARDS

The proactive management of hazards in the corporate setting is the essential first objective in any program of emergency response planning. Specifically, proactive management of workplace hazards directly impacts emergency response in two ways:

1. Lessens the likelihood that an in-plant incident will require full-fledge emergency response by both on-site first responders and community response services to protect and/or rescue personnel.
2. Lessens the likelihood that an in-plant incident will escalate into a communitywide emergency that endangers not only in-plant personnel but also the public-at-large.

It is for these reasons that *any in-plant health and safety program must be considered an essential part of both corporate and community emergency response planning.*

Deficiencies in personnel training related to human health and safety, including the health and safety not only of employees but also of the general public, clearly contribute both directly and indirectly to significant business losses, including such losses as:

- Direct health care costs for affected employees and the public
- Regulatory fines and other legal costs associated with civil and criminal proceedings related to environmental and workplace incidents
- Insurance premiums that reflect the degree of health and safety risk containment and management
- Facility audit costs associated with enforcement efforts of regulatory agencies
- Remediation costs associated with the clean-up of contaminated sites and environmental resources
- Loss of accreditation by national and international business and marketing associations, with consequent adverse impact on competitive standing within a global market
- Loss of market share due to adverse publicity generated by health and safety incidents or conditions
- Increased administrative costs due to incident reporting and follow-up, as well as recruitment and training of appropriate personnel

In light of these considerations, it is clear that an in-plant health and safety training program must first be integrated with an overall business ethos that gives the highest priority to effective health, safety, and environmental management practices—an ethos that today is rapidly becoming the essential managerial hallmark of any globally competitive enterprise and, consequently, a touchstone in modern graduate education programs in business management.

The fact that, more than 150 years after the advent of the industrial revolution, "the marketplace" has finally discovered the importance of human health and safety should not diminish the key relevance of legally enforceable and technically complex regulations. Notwithstanding the persistent debate regarding the pros and cons of governmental intrusion into boardroom deliberation, the elevation of good health and safety practices to good business practices has occurred, in fact, only after regulatory agencies caught the serious attention of business.

The number of these regulations and the range of workplace standards they establish clearly define a range of potential health and safety hazards to both employees and the public that few if any would dare refute before any objective audience.

With regard to diverse workplace health and safety standards, OSHA clearly has circumscribed certain issues that must be addressed by personnel training:

1. *Responsibility and Accountability in the Design and Day-to-Day Management of the Corporate Health and Safety Program.*

   Personnel training that does not clearly identify functional responsibilities and specific means for establishing and maintaining accountability for all policies, practices, and procedures regarding the safety of the workplace environment (as well as of environmental resources that link the workplace to the community) cannot be condoned in any circumstance and must be viewed as *prima facie* evidence that the corporation primarily is concerned with "paper compliance" with health and safety standards as opposed to the actual health and safety of its employees and the public-at-large.

2. *Behavioral Measurements of the Efficacy and Adequacy of Health and Safety Policies and Procedures.*

   The objective of any training must be objective-oriented communication, where there is always a two-way flow of information between the trainer and the persons being trained. The one-way flow of information from an instructor or, as is more commonly the case, from videotapes, from canned, computerized programs, or from pamphlets to a silent student is neither communication nor training. The only meaningful health and safety training is that which actually affects workplace behavior, and this can occur only when the training actively involves employees in the discussion of information related to their specific work-related activities and responsibilities.

3. *Active Employee Participated in All Decision-Making Regarding Health and Safety.*

   Effective personnel training must be based on the premise that health and safety are a joint objective and responsibility of both management and labor. Where health and safety practices and procedures (or the lack thereof) are perceived as emanating solely at the discretion of corporate management, it is unlikely that any personnel training program can have any measurable influence on workplace-related health or safety.

4. *The Importance of Personnel Training as a Prerequisite to Undertaking Job Assignment.*

Personnel training in health and safety practices and requirements is today an essential component of the initial in-plant processing of new employees. Although it is neither possible nor desirable to attempt to complete all health and safety training prior to undertaking actual job assignments, the company must ensure that initial training is sufficient to ensure that workers are not at special health or safety risk simply because of their status as newly assigned personnel, and that the public-at-large as well as community emergency response personnel are not at risk simply because workplace personnel are incompetent in managing in-plant operational hazards. This requires that the corporate health and safety training program be appropriately tiered or staggered to meet the needs of personnel at various stages of their employment, including the categories of newly hired, newly assigned, and temporary personnel, as well as personnel in need of refresher or advanced training, or additional training due to the implementation of new production processes or procedures.

Corporate Training and Regulatory Compliance

Health and safety training objectives too often become confused with regulatory compliance objectives—a confusion that typically reflects a misguided corporate preoccupation with doing as little as possible to comply with specific regulations, which, in turn, is an attitude that reflects an hierarchical isolation of upper level management from the realities of the modern workplace.

In the United States, for example, many executives would be surprised to learn that the Williams Steiger Occupational Safety and Health Act of 1970, which is the congressional authority for OSHA, requires "that every employer covered under the Act furnish to his employees employment and a place of employment which are free from recognized hazards that are causing or are likely to cause death or serious physical harm to his employees (29 CFR 1903.1)." Thus, *even in the absence of specific regulatory workplace standards* (e.g., Lockout/Tagout, Confined Spaces, Hazard Communication), *OSHA has the statutory authority to act to protect the health and safety of workers.*

In some jurisdictions, broad authority to ensure the health and safety of the citizen-worker is accomplished not only by legislative but also by constitutional means, as in India, where the Supreme Court in 1983 interpreted the constitutionally guaranteed right to life as requiring a healthy and safe environment, and in South Africa, where the newly elected democratic government included in its constitution the right of every citizen to an environment that is not detrimental to health and well-being. To these examples of the increasingly broad national and international mandate on behalf of human health and safety (which also presumes environmental quality) must be added the directives of the European Union, which are legally binding on its member states and which, since 1973, have specifically focused on the rights of citizens to a healthful and safe environment.

Where corporate executives understand that human health and safety and environmental quality are essential corporate objectives in an interactive and interdependent global economy, it is well established that health and safety training of personnel must ensure regulatory compliance, but must not be solely defined or constrained by (or otherwise limited to) specific regulatory requirements. In short, *regulatory requirements are best viewed as de minimus requirements that apply in all circumstances.* However, to ensure both employee and public health and safety, it is typically necessary to go well beyond published regulatory standards. To effectively integrate what may be required by written law and what is required by actual workplace (and environmental) circumstance to protect human health and safety is, accordingly, the fundamental objective of any health and safety training program.

Training Policy Document

Historically, companies have devised separate training programs to meet the legal requirements of individual regulations regarding workplace health and safety, including specific requirements for personnel training. Given the number of such regulations as well as the need for health and safety training beyond *de minimus* regulatory requirements, corporations are well advised to develop a comprehensive policy document as a basic management tool for the design, implementation, and quality control of all corporate health and safety training. Key elements of such a policy document include:

- Programs and responsibilities
- Training methods
- Scheduling constraints
- Presenters
- Training records
- General policies
- Specific programmatic requirements
- Training documentation

Programs and Responsibilities

The objectives of this section are (a) to identify precisely the individual training programs that fall within the purview of this policy document, and (b) to assign specific responsibilities for the design, content, conduct, and quality control of each program.

Programs to be included are those required by specific regulations (e.g., respiratory protection, confined spaces, hot work, blood-borne pathogens), and (b) those deemed by corporate officials and employees as appropriate to workplace and environmental circumstances or otherwise desirable but not specifically addressed by existing workplace regulations (e.g., personal hygiene and carry-home contamination, water-borne diseases).

Assigned responsibilities should include specific requirements regarding the development, review and substantive revision, and quality control of each program. Provision should also be made for the timely addition of new programs, including new topics and additional levels of training within the various programs.

Training Methods

For each training program, specific training methods should be identified on the basis of which method or combination of methods is most likely to achieve behavioral and informational objectives (see Image 13.1). Regardless of personal preferences, a comprehensive range of methods should be evaluated for efficacy, including (but not limited to):

- Classroom style lectures
- Demonstrations
- Roundtable workshops or problem solving sessions
- Seminars
- Audio-visual programs
- Topical discussions
- On-the-job practicums
- Table-top or field exercises
- Site visits to other facilities

Although on-the-job training is a valuable approach, it must be emphasized that this approach must carefully be evaluated with regard to (a) relevant regulatory requirements, (b) the risk to which the person being trained will be exposed, and (c) the risk to which the surrounding community will be exposed because of the incompetence of employees who are yet still the process of being trained.

Scheduling Constraints

Training schedules that are determined solely by routine work schedules are typically irrelevant to training objectives. The time required for a particular training session is precisely the time required to achieve specifically stated (and monitored) behavioral and informational objectives and should not be determined by any other factor. For example, although it may be convenient to train employees at the end of an eight-hour shift, it is hardly surprising that such training is most frequently a waste of time and effort. The schedule for training in each program should be established to ensure the most meaningful involvement of employees with the training exercise—an objective that can be met only by considering the type of information to be discussed, the nature of the exercise, and the mental and physical condition of the workers to be trained.

Presenters

Though many companies have tended to use consultants as trainers, the range of health and safety training is today sufficiently broad that both in-house

IMAGE 13-1

EMMITSBURG, MD, MARCH 10, 2003: EMERGENCY EDUCATION NETWORK, ONE OF THE COUNTRY'S PIONEERS IN DISTANCE LEARNING, IS LOCATED AT FEMA'S NATIONAL EMERGENCY TRAINING CENTER

Source: Jocelyn Augustineo/FEMA News Photo

personnel and external consultants should be considered for the presentation of training programs. The actual selection, of course, depends upon the type of information to be discussed and the relevance of the presenter's credentials to that type of information. In some instances, priority must be given to academic or professional credentials and, in some, to practical experience. The types and balance of the presenter's academic, professional, and experiential credentials should be specified for each training program, as well as those personal skills and attributes that are considered essential for the achievement of specific training objectives.

All presenters of personnel training programs should provide the company with a detailed resume of relevant experience as well as a syllabus for the training program and a copy of any training materials used during the presentation. It is recommended that the corporation always reserve the right to make an audio-visual recording of any health and safety training program presented by either in-house personnel or consultants, as well as the right to use any recording for purposes of documentation, quality control, and/or subsequent training purposes.

Training Records

In addition to the documents provided by each presenter (i.e., resume, syllabus, and course materials), the safety officer should maintain (at least) the following documents for each training session:

- *Training attendance form*: including the name of the program, the name of the presenter, the date of presentation, and the printed name and signature of training participants
- *Employee's training evaluation form*: to be submitted by each program participant upon completion of the training; includes a detailed assessment of the content of the training, the quality of the presentation, and the practical usefulness of the training
- *Monitor's training evaluation form*: to be completed by a designated company employee who attends the training session for the express purpose of evaluating the content and presentation of the training; usually a person with administrative or upper managerial authority

The increasing use of training evaluation forms, whether completed by training participants or by specially designated monitors, requires appropriate documentation regarding actions subsequently taken in response to those evaluations, including any revision of training session contents and the replacement of presenters. At least an annual review of all training evaluations should be conducted with appropriate documentation of findings and consequent actions.

Additional documents also may be required, such as the results of written examinations or exercises that many companies use to measure and document the efficacy of in-house training. In some instances, companies also include post-training evaluation forms that document the assessment of workplace behavior of individuals who have completed various stages of training. Documentation of *personnel actions* undertaken by the human resource department due to inappropriate employee behavior or activity specifically addressed in previous health and safety training is also often included as part of the documentation associated with that training.

General Policies

This section is devoted to those policies that must guide and inform the overall training effort, such as:

- Assessment of efficacy of training
- Programmatic review and revision
- Availability of resource information on health and safety issues
- Relationship between workplace health and safety and general lifestyle
- State-of-the-art standards and procedures

In developing these policies, the company must understand that it is increasingly subject to external legal scrutiny, especially with regard to the correspondence between written policies and the manner in which they actually are executed (or ignored) in the workplace. The basic rule to follow is the adage: "Say what you mean; mean what you say!"

Specific Programmatic Requirements

In this section, all requirements for each health and safety program (see Table 13.1) are collated, with particular emphasis given to the following:

- Regulatory reference (if any) for program
- Behavioral and informational objectives
- Personnel to be trained (by job category and work status, as in new employees, office personnel, temporary laboratory personnel)
- Frequency of presentation

TABLE 13-1

RECOMMENDED TRAINING TOPICS AND EMPHASIS (ADAPTED FROM NIOSH, USCG, AND EPA, 1985: OCCUPATIONAL SAFETY AND HEALTH GUIDANCE MANUAL FOR HAZARDOUS WASTE ACTIVITIES)

Training Topic	Emphasis of Training
Biology, Chemistry, and Physics of Hazardous Materials	Chemical and physical properties; chemical reactions;chemical compatibilties
Toxicology	Dosage; routes of exposure; toxic effects; immediately dangerous to life or health (IDLH) values; permissible exposure limits (PELs); recommended exposure limits (RELs); threshold limit values (TLVs)
Industrial Hygiene	Selection and monitoring of personal protective clothing and equipment; calculation of doses and exposure levels; evaluation of hazards; selection of worker health and safety protective measures
Rights and Responsibilities of Workers under OSHA	Applicable provisions of Title 29 of the Code of Federal Regulations
Monitoring Equipment	Functions; capabilities; selection; use; limitations; maintenance
Hazard Evaluation	Techniques of sampling assessment; evaluation of field and lab results; risk assessment

Continued

TABLE 13-1—*Continued*

RECOMMENDED TRAINING TOPICS AND EMPHASIS
(ADAPTED FROM NIOSH, USCG, AND EPA, 1985:
OCCUPATIONAL SAFETY AND HEALTH GUIDANCE
MANUAL FOR HAZARDOUS WASTE ACTIVITIES)

Training Topic	Emphasis of Training
Site Safety Plan	Safe practices; safety briefings and meetings; standard operating procedures; site safety map
Standard Operating Procedures (SOPs)	Hands-on practice; development and compliance
Engineering Controls	The use of barriers, isolation, and distance to minimize hazards
Personal Protective Clothing and Equipment (PPC & PPE)	Assignment; sizing; fit-testing; maintenance; use; limitations; hands-on training; selection of PPC and PPE; ergonomics
Medical Program	Medical monitoring; first aid; stress recognition; advanced first aid; cardiopulmonary resuscitation (CPR); emergency drills; design, planning and implementation
Decontamination	Hands-on training using simulated field conditions; design and maintenance
Legal and Reulatory Aspects	Applicable safety and health regulations (OSHA, EPA, etc.)
Emergencies	Emergency help and self-rescue; emergency drills; response to emergencies; follow-up investigation and documentation

- Method of evaluation of effectiveness
- Responsibility for design, implementation, review, and revision

Corporate Training: Special Issues

Regardless of the size of a company, the management of personnel training related to health and safety (and environmental quality) demands an important investment of time and effort that, though arguably a necessary insurance against regulatory, criminal, and civil law proceedings, is subject to numerous factors that easily can overcome the best of intent.

Some of these stubbornly difficult facts are related directly to the reality that the act of training is inextricably connected to the act of learning. Although the failure to train is very often the failure to learn, in matters related to workplace health and safety it is the corporation's responsibility to train that receives primary attention, with the consequence that an employee who refuses to learn or to change workplace behavior in accordance with good health and safety practices and who thereby suffers an injury is likely to benefit economically at the expense of the company.

It is therefore clearly incumbent upon a company not only to devise competent training programs, but also to implement stringent personnel actions whenever employees who have completed that training nonetheless fail to translate training lessons into workplace behavior. Yet, even then the company typically is constrained by a wide range of legal and societal standards that may often serve to protect a worker from the consequences of his own intransigence or incompetence—intransigence or incompetence, it must be emphasized, that place at risk not only that employee but also fellow workers and, possibly, the public at large.

Certainly we can empathize with a business manager who, unlike a teacher in a college or university, typically is blamed for the failure of someone else to learn. However, that same manager should understand that empathy is not necessarily the guarantor of sympathy. The fact remains that, in the modern world, a business does have the responsibility to make every reasonable effort to inform and instruct its employees as to the proper means for working safely—and, by proper monitoring of personnel, to ensure that they translate training into appropriate workplace behavior. Regardless of the attendant difficulties and frustrations, health and safety training and all that it implies is a basic cost of doing business. In light of the clearly dismal history of worker health and safety throughout most of the industrial revolution, we might reasonably add, "Finally!"

In undertaking its admittedly burdensome and difficult task of translating training into safe work-related behavior, any business must come to grips with two key issues that, regardless of a company's size or geographic location or industrial code, typically demand particular attention: managerial skills and the process of communication.

Professional Managerial Skills

The overall responsibility for personnel training in health and safety matters is most often given to a safety or training officer or other persons who, regardless of the extent of their technical, scientific, or other skills, are not professionally trained managers. What managerial skills they do posses typically have been obtained through limited on-the-job experience, with little if any guide or instruction by professional managers. Perceived as essentially technicians, they occupy relatively low-level and low-status positions in a corporate hierarchy that, minimizing their authority even while expanding their responsibility, effectively defines their contribution as a white-collar service function that, at best, often is seen as subservient to both mainline corporate managerial and production tasks.

More sophisticated corporations have in recent years begun to elevate the status of personnel training by assigning this function to higher level departments, such as a human resource or loss control department, or even, in a very few cases, to executive level officers, but the vast majority of companies persist in marginalizing personnel training. The consequence is that the typical safety or training officer is essentially ignorant of basic managerial skills, especially those related to the management of information, quality control, and objective-oriented systems analysis.

Consider the fact, for example, that even a small manufacturing company having on the order of 40 employees may be legally required to comply with a dozen or more relatively complex health, safety, and environmental regulations that serve not only to protect the workplace employee but also to protect the community surrounding that workplace. In addition to these regulations, the same company may have a variety of additional health, safety, and environmental training requirements imposed by the concerns of corporate executives, insurance carriers, corporate owners, and unions. In this rather common situation, which specific employees must be trained in what, to what degree or level of competence, how often, and with what measure of success or failure are fundamental questions—yet few safety officers who have mainline training responsibility can immediately provide the answers or even know how to organize a relevant database or computerize a database to generate the answers.

The typical ignorance of safety officers with respect to basic managerial skills, and the consequent ineffectiveness of much of health, safety, and environmental training programs conducted within corporations cannot be blamed on the safety officer, but rather, should be attributed to that corporate executive who considers the management of finances, productivity, raw materials, and product distribution to be inestimably more important than the management of human health and safety—that corporate executive who, despite a long reign in the history of corporations, is well poised to become an endangered species throughout the world.

The Realities of Communication

That there can be no effective training without effective communication is a bromide so logically soporific it is usually ignored in practice, especially in the United States where the Americanized English language is consider the *lingua franca* that not only overcomes all linguistic and cultural barriers but also obviates any and all distinctions imposed by diverse personal experience and values. The perception is, of course, quite wrong—as evidenced in the United States by the rapidly expanding influx of non-English speaking persons into the work force as well as by the tardy and painful recognition that many of our English-speaking fellow citizens (including some with college degrees) are in fact functionally illiterate.

The political rhetoric that bemoans this real situation as well as regulations requiring the use of English in warning signs and labels are, however, absolutely irrelevant to the fact that, for now and for the foreseeable future, corporate health and safety training must effectively confront the linguistic pluralism of the American work force, whether that pluralism derives from differences in primary language, from differences in language skills, or, for that matter, from differences in linguistic expression and cognition imposed by personal experience. To do otherwise is essentially equivalent to defining worker health and safety as a reward for social conformity rather than as a right regardless of human diversity.

The enormous difficulty inherent in the act of communication within an actual linguistic, cultural, and experiential pluralism cannot be made any the less, of course, simply by extolling the importance of the common objective of human health and safety, nor is the American experiment in linguistic diversity yet so far progressed as to give universally relevant clues as to the most efficient strategies for dealing with that difficulty. However, we do know that one does not overcome it simply by speaking English more loudly and more slowly!

We also know that the American business community, which is increasingly dependent for its very livelihood upon communication across cultural and linguistic barriers, has had to begin to divest itself of its traditional linguistic and cultural isolationism and to experiment with practical means of fostering cross-cultural and linguistic fluency.

Finally, we know that computer technology has only begun to be tapped for its contribution to human communication, whether in the university, at home, or in the business world. With a realistic understanding of the limitations of any language, with an experimental ethos directed toward achieving business objectives despite those limitations, and with a sophisticated electronic technology simply waiting to be exploited, we already perceive that perhaps our long-trusted approach to education and training is already grossly outdated and in need of drastic revision.

INCIDENT RESPONSE PERSONNEL

Except in special circumstances, it cannot be expected that in-plant first responders will be trained to the same level of expertise and competence as professional community and governmental response services. However, because of the critical role that first responders play, it is necessary that industry become more knowledgeable of the guidance and training materials available through professional response services (see Image 13.2) and take specific steps to integrate that guidance and training experience into the corporate training of, at least, in-plant initial responders and, preferably, all in-plant managers and other personnel who have mainline responsibility for materials and operations that present risk to facility personnel, environmental resources, and the surrounding community.

Of particular importance is the experience of professionals with regard to:

- Designing a comprehensive emergency response training program
- Identifying those types of training that have been proven to be highly effective for ensuring the development of emergency response skills
- Utilizing professional emergency response services, resources, and information to meet specific training objectives

DESIGN OF COMPREHENSIVE TRAINING PROGRAM

The comprehensiveness of an adequate training program in emergency planning and response must not be defined solely in terms of corporate experience with so-called "personnel training," which, after all, being inclusive of all aspects of business operations, is far more focused on normal procedures and SOPs than life-threatening incidents. The comprehensiveness of training that focuses on emergency planning and response must be based on what the professionals in emergency response deem important.

IMAGE 13-2

FEMA'S PUBLICATIONS WAREHOUSE CONTAINS BOOKS, PAMPHLETS, CD-ROMS, AND VIDEOS ON A WIDE RANGE OF EMERGENCY PREPAREDNESS TOPICS. PUBLICATIONS ARE AIMED AT BOTH PROFESSIONALS AND THE GENERAL PUBLIC

Source: Bill Reckert/FEMA News Photo

For example, Figure 13.1 is a partial listing of diverse resources provided by the U.S. National Response Team (http://www.epa.gov/superfund/programs/er/nrs/nrsnrt.htm), including handbooks, reports, bulletins, courses, videos, model exercise plans, and other documents specifically designed by professionals as effective tools for training related to emergency response.

In some instances, guidance for emergency response training is provided across a broad spectrum of types of emergencies (as in Figure 13.1); in others, detailed guidance is provided with respect to particular types of emergencies. An example of the latter type of resource that provides invaluable assistance to industries using, storing, or producing petrochemicals is the National Preparedness for Response Exercise Program (PREP), developed by the U.S. Coast Guard, EPA, the Research and Special Programs Administration in the U.S. Department of Transportation, and the Minerals Management Service. This program (see Figure 13.2) provides special guidelines for conducting response training exercises in compliance with the Oil Pollution Act of 1990.

Of course, the comprehensiveness of emergency response training is not to be determined simply by the substantive content of training related to response procedures and techniques, but also by the substantive content related to the overall design and quality control of the training effort itself. A valuable

FIGURE 13-1

EXAMPLE OF TRAINING RESOURCES AVAILABLE THROUGH U.S. NATIONAL RESPONSE TEAM (NATIONAL RESPONSE TEAM, INTERNET HOME PAGE)

National Response Team

http://www.nrt.org/nrt/home

RESOURCES

The following resources can help you to design, conduct, and evaluate exercises that test your emergency response procedures. If you are aware of additional materials that could be included in future issues, please send a complete description, including contact information, to the NRT Preparedness Committee, Mail Code 5101, U.S. Environmental Protection Agency, Washington, DC 20460.

Developing a Hazardous Materials Exercise Program: A Handbook for State and Local Officials (NRT-2). The NRT developed NRT-2 to provide guidance for the initial development of (or refinement of an existing) exercise program. It also identifies federal-level resources available to state and local officials to assist in the implementation of comprehensive exercise programs to assess their hazardous materials plans and annexes. Contact EPA's EPCRA Hotline at (800) 535-0202 / FAX: (703) 412-3333 to obtain a free copy.

The Exxon Valdez Oil Spill: A Report to the President (NRT). This report addresses the preparedness for, response to, and early lessons learned from the Exxon Valdez oil spill in Prince William Sound, Alaska. Exercise planners should review the document when developing exercises that test response procedures for major incidents. Contact EPA's EPCRA Hotline at (800) 535-0202/ FAX: (703) 412-3333 to obtain a free copy of this report.

Guide to Exercises in Chemical Emergency Preparedness Programs (EPA) - This series of three bulletins provides an overview of the major types of exercises and describes some resources currently available for conducting exercises. You can obtain free copies of Introduction to Exercises in Chemical Emergency Preparedness Programs, A Guide to Planning and Conducting Table-Top Exercises, and A Guide to Planning and Conducting Field Simulation Exercises by contacting EPA s EPCRA Hotline at (800) 535-0202 / FAX: (703) 412-3333.

Exercise Design Course (FEMA) - This course develops skills that will enable participants to train staff and to conduct an exercise that will test a community's plan and its operational response capability. Contact the Emergency Management Institute at (301) 447-1286.

EPA Region 2 Exercise Video and Manual (EPA) - A video and manual demonstrating field exercises held in the state of New Jersey features comments from Federal On-Scene Coordinators (FOSCs). Contact John Ulshoefer, EPA Region 2 at (908) 321-6620 for more information.

GSA Exercise Diskettes (GSA) - Government employees can review GSA's model exercise plan that includes handbooks for both exercise players and controllers, an exercise evaluation plan, a master scenario events list, and sample standard operating procedures for an emergency operations center. Contact Gordon Tassi, GSA Emergency Management Department at (202) 501-0900 for more information.

Hazardous Materials Exercise Evaluation Methodology (HM-EEM) and **Manual** (FEMA) - This document was designed to assist state and local governments in the comprehensive evaluation of hazardous materials exercises. It contains a series of modules prepared to evaluate major exercise objectives, so that emergency plans can be amended to reflect the lessons learned. Contact FEMA s Publications Management Office, 500 C Street, SW, Washington, DC 20472 to obtain a copy.

---

resource to industry at large is a handbook prepared by the U.S. Department of Energy (see Figure 13.3), which emphasizes alternative approaches to training and techniques for evaluating both their selection and success—precisely those elements of training that are so crucial to developing an effective in-plant emergency response capability.

Any training program for incident response personnel should, of course, consider the training materials and resources made available through the Federal Emergency Management Agency (FEMA; http://www.fema.gov/tab_education.shtm).

Proven Training Methods

The general literature is replete with learned discussions and assessments of a plethora of alternative training methods, and there can be no doubt that new methods will continue to be developed, especially with regard to the use of rapidly evolving computer and communication technologies. Given the seemingly limitless plenitude of training methods, it is understandable that trainers tend to have their favorite few. Nor is it surprising that, in industry at large, such favorites are generally those (e.g., brief lectures and video presentations) that minimize the absence of attending personnel from their main productive work.

Professional emergency response personnel also utilize lectures and video presentations in the process of their training, although the emphasis of the professional is on exercises and drills (see Table 13.2)—on their actual practice of specific skills, on the personal doing rather than on a safely removed viewing, on being where the real action is rather than on seeing someone else perform, on playing an actual role rather than on simply sitting in an audience.

There are standard distinctions made among the various types of drills and exercises by professional response services; however, it should be noted that different organizations (including industrial companies) often experiment with

---

FIGURE 13-2

OVERVIEW OF NATIONAL PREPAREDNESS FOR RESPONSE EXERCISE PROGRAM FOR CONDUCTING RESPONSE TRAINING EXERCISES IN COMPLIANCE WITH THE OIL POLLUTION ACT OF 1990 (U.S. EPA, ELECTRONIC REFERENCE LIBRARY)

U.S. EPA Emergency Response Program

**Preparedness for Response
Exercise Program**

Overview

The U.S. Coast Guard, EPA, the Research and Special Programs Administration in the U.S. Department of Transportation, and the Minerals Management Service developed the National Preparedness for Response Exercise Program (PREP) to provide guidelines for compliance with the Oil Pollution Act of 1990 (OPA) pollution response exercise requirements. These guidelines are voluntary in nature. While plan holders are not required to follow the PREP guidelines, they are still bound by the regulatory exercise requirements of the OPA and may develop their own exercise program in order to comply.

Consisting of periodic unannounced drills as required by OPA, the PREP guidelines also recommend announced drills. The guiding principles for PREP establish both internal exercises, which are conducted within the plan holder's organization, and external exercises, which extend beyond the plan holder's organization to involve other members of the response community. External exercises are separated into two categories: industry-led Area Exercises and government-initiated unannounced exercises. These exercises are designed to evaluate the entire response mechanism in a given Area to ensure adequate pollution response preparedness. The goal of PREP is to conduct approximately 20 Area exercises per year, with the intent of exercising most Areas of the country over a three-year period.

Continued

FIGURE 13-2—*Continued*

OVERVIEW OF NATIONAL PREPAREDNESS FOR RESPONSE EXERCISE PROGRAM FOR CONDUCTING RESPONSE TRAINING EXERCISES IN COMPLIANCE WITH THE OIL POLLUTION ACT OF 1990 (U.S. EPA, ELECTRONIC REFERENCE LIBRARY)

U.S. EPA Emergency Response Program
Oil Spill Prevention, Preparedness & Response

Oil Spill Training

Oil spill training is an important element in EPA's oil spill prevention and preparedness efforts. Studies indicate that a significant number of oil spills at fixed facilities are caused by operator error, such as failing to close valves or overfilling tanks during transfer operations. Because operator error is more likely to be a factor in causing spills, training and briefings are critical for the safe and proper functioning of a facility.

Training provides a number of benefits in the area of oil spill preparedness. Proper training of facility personnel can reduce the occurrence of operator-related spills and reduce the severity of impacts when a spill does occur. Training encourages up-to-date planning for the control of, and response to, an oil spill, and also helps to sharpen operating and response skills, introduces the latest ideas and techniques, and promotes interaction with the emergency response organization and familiarity with the facility's SPCC Plan. EPA offers training courses for conducting proper response measures in cases of inland oil spills as well as a drill/exercise program for oil-storage facilities.

EPA Training Requirements

EPA requires owners and operators of facilities subject to the Oil Pollution Prevention regulations to conduct training on facility-specific oil spill prevention and response measures. Under the Oil Pollution Prevention regulation, EPA requires owner/operators to instruct their personnel on the operation and maintenance of equipment to prevent discharges of oil. In addition, regulated facilities should have a designated person who is accountable for oil spill prevention and who reports to line management. The current regulations also compel facility owners or operators to conduct spill prevention briefings for their operating personnel as often as needed to ensure an adequate understanding of the SPCC Plan for that facility.

In 1994, EPA added requirements for oil spill response training for facilities that are required to prepare a facility response plan. Specifically, facility owner or operators are required to develop and implement a facility response training program if their facility is determined to pose substantial harm to the environment. According to the rule, training must be specific in nature and scope to the responsibilities of facility personnel identified in the facility response plan. In addition, facilities are required to develop and implement an oil spill drill/exercise program. The drill/exercise program is comprised of tabletop and deployment exercises that are both announced and unannounced, as well as participation in larger area drills and exercises. To satisfy the drill/exercise program, facilities may participate in the federal government's Preparedness for Response Exercise Program.

In 1991, EPA proposed revisions to the SPCC regulations to clarify the mandatory nature of the oil spill prevention training requirements and proposed several additional requirements. Specifically, EPA proposed the following spill prevention training requirements:

- All employees who are involved in oil-handling activities would be required to receive 8 hours of facility-specific training within one year of the final regulations.
- In subsequent years, employees would be required to undergo 4 hours of refresher training.
- Employees hired after the training program has been initiated would be required to receive 8 hours of facility-specific training within one week of starting work and 4 hours each subsequent year.

EPA currently is reviewing and evaluating comments received from the public on these proposed revisions.

---

and mix elements of different techniques. For example, whereas a table-top exercise usually is considered to be more of a conference than an actual drill, many organizations have developed table-top exercises into simulations that others would describe as walk-through or functional drills. The categorical name is not important—the objective is! And the objective is to practice what has been learned. . . and, then, to practice again and again, being always mindful

FIGURE 13-3

U.S. DEPARTMENT OF ENERGY HANDBOOK ON ALTERNATIVE APPROACHES TO TRAINING AND TECHNIQUES FOR EVALUATING TRAINING PROGRAMS (U.S. DEPARTMENT OF ENERGY, ELECTRONIC REFERENCE LIBRARY)

DOE HANDBOOK

Alternative Systematic Approaches to Training

U.S. Department of Energy, Washington, D.C. 20585   FSC 6910

http://www.osti.gov/html/techstds/standard/hdbk1074/hdb1074.html

Disclaimer

This document is an electronic representation of the official, printed standard. The printed document takes precedence and is available as follows:

DOE and DOE contractors: Contact Office of Scientific and Technical Information, P.O. Box 62, Oak Ridge, TN 37831; prices available from (423) 576-8401.

Public Contact: U.S. Department of Commerce, Technology Administration, National Technical Information Service, Springfield, VA 22161; (703) 487-4650. Order No. DE95006851

Table of Contents

Continued

that the next emergency (e.g., another Hurricane Katrina) may nevertheless overwhelm even the most thoroughly prepared (see Image 13.3).

Various types of training, of course, are mandated by specific regulations (see Table 13.3). However, even when not mandated, training may in fact be necessary to achieve the objective of effective emergency response. In such

FIGURE 13-3—*Continued*

U.S. DEPARTMENT OF ENERGY HANDBOOK ON ALTERNATIVE
APPROACHES TO TRAINING AND TECHNIQUES FOR EVALUATING
TRAINING PROGRAMS (U.S. DEPARTMENT OF ENERGY, ELECTRONIC
REFERENCE LIBRARY)

3.5.4 Training Program Description

3.5.5 Facility Involvement

3.6 Documentation

4. TRAINING DEVELOPMENT

4.1 Purpose

4.2 Techniques for Development

4.3 Elements of Development

4.4 Products of Development

4.5 Application

4.6 Documentation

5. TRAINING IMPLEMENTATION

5.1 PURPOSE

5.2 Techniques for Implementation

5.2.1 On-the-Job Training

5.2.2 Classroom Training

5.2.3 Individualized Instruction

5.2.4 Laboratory Training

5.2.5 Simulator Training

5.3 Elements of Implementation

5.4 Products of Implementation

5.5 Application

5.6 Documentation

6. TRAINING EVALUATION

6.1 Purpose

6.2 Methods of Evaluation

6.3 Elements of Evaluation

6.4 Products of Training Evaluation

6.5 Application

6.6 Conducting Training Evaluations

6.6.1 In-Training Evaluations

6.6.2 Training Delivery Evaluations

6.6.3 Post-Training Evaluations

6.6.4 Change Actions

6.6.5 Evaluating Facility and Industry
Operating Experience

6.6.6 Comprehensive Training
Program Evaluation

6.7 Documentation

6.7.1 Approval and Tracking of
Changes/Improvements

6.7.2 Updating Analysis Data

**APPENDIX A, FACTORS AFFECTING
TECHNIQUE SELECTION**

General Guidance - Grading Based on Hazard

Nuclear Hazard Category 1 (High-Hazard) and
2 (Moderate-Hazard) Facilities

Nuclear Hazard Category 3 (Low-Hazard) Facilities

General Guidance - Technique Selection Considerations

Key Factors

**APPENDIX B, SAMPLE TEMPLATE FOR DETERMINING
SYSTEM KNOWLEDGE AND SKILLS**

APPENDIX C, ON-THE-JOB TRAINING GUIDANCE

OJT INSTRUCTOR

GENERAL GUIDANCE

Preparation

Conduct

CONDUCT OF OPERATIONS GUIDELINES

OJT INSTRUCTIONS TO THE TRAINEE

EVALUATOR INSTRUCTIONS

PERFORMANCE EVALUATION

INSTRUCTIONS TO THE TRAINEE

instances, it is necessary to define individual training topics precisely and then to determine the most appropriate training technique (or combination of techniques) for each topic. In undertaking these tasks, corporate training personnel should be guided by the examples of professional organizations and agencies that have had to meet the same need (see Tables 13.4 and 13.5).

TABLE 13-2

BASIC FORMS OF TRAINING (ADAPTED FROM FEMA, 1996: EMERGENCY MANAGEMENT GUIDE FOR BUSINESS. FEMA ELECTRONIC LIBRARY)

Orientation & Education Session	Regularly scheduled discussion session to provide information, answer questions, and Identify needs and concerns.
Table-top Exercise	Members of the emergency management group meet in a conference room setting to discuss their responsibilities and how they would react to emergency scenarios. This is a cost-effective and efficient way to identify areas of overlap and confusion before conducting more demanding training activities.
Walk-through Drill	The emergency management group and response teams actually perform their emergency response functions. This activity generally involves more people and is more thorough than a table-top exercise.
Functional Drill	This drill tests specific functions, such as medical response, emergency notification, warning and communications procedures and equipment, though not necessarily at the same time. Personnel are asked to evaluate the systems and identify problem areas.
Evacuation Drill	Personnel walk the evacuation route to a designated area where procedures for accounting for all personnel are tested. Participants are asked to make notes as they go along of what might become a hazard during an emergency (e.g., stairways cluttered with debris; smoke in the hallways). Plans are modified accordingly.
Full-scale Exercise	A real-life emergency situation is simulated as closely as possible. This exercise involves company emergency response personnel, employees, and management and community response organizations.

Professional Emergency Response Training Services, Resources, and Information

One of the most important advantages of using the training services, resources, and information of professional response services is that such services, which must devote significant effort to evaluating the effectiveness of their own efforts, typically make these evaluations available so that others can learn from their experience. For example, the U.S. National Response Team (NRT) Preparedness Committee developed an *information exchange*, which is intended to share NRT experience (via the Internet) with both training exercises and actual incidents with the broad emergency response community (see Figure 13.4). Another example of an excellent training resource for general industry as well as governmental agencies is the Learning Resource Center (LRC) maintained by the U.S. Fire Administration (see Figure 13.5).

These and other similar services that are easily available through governmental agencies are invaluable resources and should be regularly consulted by emergency response trainers and managers throughout industry. Other gov-

IMAGE 13-3

NEW ORLEANS, LA, AUGUST 28, 2005. RESIDENTS BRINGING THEIR BELONGINGS AND LINING UP TO GET INTO THE SUPERDOME, WHICH WAS OPENED AS A HURRICANE SHELTER IN ADVANCE OF HURRICANE KATRINA

Source: Marty Bahamonde/FEMA News Photo

ernmental training services are available that meet more narrowly defined needs and/or focus primarily on the training needs of governmental agencies.

For example, the U.S., National Institute for Occupational Safety and Health (NIOSH; http://www.cdc.gov/niosh/homepage.html) has developed a computer program that simulates a major mine emergency. Known as the *Mine Emergency Response Interactive Training Simulations* (MERITS), this program serves as an important training device for command center personnel, including personnel from mining companies, labor organizations, and governmental agencies. MERITS is a highly interactive program that provides simulated emergency-related data and information to users via personal computers and the Internet, and requires input regarding both underground and surface operations.

TABLE 13-3

SELECTED STANDARDS THAT AFFECT TECHNICAL RESCUE TRAINING (ADAPTED FROM U.S. FIRE ADMINISTRATION, 1995: TECHNICAL RESCUE PROGRAM DEVELOPMENT MANUAL [FA-159])

Rescue Discipline	OSHA Standard	NFPA Standard	Comment
Confined Space	29 CFR 1910.146	None	Training requirements mandate annual entry training at a representative permit space and basic first aid training, but do not specify levels of training or minimum training proficiencies. A separate OSHA standard on hazmat operations training (29 CFR 1910.120) affects training for operations in confined space with IDLH (toxic or oxygen deficient) environments.
Collapse	None	NFPA 1470	
Water/Diving	None	None	Professional Association of Diving Instructors (PADI) and other dive organizations have standards for dive training. The American Red Cross also has water rescue training standards.
Trench	29 CFR 1926.650-.652	None	Mandates training on hazards of trench activities, including proper use of shoring, but does not establish operational training levels.
Rope	None	None	NFPA 1983 is the standard for rope to be used for rescue but does not discuss training.

TABLE 13-4

EXAMPLES OF APPROPRIATE TRAINING TOPICS FOR VARIOUS TYPES OF RESCUE (ADAPTED FROM U.S. FIRE ADMINISTRATION, 1995: TECHNICAL RESCUE PROGRAM DEVELOPMENT MANUAL [FA-159])

Rope Rescue
- Types of rope
- Types of equipment
- Types of hardware and technical gear
- Communications
- Knots, hitches, and anchors
- Lashing and picketing techniques
- Simple and complex mechanical advantage systems
- Belay techniques
- Litter rigging and evacuation techniques
- Low angle rescue
- High angle rescue
- Urban rescue operations
- Traverse techniques
- Incident command
- Self rescue techniques
- EMS considerations
- Helicopter operations

Equipment: Helmet; Boots; Leather gloves; Harness; Clothing

Continued

TABLE 13-4—*Continued*

EXAMPLES OF APPROPRIATE TRAINING TOPICS FOR VARIOUS TYPES OF RESCUE (ADAPTED FROM U.S. FIRE ADMINISTRATION, 1995: TECHNICAL RESCUE PROGRAM DEVELOPMENT MANUAL [FA-159])

Confined Space Rescue
- Types of confined spaces
- OSHA rules
- Hazard recognition
- Securing the scene
- Resources

- Atmospheric monitoring
- Incident Command
- Rescuer entry techniques
- Retrieval systems
- Rope and hardware and technical equipment

- Lockout/Tagout procedures
- Breathing apparatus equipment
- EMS and patient care considerations
- Safety and survival

Equipment: Helmet; Gloves; Boots; Clothing; Harness; Knee pads/elbow pads; Eye protection; SCBA

Trench Rescue
- Trench hazards
- Securing the scene
- Safety

- Incident command
- Equipment and Resources
- Department SOPs
- Shoring techniques

- Rigging
- EMS care
- Entry and patient removal techniques

Equipment: Helmet; Gloves; Boots; Clothing; Harness Knee pads/elbow pads; Eye protection; SCBA; Folding shovel

Structural Collapse
- Size up and command considerations
- Construction types
- Types of collapses
- Initial actions
- Dangers to rescuers
- Basic search techniques
- Advanced search techniques

- Shoring and stabilizing techniques
- Equipment and technologies for collapse rescue
- EMS and patient considerations
- Safety and psychological impact

- Critical incident stress debriefing
- Breaching concrete and steel and other barriers
- Tunneling and excavation techniques
- Hazards to rescuers
- Heavy construction equipment operations

Equipment: Helmet; Gloves; Boots; Clothing; Harness; Knee pads/elbow pads; Eye protection; SCBA; Folding shovel

Water Rescue
- Water hazards
- Ice characteristics and dangers
- Swift water hazards and hydraulic characteristics
- Reach techniques
- Throw techniques
- Row techniques
- Go techniques
- Helicopter uses
- Cold water drowning and Hypothermia

- Self rescue and survival techniques
- Rescue *vs.* recovery
- Diver support
- Search patterns and techniques
- Safety
- Incident command
- Boat operations
- Flash flood and rising water

- Contaminated bodies of water
- Ice rescue equipment and techniques
- Swift water rescue equipment and techniques
- Surf rescue equipment and techniques
- Basic water safety
- Swimming test

Equipment: Personal floatation device/life vest; Whistle; Knife or shears; Flashlight; Rope throwbag; Helmet; Gloves; Goggles/eye protection; Wet or dry suit; Suitable footwear; SCUBA gear (dive team only)

TABLE 13-5

MAJOR CATEGORIES OF TRAINING FOR DISPATCHERS
(ADAPTED FROM U.S. FIRE ADMINISTRATION, 1995:
FIRE DEPARTMENT COMMUNICATIONS MANUAL:
A BASIC GUIDE TO SYSTEM CONCEPTS AND EQUIPMENT)

Elements of Dispatcher Training
Communication Skills:
Includes the ability to listen intently, speak clearly and accurately, not jump to conclusions,
    and obtain the correct information.
Departmental Procedures:
Includes basic types of fire incidents, emergency medical incidents, responses to other
    emergencies, administrative notifications, and interagency communications.
Equipment Use:
Covers the gamut from telephones to sophisticated computer systems; today's dispatcher
    must be skilled in the use of the latest communications equipment available to department.
Interpersonal Communications and Understanding:
Addresses the need to be aware of the community; differences in verbal mannerisms can
    prevent accurate information from being communicated.

Another such program is the Comprehensive Exercises Program (CEP), developed by the Federal Emergency Management Agency (FEMA; http://www.fema.gov/), which is intended to be inclusive of comprehensive, all-hazard, risk-based multiscenario training exercises that test and evaluate the effectiveness of plans, policies, procedures, systems, and facilities used to respond to diverse emergency situations. The CEP, which is intended primarily to contribute to the development of compatible federal, state, and local Emergency Operation Plans (EOPs), is based on the concept of an emergency response partnership among federal, state, and local governmental authorities, as well as volunteer and private sector organizations. One of the major objectives of this partnership is to provide a means for sharing response-related information throughout the emergency management community.

The U.S. Fire Administration (U.S. FA) provides a wide range of training materials and information on fire response, including:

- Training courses released through U.S. FA's National Fire Academy and available from the National Technical Information Service (NTIS; http://www.ntis.gov/); industrial trainers may also be able to review course materials at state fire training schools and metropolitan fire departments.
- Other training materials and packages developed by he U.S. FA and available through NTIS
- Documents and reports that are easily integrated into a corporate training program on emergency response and that are available in hard copy and by download from U.S. FA (http://www.usfa.fema.gov/pubs)

Finally, it must be emphasized that many training resources are available not only through national, regional, and local emergency response organizations,

FIGURE 13-4

INFORMATION EXCHANGE PROGRAM FOR SHARING NATIONAL
RESPONSE TEAM EXPERIENCE WITH TRAINING EXERCISES (U.S.
NATIONAL RESPONSE TEAM, ELECTRONIC REFERENCE LIBRARY)

**September 1995
Volume 3, Issue 1**

*Prepared by the
National Response Team
Preparedness Committee*

Lessons Learned From Exercises and Incidents

A WORD FROM THE NRT

This edition of the NRT/RRT Information Exchange, *Lessons Learned from Exercises and Incidents*, was developed by the NRT Preparedness Committee as a vehicle for sharing the lessons learned from recent exercises and from actual incidents. The insight gained from one exercise can help other planners and responders further develop their own emergency response plans, and ultimately improve preparedness and response capabilities.

Agency representatives serving on the NRT Preparedness Committee collected and submitted the lessons learned described in this issue - each agency is identified in the text. In order to publish this document annually and to keep the information timely, we have developed a SURVEY FORM to collect lessons learned from recent exercises and incidents. Please take a few moments after you have been involved in an exercise to fill out this form and provide us with the lessons you have learned so that others may benefit from your experience. Use the summary of exercise objectives and explanatory notes listed on the back of the form to help us categorize the information. The exercise objectives, adapted from those found in *Developing a Hazardous Materials Exercise Program: A Handbook for State and Local Officials (NRT 2)*, correspond to the sections of this document. You will notice that there are more objectives on the form than appear in this issue.

While this effort is certainly voluntary, your input is needed for effective information exchange between and among all levels of government. The NRT Preparedness Committee encourages you to copy this document and distribute it to those likely to plan and conduct exercises in your jurisdiction. Should you have any questions or if you would like to provide feedback on this project, please contact **NRT** by fax at: **(202) 260-0154.**.

http://www.nrt.org/nrt/home

but also through the international network of practicing professionals dedicated to comprehensive and effective emergency response. For example, Figure 13.6 includes a summary of a table-top exercises developed by the Director General of the Malaysian Federal Fire and Rescue Services (Dato' Soh Chai Hock) for use in a three-day training program presented to industrial managers throughout Malaysia, but also made freely available by him to anyone who could use it in their own emergency response training program.

The broad sharing of such information and resource training materials, as exemplified by individuals, governmental agencies, and professional associations throughout the world, underscores the fact that no corporate trainer having responsibility for training in-plant emergency response personnel lacks direct, readily available (and, in most cases, free) access to extensive professional guidance and training materials.

FIGURE 13-5

LEARNING RESOURCE CENTER RESOURCES AVAILABLE FOR TRAINING PURPOSES (U.S. FIRE ADMINISTRATION, ELECTRONIC REFERENCE LIBRARY)

Learning Resource Center

United States Fire Administration

The Learning Resource Center (LRC) provides current information and resources on fire and emergency management subjects. With its collection of more than 50,000 books, reports, periodicals, and audiovisual materials, the LRC facilitates and supports student and faculty research and supplements classroom lectures and course materials.

The LRC routinely answers simple requests; e.g., an organization's telephone number and address. In response to more complex inquiries on specific subjects, the LRC will do literature searches, compile bibliographies, and, depending on length and copyright restrictions, provide documentation in the form of reports and articles. Users of this website may access the LRC's Online Card Catalog to perform their own literature searches.

Only FEMA personnel and National Emergency Training Center (NETC) students may borrow materials from the LRC. However, via interlibrary loan through local libraries, the general public can access the LRC's collection of books and research reports. Audiovisuals, magazines, and general reference materials are noncirculating.

Call (800) 638-1821 outside of Maryland or (301) 447-1030 for information on how to borrow materials from the LRC. Our email address is netclrc@fema.gov. Please include your mailing address and phone number in all email messages. Or, write to us at:

National Emergency Training Center
Learning Resource Center
16825 South Seton Avenue
Emmitsburg, MD 21727

Updated September 8, 1997

STUDY GUIDE

True or False

1. Corporate training of personnel in health and safety is best viewed as a key element of corporate risk management, which is inclusive of all corporate effort to control losses in productivity, capital resources, human resources, and market performance.
2. Personnel training must clearly identify functional responsibilities and specific means for establishing and maintaining accountability for all policies, practices, and procedures regarding the safety of the workplace.
3. The only meaningful health and safety training is that which actually affects workplace behavior.
4. OSHA has statutory authority only to enforce specific health and safety standards or regulations.
5. There is little regulatory/governmental authority regarding the content of health and safety training.

6. It is essential to connect corporate training programs with appropriate corporate personnel actions.
7. The provision of information regarding emergency procedures and policies is only one form of training and must be integrated with a variety of drills and practical exercises.
8. Training manuals and other materials generally are not available through governmental/professional organizations.

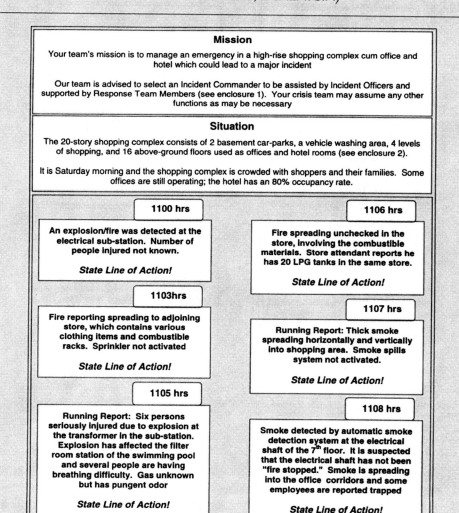

FIGURE 13-6

PORTION OF TABLE-TOP EXERCISE MATERIALS SHOWING TIMED UNITS OF INFORMATION SEQUENTIALLY PROVIDED TO CRISIS TEAM MEMBERS (ADAPTED FROM SCENARIO DESIGNED AND PROVIDED BY DATO'SOH CHAI HOCK, DIRECTOR GENERAL OF FIRE AND RESCUE SERVICES, MALAYSIA)

Mission

Your team's mission is to manage an emergency in a high-rise shopping complex cum office and hotel which could lead to a major incident

Our team is advised to select an Incident Commander to be assisted by Incident Officers and supported by Response Team Members (see enclosure 1). Your crisis team may assume any other functions as may be necessary

Situation

The 20-story shopping complex consists of 2 basement car-parks, a vehicle washing area, 4 levels of shopping, and 16 above-ground floors used as offices and hotel rooms (see enclosure 2).

It is Saturday morning and the shopping complex is crowded with shoppers and their families. Some offices are still operating; the hotel has an 80% occupancy rate.

1100 hrs

An explosion/fire was detected at the electrical sub-station. Number of people injured not known.

State Line of Action!

1103hrs

Fire reporting spreading to adjoining store, which contains various clothing items and combustible racks. Sprinkler not activated

State Line of Action!

1105 hrs

Running Report: Six persons seriously injured due to explosion at the transformer in the sub-station. Explosion has affected the filter room station of the swimming pool and several people are having breathing difficulty. Gas unknown but has pungent odor

State Line of Action!

1106 hrs

Fire spreading unchecked in the store, involving the combustible materials. Store attendant reports he has 20 LPG tanks in the same store.

State Line of Action!

1107 hrs

Running Report: Thick smoke spreading horizontally and vertically into shopping area. Smoke spills system not activated.

State Line of Action!

1108 hrs

Smoke detected by automatic smoke detection system at the electrical shaft of the 7th floor. It is suspected that the electrical shaft has not been "fire stopped." Smoke is spreading into the office corridors and some employees are reported trapped

State Line of Action!

Continued

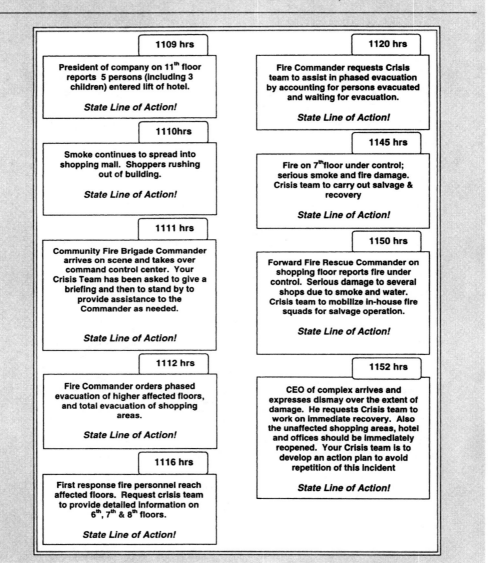

FIGURE 13-6—*Continued*

PORTION OF TABLE-TOP EXERCISE MATERIALS SHOWING TIMED
UNITS OF INFORMATION SEQUENTIALLY PROVIDED TO CRISIS TEAM
MEMBERS (ADAPTED FROM SCENARIO DESIGNED AND PROVIDED BY
DATO'SOH CHAI HOCK, DIRECTOR GENERAL OF FIRE AND
RESCUE SERVICES, MALAYSIA)

1109 hrs

President of company on 11th floor reports 5 persons (including 3 children) entered lift of hotel.

State Line of Action!

1110hrs

Smoke continues to spread into shopping mall. Shoppers rushing out of building.

State Line of Action!

1111 hrs

Community Fire Brigade Commander arrives on scene and takes over command control center. Your Crisis Team has been asked to give a briefing and then to stand by to provide assistance to the Commander as needed.

State Line of Action!

1112 hrs

Fire Commander orders phased evacuation of higher affected floors, and total evacuation of shopping areas.

State Line of Action!

1116 hrs

First response fire personnel reach affected floors. Request crisis team to provide detailed information on 6th, 7th & 8th floors.

State Line of Action!

1120 hrs

Fire Commander requests Crisis team to assist in phased evacuation by accounting for persons evacuated and waiting for evacuation.

State Line of Action!

1145 hrs

Fire on 7th floor under control; serious smoke and fire damage. Crisis team to carry out salvage & recovery

State Line of Action!

1150 hrs

Forward Fire Rescue Commander on shopping floor reports fire under control. Serious damage to several shops due to smoke and water. Crisis team to mobilize in-house fire squads for salvage operation.

State Line of Action!

1152 hrs

CEO of complex arrives and expresses dismay over the extent of damage. He requests Crisis team to work on immediate recovery. Also the unaffected shopping areas, hotel and offices should be immediately reopened. Your Crisis team is to develop an action plan to avoid repetition of this incident

State Line of Action!

9. A walk-through drill generally involves more people and is more thorough than a table-top exercise.
10. A full-scale exercise is essentially a simulated real-life emergency situation involving all appropriate in-house and community response authorities and personnel.

Multiple Choice

1. The scheduling of in-plant personnel training must be based on
   A. production schedules
   B. type of information to be discussed in training
   C. the nature of the training exercise
   D. the mental and physical condition of the workers to be trained
   E. A and C
   F. B through D
2. Appropriate trainers for a comprehensive in-plant training program may include
   A. consultants
   B. external experts
   C. local emergency authorities
   D. academics
   E. all of the above
3. In the Training Policy Document, particular attention must be given to General Policies, because
   A. they demonstrate corporate intent toward compliance with regulatory authority
   B. they establish areas of nonliability for the corporation
   C. they may be used by external authority to define actual corporate responsibility regardless of regulatory authority

Essays

1. Table 13.2 and Figure 13.6 present two alternative descriptions/forms of a table-top exercise. Compare and contrast the two, and discuss how they both play appropriate roles in an effective training program.
2. Refer to Figure 13.3. (Note: the Internet address for this table of contents is included in the figure.) Select either Chapter 5 (training implementation) or Chapter 6 (training evaluation) and, using additional information obtained from the included Internet address, relate the additional information to appropriate sections of this chapter.

Case Study

You are the Fire Response Controller for a major international oil company with import facilities located in Port of Singapore. There are eight other international oil companies with dock facilities in the same vicinity. Based on your reading and discussions of this chapter, what might be some key advantages to your integrating the emergency response training of your personnel with the training efforts of these other companies? Be very specific.

GLOSSARY

A

Abrasion   The wearing away of any part of a material by rubbing against another surface.

AC   Blood agent hydrogen cyanide.

Acclimatization   Process whereby worker is conditioned to work under high temperature and relative humidity; usually accomplished over a period of six days by gradually increasing work load and time of exposure to heat.

ACGIH   American Conference of Government Industrial Hygienists. A private organization of occupational safety and health professionals; recommends occupational exposure limits for toxic substances. ACGIH limits are not legally enforceable.

ACP   See *Area contingency plan*.

Acquired immune deficiency syndrome   A communicable disease caused by human immunodeficiency virus (HIV).

Action plan   See *Incident action plan*.

Acute   As used regarding a disease, of short duration, usually with an abrupt onset. As used regarding an illness due to chemical exposure, developing within a short time period (seconds, minutes, hours) after the exposure.

Administrative control   Any administrative procedure, policy, or protocol that results in a reduction of risk to personnel.

Advanced life support   Emergency medical treatment at an advanced level, usually provided by paramedics, and including use of drugs, cardiac monitoring/intervention, and intravenous fluids.

Aerial torch   An ignition device suspended under a helicopter, capable of dispensing ignited fuel to the ground for assistance in burnout or backfiring.

Afterflame time   The length of time for which a material (fabric) continues to flame after the ignition source has been removed.

Afterglow   Glow that persists in the material after the removal of an external ignition source or after the cessation (natural or induced) of flaming of the material.

AHE   Acute hazardous event.

AIDS   See *Acquired immune deficiency syndrome*.

AIDS related complex (ARC)   An outdated term used to describe symptoms of HIV infection in patients who have not developed AIDS. These include fatigue, diarrhea, night sweats, and enlarged lymph nodes. ARC is not included in the current Centers for Disease Control classification of HIV infection.

Airborne pathogen   Pathologic microorganism spread by droplets expelled into the air, typically through a productive cough or sneeze.

Air permeability   The rate of air flow through a material under a differential pressure between two fabric surfaces.

Air tanker   Any fixed wing aircraft certified by the U.S. Federal Aviation Administration as being capable of transport and delivery of fire retardant solutions.

Alarm signal   An identifiable audible warning that indicated that a firefighter is in need of assistance.

Allocated resources   Resources dispatched to an incident.

Alpha radiation   The least penetrating type of nuclear radiation; not considered dangerous unless alpha-contaminated particles enter the body (e.g., through inhalation); indistinguishable from the nucleus of helium atom (2 protons and 2 neutrons).

ALS   See *Advanced life support*.

Alternative response technology   Response method or technique other than mechanical containment or recovery; may include use of chemical dispersants, in-situ burning, bioremediation, or other alternatives.

Ambient concentration   The concentration of a chemical or material (e.g., dust, vapor, mist) in a volume of air or water.

Amniotic fluid   The watery fluid that surrounds the fetus or unborn child in the uterus.

Annunciator   The device of a PASS unit designed to emit the alarm signal.

ANSI   American National Standards Institute.

Anthrax   Contagious disease of warm-blooded animals caused by *Bacillus anthracis* bacterium; characterized by fever, prostration, malignant pustules on exposed skin, and internal hemorrhage.

Antibody   A component of the immune system that eliminates or counteracts a foreign substance (antigen) in the body.

Antibody positive   The result of a test or series of tests to detect antibodies in blood; a positive result means that a particular antibody is present.

Antigen   A foreign substance (including pathogens) that stimulates the production of antibodies in the immune system.

Antiviral drug   A drug that can interfere with the lifecycle of a virus.

APELL   See *Awareness and preparedness for emergencies at local level*.

Approach clothing   Protective clothing designed to provide protection from radiant heat.

ARC   See *AIDS related complex*.

Area command   In ICS, an organization established to (1) oversee the management of multiple incidents that are each being handled by an Incident Command System organization; or (2) to oversee the management of a very large incident that has multiple Incident Management Teams assigned to it. Area Command has the responsibility to set overall strategy and priorities, allocate critical resources based on priorities, ensure that incidents are properly managed, and ensure that objectives are met and strategies followed.

Area committee   In the United States, committee appointed by the President for each national area that is designated by the President under the Oil Pollution Act of 1990.

Area contingency plan   An incident response plan prepared, under authority of the U.S. Oil Pollution Act of 1990, by a presidentially designated Area Committee, to be implemented in conjunction with the National Contingency Plan to remove a worst case discharge, and to mitigate or prevent a substantial threat of such a discharge, from a vessel, offshore facility, or onshore facility operating in or near the affected area.

Area of probability   Limited area where a lost object or person is most likely to be found.

ARIP   Accidental Release Information Program

ART   See *Alternative response technology.*

Asphyxiant   Any vapor that can displace air and thereby cause suffocation.

Aspiration hazard   The danger of drawing a fluid into the lungs and causing an inflammatory response to occur.

Assigned resources   Resources checked in and assigned work tasks on an incident.

Assignments   Tasks given to resources to perform within a given operational period, based upon tactical objectives in the incident action plan.

Assisting agency   In ICS, any agency or organization that directly participated (i.e., provides tactical or service resources) in the emergency response effort; a wide range of different agencies may provide such assistance in a multi-jurisdiction or multiagency incident.

ASTM   American Society for Testing and Materials.

Asymptomatic   Having a disease-causing agent in the body but showing no outward signs of disease.

Asymptomatic HIV seropositive   The condition of testing positive for HIV antibody without showing any symptoms of the disease; a person who is HIV-positive, even without symptoms, is capable of transmitting the virus to others.

Audiogram   Hearing test to determine normal hearing capacity of individual.

Autoignition temperature   The lowest temperature at which combustion occurs in the bulk gas in a heated gas-air mixture.

Available resources   Incident-based resources that are ready for deployment.

Awareness and preparedness for emergencies at local level   A process for responding to technological accidents; a component of the United Nations Environmental Program.

AZT   The first FDA-approved drug used to treat AIDS.

B

Bacterium   A type of microorganism; some can produce disease in a suitable host. Bacteria can self-reproduce, and some forms may produce toxins harmful to their host.

Base   In ICS, the location at which primary logistics functions for an incident are coordinated and administered; there is only one base per incident. The incident command post may be collocated with the base.

Baseline audiogram   An audiogram performed to determine normal hearing capacity in the absence of any work-related hearing impairment; used to compare with results of subsequent audiograms and thereby determine any development of hearing impairment.

Basic life support   Emergency medical treatment at a level authorized to be performed by emergency medical technicians as defined by the medical authority having jurisdiction; generally refers to treatment provided at EMT-A level.

BATF   Bureau of Alcohol, Tobacco, and Firearms (U.S.).

Beta radiation   A type of nuclear radiation (fast moving electrons) that is more penetrating than alpha radiation and can damage skin tissue and harm internal organs.

Binary device   Any CW device composed of two or more chemical components (i.e., precursors) that, when mixed, produce a toxic gas or liquid; mixing may be accomplished by a wide range of mechanical or electrical forces.

Biological agents   Living organisms, or the materials derived from them, that cause disease (or harm) in humans, animals, or plants, or cause deterioration of materials. Biological agents may be found as liquid droplets, aerosols, or dry powders. A biological agent can be adapted and used as a terrorist weapon, such as anthrax, tularemia, cholera, encephalitis, plague, and botulism.

Bioremediation   Technology in which living organisms (e.g., bacteria, yeasts) are used to effect chemical changes in hazardous chemicals; can be used in both water and soil.

Bleve   A boiling liquid expanding vapor explosion, with the sudden release and ignition of a great mass of pressurized liquid into the atmosphere.

Blister agent   A chemical agent (also called a *vesicant*) that causes sever blistering and burns to eyes, skin, and tissues of the respiratory tract; exposure is through liquid or vapor contact; also referred to as mustard agent (e.g., mustard, lewisite).

Blizzard warning   Condition of alert that severe winter weather with sustained winds of at least 35 mph is expected.

Blood agent   A chemical agent that interferes with the ability of blood to transport oxygen and causes asphyxiation (e.g., hydrogen cyanide, cyanogens chloride).

BLS   See *Basic life support*.

Blood-borne pathogen   Pathologic microorganisms that are present in human blood and that can cause disease in humans; the term "blood" includes blood, blood components, and products made from human blood.

Boat-directed search   Offshore search pattern that is surface-directed by tender in boat with on-board safety diver equipped and standing by.

Body Fluid   Fluid that has been recognized by the U.S. Center for Disease Control as directly linked to the transmission of HIV and/or HBV and/or to which universal precautions apply; includes blood, semen, blood projects, vaginal secretions, cerebrospinal fluid, synovial fluid, pericardial fluid, amniotic fluid, and concentrated HIV or HBV viruses.

Body substance isolation An infection control strategy that considers all body substances potentially infectious; more stringent than universal precautions, the strategy used to control exposure to blood-borne pathogens.

Boiling point The temperature at which a liquid's vapor pressure equals the atmospheric pressure, resulting in rapid vaporization of the liquid.

Botulin toxin Extremely potent toxin typically associated with food poisoning; produced by *Clostridium botulinum* bacterium. Infection is characterized by disturbances in vision, speech, and swallowing and, within a few days, paralysis of respiratory muscles and death by suffocation.

Branch The organizational level (in ICS) having functional or geographic responsibility for major parts of incident operations; the branch level is organizationally between section and division/group in the operations section, and between section and units in the logistics section. Branches are identified by the use of roman numerals or by functional name (e.g., medical, security).

Breathability The capacity of a material to allow air permeability and water vapor transmission.

BSI See *Body substance isolation*.

Bubonic plague Contagious, often fatal epidemic disease caused by *Yersinia pestis* bacterium; transmitted from person to person by the bite of fleas from an infected host, especially a rat; characterized by chills, fever, vomiting, diarrhea, and the formation of buboes (swellings of lymph nodes).

Bursting strength The force of pressure required to rupture a textile by expanding it with a force applied perpendicular to the fabric surface.

Business Continuity Program An ongoing process supported by senior management and funded to ensure that the necessary steps are taken to identify the impact of potential losses, maintain viable recovery strategies and recovery plans, and ensure continuity of services through personnel training, plan testing, and maintenance.

BW Biological warfare.

C

Cache A predetermined complement of tools, equipment, and/or supplies stored in a designated location, available for incident use.

CAER Community awareness and emergency response.

CAMEO Computer-aided management of emergency operations; a suite of programs developed by U.S. EPA and NIOSH for management of chemical incidents.

Camp In ICS, a geographical site, within the general incident area, separate from the incident base, equipped and staffed to provide sleeping, food, water, and sanitary services to incident personnel.

Carcinogen Any material that causes cancer.

Carpal tunnel syndrome Entrapment of the median nerve of the hand and wrist in the tunnel through the carpal bones of the wrist; symptoms include finger numbness and pain upon gripping.

Carrier A person who apparently is healthy, but who is infected with some disease-causing organism that can be transmitted to another person.

Carries   Means of moving a drowning victim, under control of the rescuer, to a place of safety.

CAS   Chemical abstract service; an organization operated by the American Chemical Society that indexes information about chemicals.

Catalyst   A substance that promotes or increases the rate of a chemical reaction, but is not a reactant and is not consumed by the reaction.

Catastrophic release   A major uncontrolled emission, fire, or explosion, involving one or more highly hazardous chemicals.

Cause-consequence analysis   A hazard analysis method that combines fault tree and event tree analysis, displaying the relationships between the accident's consequences (modeled by event tree analysis) and the basic causes (modeled by fault tree analysis).

CBW   Chemical/biological warfare.

CDC   See *Centers for Disease Control*.

Center punch   Device for breaking side and rear windows of a vehicle.

Centers for Disease Control   A branch of the U.S. Public Health Service, Department of Health and Human Services, concerned with communicable disease tracking and control.

CEPP   Chemical Emergency Preparedness program.

CERCLA   Comprehensive Emergency Response, Compensation, and Liability Act (U.S.).

CFR   U.S. Code of Federal Regulations.

Chain of command   A series of management positions in order of authority.

Char length   The distance a material burns after being exposed to a flame as measured using a weight to tear the materials; specifically, the distance the material tears along the burned area is the char length.

Charring   The formation of carbonaceous residue as the result of pyrolysis or incomplete combustion.

Check-in   In ICS, the process whereby resources first report to an incident; check-in locations include incident command post (resources unit), incident base, camps, staging areas, helibases, helispots, and division supervisors (for direct line assignments).

Checklist analysis   A hazard evaluation method that uses written lists of design or operational features as a guide in assessing the process safety status of a system; can be generic in scope or specific to a type of process.

Chelation   Decontamination method involving the addition of chemical reagents that tightly bind with the target contaminant; can effectively reduce the chemical dynamics of the contaminant and/or facilitate its removal.

Chemical agent   Any of five classes of chemicals that produce incapacitation, serious injury, or death, including nerve agents, blister agents, blood agents, choking agents, and irritating agents.

Chemical asphyxiant   Known also as blood agents or blood poisons, these chemicals interrupt the flow of oxygen in the blood or the tissues in any of three ways: (1) react more readily than oxygen with the blood (e.g., carbon monoxide); (2) liberate the hemoglobin from red blood cells, resulting in a lack of oxygen transport; or (3) cause a malfunction in the oxygen-carrying capacity of red blood cells (e.g., benzene and toluene).

Chemical penetration The bulk flow of a liquid chemical through seams, closures, and openings or pores and imperfection in a clothing material.

Chemical process quantitative risk analysis A numerical evaluation of overall risk from a chemical process that combines consideration of both incident consequences and incident probabilities.

CHEMTREC Chemical Transportation Emergency Center; a nationwide service established by the Chemical Manufacturers Association to relay emergency information concerning specific chemicals that have been involved in a transportation emergency.

Cholera Infectious disease of the small intestine caused by *Vibrio cholerae* bacterium; characterized by profuse watery diarrhea, vomiting, muscle cramps, severe dehydration, and depletion of electrolytes.

Chicken pox A highly communicable disease caused by a herpes virus; commonly occurs in childhood.

Chief The ICS title for individuals responsible for command of functional sections: operations, planning, logistics, and finance/administration.

Choking agent A chemical agent (e.g., chlorine, phosgene) that causes physical injury to the lungs; in extreme cases, membranes swell and lungs become filled with liquid, which can result in asphyxiation.

Chronic As used regarding a disease, of long duration or recurring often. As used regarding an illness due to chemical exposure, developing within a long time period (years, decades) after the exposure.

Chronic carrier A person who is a long-term carrier of a disease, whether or not that person shows any symptoms of that disease; in the case of a communicable disease, often used to denote persons who can serve as the source of infection for other persons.

Circular pattern Circular or semicircular search pattern of area of probability when datum point is known.

CISD See *Critical incident stress debriefing*.

Cleaning The physical removal of dirt and debris.

Cleaning shrinkage The change in dimension of a fabric specimen after exposure to a specified cleaning process.

Clear text The use of plain English in radio communications transmissions; no ten codes or agency-specific codes are used when utilizing clear text.

Closed-circuit SCBA A recirculation-type SCBA in which the exhaled gas is rebreathed by the wearer after the carbon dioxide has been removed from the exhalation gas and the oxygen content within the system has been restored from sources such as compressed breathing gas, chemical oxygen, liquid oxygen, or compressed gaseous oxygen.

Clostridian toxin Potent toxin produced by the bacterium *Clostridium perfringens*, the causative agent of gas gangrene; characterized by slow asphyxiation and subsequent necrosis (cellular death) of living tissue.

CMV See *Cytomegalovirus*.

CNS Central nervous system (brain and spinal cord).

Colorfastness The ability of a material to retain the same color following exposure to specific physical or environmental conditions.

Combustible   Any material that burns when subjected to a temperature greater than 100°F and below 200°F.

Combustion   A chemical process of oxidation that occurs at a rate fast enough to produce heat and usually light in the form of either flames or glow.

Command post   See *Incident command post*.

Command staff   In ICS, the command staff consists of the information officer, safety officer, and liaison officer, who report directly to the incident commander; in minor incidents, the incident commander may choose not to establish a separate command staff, in which case these functions become the responsibility of the incident commander.

Communicable disease   A disease that can be transmitted from one person to another; also known as contagious disease.

Compliance audit   A documented evaluation and assessment of a facility's compliance with regulatory provisions.

Composite   The layer or combination of layers for a protective clothing item.

Compressed breathing gas   Oxygen or a respirable gas mixture stored in a compressed state and supplied to the user in gaseous form.

Conduction   For thermal protection, refers to heat transfer through a material by direct contact with another material or surface.

Confined space   An enclosed space large enough for a person to enter and perform work but which has restricted means of entry and exit (e.g., tanks, vessels, vaults).

Consequence management   As described in (U.S.) Presidential Decision Directive 39, the response to a terrorist incident, with focus on the alleviation of damage, loss, hardship, or suffering; the Federal Emergency Management Agency is the lead agency for consequence management.

Contaminant   A substance or microbial organism that poses a threat to human health or life or to the environment.

Contamination reduction corridor   Designated area of hazardous waste site within the contamination reduction zone (CRZ) where decontamination procedures are performed.

Contamination reduction zone   At a hazardous waste site, the transition area between the contaminated area and the clean area.

Contraindicated   In reference to any medicine or vaccine, determined to be actually counterproductive to health and well being in certain circumstances.

Convection   For thermal protection, refers to heat transfer through a material by air transmission.

Cooperating agency   In ICS, any agency or organization that, though not directly participating in the response effort, becomes involved in the incident to provide key services (e.g., Red Cross, Red Crescent).

Corrosive   Any material that chemically burns living tissues on contact; may be either an acid or a base (alkali).

CPQRA   See *Chemical process quantitative risk analysis*.

CPR   Cardiopulmonary resuscitation.

CRC   See *Contamination reduction corridor*.

Crisis management   As described in (U.S.) Presidential Decision Directive 39, the response to a terrorist incident, with focus on the criminal aspects of the incident; the Federal Bureau of Investigation (FBI) is the lead agency in crisis management.

Critical equipment   Chemical process equipment that is deemed to be critical to process safety.

Critical incident   Any event, circumstance, or condition that overwhelms an individual's capacity to cope; an extreme situation that causes significant and potentially long-lasting psychological effects.

Critical incident stress   Stress resulting from the experience of a critical incident and having the potential to cause significant behavioral, cognitive, emotional, and physical changes in the individual who experiences the stress.

Critical incident stress debriefing   Stress reduction processes designed to address the special needs of emergency response personnel in dealing with situations that cause strong emotional reactions or interfere with the ability to function.

Critical operating parameter   A parameter that indicated imminent approach of a major process hazard when a certain measurable level of that parameter is exceeded.

Cumulative stress disorder   The cumulative wear and tear of joints and associated tissue due to mechanical injury to muscle, nerve, tendon, and/or bone tissue; also called cumulative trauma disorder.

CRZ   See *Contamination reduction zone.*

CSD   See *Critical stress disorder.*

Cut resistance   The capacity of a material to prevent cut-through by a sharp-edged blade.

CW   Chemical warfare.

CX   Choking agent phosgene oxime.

Cyalume   Chemical light stick attached to buoyancy compensator by lanyard; to be activated and displayed on all night operations.

Cytomegalovirus   A viral infection that may occur without any symptoms or result in mild flu-like symptoms; severe CMV is shed in body fluids (urine, semen, sputum, and saliva); in the presence of immune deficiency, such as AIDS, it can also affect other internal organs and vision, sometimes leading to blindness.

D

Damage assessment   An appraisal or determination of the effects of the disaster on human, physical, economic, and natural resources.

Datum point   Last known position of lost object or person.

Debilitating illness   (or injury) A condition that temporarily or permanently prevents a person from engaging in normal work or activities.

Decibel   The unit of measurement (dB) used to express sound levels.

Decontamination   The physical and/or chemical process of reducing and preventing the spread of contamination from contaminated persons or equipment.

Deflagration   A chemical reaction where the reaction or flame front travels at less than sonic velocity.

Delta hepatitis   See *Hepatitis D.*

Demand SCBA   See *Negative pressure SCBA.*

Dermal   Pertaining to the skin.

Detoxification   Decontamination method that involves a chemically, physically, or biologically mediated change in either the molecular structure of a contaminant molecule or the molecular dynamics (e.g., acidity, alkalinity) in which the contaminant plays a role; used to reduce the hazard associated with a contaminant.

Diphtheric toxin   Potent toxin produced by the bacterium *Corynebacterium diphtheriae*, which causes tissue destruction and the formation of a gray membrane in the upper respiratory tract that can detach to cause asphyxiation; toxin may also enter into blood and subsequently damage tissues elsewhere in the body.

Direct disease transmission   The transmission of a disease from one person to another due to direct contact with infected blood, body fluids, or other infections materials.

Disease   An alteration of health, with a characteristic set of symptoms, that may affect the entire body or specific organs. Diseases have a variety of causes and are known as infectious diseases when they are due to pathogenic microorganisms such as bacteria, viruses, or fungi.

Disease vector   An organism (e.g., mosquito, tick) that plays an intermediary role in the transmission of a pathogenic organism from an infected person to a noninfected person.

Disinfection   A procedure that inactivates virtually all recognized pathogenic microorganisms, but not necessarily all microbial forms (e.g., bacterial endospores) on inanimate objects.

Dispatch center   In ICS, a facility from which resources are directly assigned to an incident.

Dispersant   A chemical agent used to break up concentrations of a hazardous material (e.g., oil spill).

Dispersion model   A quantitative model (usually computerized) that is used to make predictions concerning the spatial distance traversed by and the changing concentration of a volume of a chemical or material released into either air or water.

Division   In ICS, divisions are used to divide an incident into geographical areas of operation; a division is located between the branch and the task force/strike team; divisions are identified by alphabetic characters for horizontal applications and, often, by floor numbers when used in buildings.

Documentation unit   In ICS, functional unit within the planning section responsible for collecting, recording, and safeguarding all documents relevant to the incident.

Dose   The amount of a chemical that enters into the body relative to the weight of that body; usually measured in terms of milligrams (of chemical) per kilogram (of body weight).

Dose-effect   See *Dose-response*.

Dose-response   The direct mathematical relationship between the dose of a chemical administered to an individual or group and the probability that the individual or group will experience a defined response (effect) to the chemical.

DOT   U.S. Department of Transportation.

DWW   Dangerous when wet.

E

EAP See *Member assistance program.*

EC_{50} Effective concentration of a chemical (in ambient air or water) for 50% of the population (or the ambient concentration that presents an individual with a 0.5 probability of experiencing the defined effect).

ED_{50} Effective dose of a chemical for 50% of the population (or the dose that presents an individual with a 0.5 probability of experiencing the defined effect).

ELISA See *Enzyme-linked immunosorbent assay.*

Emergency management coordinator/director The individual within each political subdivision that has coordination responsibility for jurisdictional emergency management.

Emergency medical care The provision of treatment to patients, including first aid, cardiopulmonary resuscitation, basic life support (EMT level), advanced life support (paramedic level), and other medical procedures that occur prior to arrival at a hospital or other health care facility.

Emergency medical operations Delivery of emergency medical care and transportation prior to arrival at a hospital or other health care facility.

EMS See *Emergency medical services.*

Emergency medical services A group, department, or agency that is trained and equipped to respond in an organized manner to any emergency situation where there is the potential need for the delivery of prehospital emergency medical care and/or transportations; EMX can be provided by fire department, private, third service, or hospital-based systems or any combination thereof.

EMT See *Emergency medical technician.*

Emergency medical technician A healthcare specialist with particular skills and knowledge in prehospital emergency medicine.

Emergency operations center A predesignated facility established by an agency or jurisdiction to coordinate the overall agency or jurisdictional response and support to an emergency.

Emergency operations plan A document that (a) assigns responsibility to organizations and individuals for carrying out specific actions at projected times and places in an emergency that exceeds the capability or routine responsibility of any one agency; (b) sets forth lines of authority and organizational relationships, and shows how all actions will be coordinated; (c) describes how people and property will be protected in emergencies and disasters; (d) identifies personnel, equipment, facilities, supplies, and other resources available for use during response and recovery operations; and (e) identifies steps to address mitigation concerns during response and recovery activities.

End-of-service-time indicator A warning device on a SCBA that warns the user that the end of the service time of the SCBA is approaching.

Endothermic A chemical reaction that required the addition of heat.

Engineering control Any method of controlling exposure to a hazard by modifying the physical space in which the hazard is contained (e.g., barrier, ventilation).

Enteric precautions   A system of precautions to prevent transmission of disease by the oral/fecal route.

Entry clothing   Protective clothing that is designed to provide protection from conductive, convective, and radiant heat, and permit entry into flames.

Environmental fate   The totality of translocations (movements) and transformations that a chemical undergoes upon release to the environment.

Enzyme-linked immunosorbent assay   A test used to detect antibodies to the AIDS virus, indicating infection. For accuracy, a positive ELISA test is always repeated; if still positive, a Western blot test is then performed to confirm the diagnosis. The sensitivity and specificity of a properly performed ELISA test 12 weeks after exposure is at least 99%.

EOC   See *Emergency operations center*.

EOP   See *Emergency operations plan*.

EPA   U.S. Environmental Protection Agency.

EPCRA   Emergency Planning and Community Right-to-Know Act (U.S.).

Epidemic typhus   Any of several infectious diseases caused by *rickettsia* (e.g., *Rickettsia prowasecki*); typically transmitted by fleas, lice, or mites; characterized by severe headache, sustained high fever, depression, delirium, and the eruption of red skin rashes.

Epidemiology   The study of the incidence, distribution, and control of a disease in a population.

Ergonomics   Discipline concerned with designing plant, equipment, operations, and work environments so that they match human capabilities.

Ergonomic stress   Any stress associated with mechanical tensions in the musculoskeletal system.

Etiologic agent   A living organism (or its toxins) that may cause human disease.

Etiology   The cause(s) or origin(s) of a disease.

Evacuation drill   A training exercise in which personnel walk the evacuation route to a designated area where procedures for accounting for all personnel are tested; participants are asked to make note of what might become a hazard during an actual emergency (e.g., stairways cluttered with debris, smoke in hallways) so that appropriate modifications of the evacuation plan may be made.

Evaporative heat transfer   The process by which heat is removed from a surface by the evaporation of liquid; as applied to protective clothing, this measure refers to how easily body heat can be released from inside the clothing to the outside environment by the process of sweating, which involves both permeation of air and transmission of water vapor as fluid for transferring heat.

Event   Any planned, nonemergency activity (e.g., parades, concerts).

Event tree analysis   One of several methods of hazard analysis, which involves inductive determination of pathways of disturbances having led to a hazardous situation.

Exclusion zone   Area at hazardous waste site where contamination by hazardous chemicals has already occurred or might occur.

Exothermic   A term used to characterize the evolution of heat (e.g., from a chemical process).

Expanding square search   Compass-directed search pattern for offshore use.

Explosive   Any material that suddenly releases pressure, gas, and heat when ignited.

Explosive (Class A)   U.S. Department of Transportation classification for those substances that pose maximum explosion hazard through detonation.

Explosive (Class B)   U.S. Department of Transportation classification for those substances that ignite by rapid combustion rather than by detonation; includes fireworks, flash powders.

Explosive (Class C)   U.S. Department of Transportation classification for those substances that contain restricted quantities of Class A and/ or Class B explosives; minimum explosion hazard.

Explosive Limits   See *Upper and Lower Explosive Limit.*

Exposure   Eye, mouth, mucous membrane, skin, or parenteral contact with blood, or other body fluids, or other potentially infections material.

Extremely hazardous substance   Any of more than several hundred chemicals listed by the U.S. Environmental Protection Agency to provide a focus for state and local emergency planning.

F

Face-piece   The component(s) of a SCBA that covers, at a minimum, the wearer's nose, mouth, and eyes; also known as a facemask.

Face-shield   A transparent shield on the protective helmet that provides protection for the face and supplements primary eye protection.

Facilities unit   In ICS, a functional unit within the support branch of the logistics section that provides fixed facilities for the incident; these facilities may include the incident base, feeding areas, sleeping areas, and sanitary facilities.

False negative   Incorrect test result indicating that no antibodies or diseases are present when they are in fact present.

False positive   Incorrect test result indicating that antibodies or diseases are present when they are in fact not present.

Fan sweep   A semicircular variation of a circular search pattern, shore- or boat-based; surface-directed for down-current search use.

Fault tree analysis   One of several methods of hazard analysis, which involved deductive description of events leading from the failure of one or more components to a hazardous situation.

Federal response plan   In the United States, a federal plan that is activated when the state's resources are not sufficient to cope with a disaster and the governor has requested federal assistance.

Finance/administration section   In ICS, the section responsible for all incident costs and financial considerations; includes the time unit, procurement unit, compensation/claims unit, and cost unit.

Fire point   The lowest temperature at which a material can evolve vapors fast enough to support continuous combustion.

First-degree burn   A mild burn characterized by pain and reddening of the skin.

First responder   Person(s) who arrives first on the scene at emergency incidents and has the responsibility to act; includes fire, police, EMS, and other public safety workers.

Flame resistance   The property of a material to prevent, terminate, or inhibit combustions following application of a flaming or nonflaming source of ignition, with or without subsequent removal of the ignition source; flame resistance can be an inherent property of the material, or it may be imparted by specific treatment.

Flammable   Any material that burns when subjected to a temperature less than 100°F.

Flash fire   The combustion of a flammable vapor in which the flame front travels at less than sonic velocity, resulting in a negligible blast wave.

Flash point   The lowest temperature at which the vapor of a substance will burn or explode; not to be confused with ignition temperature.

Flood warning   Condition of alert that flooding is already occurring or will soon occur; precautions should be taken immediately; preparation should be made to move to higher ground.

Flood watch   Condition of alert that flooding is possible.

Flue gas   The air coming out of a chimney after combustion in the burner it is venting.

Fluid resistant clothing   Clothing designed and constructed to provide a barrier against accidental contact with body fluids.

FOSC   Federal on-scene commander.

FRP   See *Federal response plan.*

FSSA   Firefighter Safety Study Act (U.S.).

Fuel tender   Any vehicle capable of supplying fuel to ground or airborne equipment.

Fugitive emission   Any emission of dust, vapor, mist, or gas that is not prevented by a capture system.

Full-scale exercise   A training exercise in which a real-life emergency situation is simulated as closely as possible; involves company emergency response personnel, employees, management, and community response organizations.

Function   In ICS, function refers to the five major activities: command, operations, planning, logistics, and finance/administration. Also used when describing the activity involved, for example, the planning function.

Functional drill   A training exercise involving an intensive practicum on specific emergence response functions, such as medical response, emergency notification warning and communications procedures, and equipment; personnel generally are asked to evaluate the protocols and procedures tested during the drill and to identify any problem areas.

Fungus   A group of microorganisms that includes molds and yeasts; some fungi are pathogenic.

G

GA   Nerve agent tabun.

Gamma radiation   High-energy ionizing radiation (X-ray) that travels at the speed of light and has great penetrating power; can cause skin burns, severely injure internal organs, and have long-term physiological effects.

Gastric lavage   The washing out or irrigation of the stomach with innocuous fluids by means of a gastric tube.

GB   Nerve agent sarin.

GD   Nerve agent soman.

GEDAPER   Acronym used to describe steps in the analysis of an incident: gathering information, estimating course and harm, determining strategic goals, assessing tactical options and resources, planning and implementing actions, evaluating, and reviewing.

General exhaust   Removal of contaminated air from a large area by use of an air circulation or exchange system.

General staff   In ICS, the group of incident management personnel reporting to the incident commander, including operations section chief, planning section chief, logistics section chief, and finance/administration section chief.

Generic ICS   The description of an ICS management system that is generally applicable to any kind of incident or event.

Geographic information system   Any satellite-mediated information system that provides the user with specific on-the-ground location data.

German measles   See *Rubella*.

GF   Fluoride-containing, organophosphate nerve agent.

GIS   See *Geographic information system*.

Gonorrhea   A sexually transmitted disease caused by the bacterium *Nesseria gonorrhea*.

Ground support unit   In ICS, the functional unit within the support branch of the logistics section responsible for the fueling, maintaining, and repairing of vehicles, and the transportation of personnel and supplies.

Group   In ICS, groups are established to divide the incident into functional areas of operation; composed of resources assembled to perform a special function not necessarily within a single geographic division; located between branches (when activated) and resources in the operations section.

H

Halogen stripping   A decontamination method involving the removal of halogen atoms (e.g., chlorine, bromine, iodine) from a contaminant molecule, thus reducing the hazard associated with the contaminant.

Hanta disease   Viral infection due to any member of the genus *Hantavirus*; transmitted by rodents; characterized by flu-like symptoms and, in more severe cases, shock, kidney failure, internal bleeding, fluid accumulation in the lungs, and death.

Hazard   The potential harm or injury that may be associated with any material or situation; also defined by various governmental agencies (e.g., OSHA, USDOT) in terms of hazard classes or hazard categories (e.g., flammable, carcinogen).

Hazard analysis   Identification of individual hazards of a system, determination of the mechanisms by which they could give rise to undesirable events, and evaluation of the consequences of those events.

Hazard and operability study   One of several methods of hazard analysis carried out by application of guide words to engineering and instrument drawings to identify all deviation from design intent with undesirable effects for safety or operability, with the aim of identifying potential hazards.

Hazard class   The category used to describe the type of physical or health hazard of a chemical (e.g., flammable, mutagen); various regulatory agencies have adopted hazard classes of particular usefulness within their jurisdictional authority.

HAZOP   See *Hazard and operability study.*

HAZWOPER   U.S. OSHA safety standard for hazardous waste operations and emergency response (29 CFR 1910.120).

HBIG   See *Hepatitis B Immune Globulin.*

HBV   See *Hepatitis B.*

HCS   Hazard communication standard; U.S. OSHA regulated standard under 29 CFR 1910.1200.

HCV   See *Hepatitis C.*

HD   Blister agent distilled mustard.

HDV   See *Hepatitis D.*

Health hazard   Any property of a material or substance that either directly or indirectly can cause injury or incapacitation, either temporary or permanent, from exposure by contact, inhalation, or ingestion.

Healthcare worker   An employee of a healthcare facility including, but not limited to, nurses, physicians, dentists and other dental workers, optometrists, podiatrists, chiropractors, laboratory and blood bank technologists and technicians, research laboratory scientists, phlebotomists, dialysis personnel, paramedics, emergency medical technicians, medical examiners, morticians, housekeepers, laundry workers, and others whose work may involve direct contact with body fluids from living individuals or corpses; this definition includes firefighters, due to potential for direct contact with body fluids during firefighting, rescue, extrication, and other emergency response activities.

Health database   A compilation of records and data relating to the health experience of a group of individuals, maintained in a manner such that it is retrievable for study and analysis over a period of time.

Health promotion   Preventative health activities that identify real and potential risks in the workplace, and that inform, motivate, and otherwise help people to adopt and maintain healthy practices and lifestyles.

Heat cramps   Relatively early phase of heat stress caused by profuse sweating with inadequate replacement of electrolytes; characterized by muscle spasms and pain in hands, feet, and abdomen.

Heat exhaustion   Condition of severe dehydration and stress on body organs due to heat-induced cardiovascular insufficiency; characterized by pale, cool, moist skin, heavy sweating, dizziness, nausea, and fainting spells.**Heat resistance**   The capacity of a material to retain useful properties as measured during exposure of the material to a specified temperature and environment for a specified time.

Heat stroke   Advanced state of heat stress that requires immediate medical attention due to failure of the body's system of temperature regulation; characterized by red, hot, dry skin, lack of or reduced perspiration, nausea, dizziness or confusion, strong, rapid pulse, coma.

Helibase   In ICS, the main location for parking, fueling, maintenance, and loading of helicopters operating in support of an incident; usually located at or near the incident base.

Helicopter tender   A ground service vehicle capable of supplying fuel and support equipment to helicopters.

Helispot   Any designated location where a helicopter can safely take off and land. Some helispots may be used for loading of supplies, equipment, or personnel.

Helitack   The initial attack phase of fire suppression using helicopters and trained airborne teams to achieve immediate control of wildfires.

Helitanker   A helicopter equipped with a fixed tank or a suspended bucket type container that is used for aerial delivery of water or retardants.

Helper/suppressor T-cells   White blood cells that are part of the immune system.

Hemorrhagic fever   Type of fever characterized by profuse bleeding from internal organs and rapid wasting and death; caused by variety of viruses (e.g., Ebola virus, Marburg virus).

Hepatitis   Inflammation or swelling of the liver; can be caused by certain drugs, toxins, or infectious agents, including viruses; hepatitis caused by viruses include Hepatitis A, B, and D, and non-A non-B. Non-A non-B hepatitis includes Hepatitis C, Hepatitis E, and other, as yet unclassified types of hepatitis.

Hepatitis A   Infections hepatitis; a viral form of hepatitis normally spread by fecal contamination and generally not a significant risk for emergency care providers.

Hepatitis B (HBV)   Serum hepatitis; a viral form of hepatitis spread through blood contact, and also as a sexually transmitted disease; a significant risk for emergency care workers; infection may result in death, chronic hepatitis, liver cancer, or cirrhosis of the liver; vaccine to prevent spread of hepatitis B is available.

Hepatitis B immune globulin   A preparation that provides some temporary protection following exposure to BHV if given within seven days after exposure.

Hepatitis C (HCV)   A viral form of hepatitis spread via blood contact.

Hepatitis D (HDV)   Delta hepatitis; a viral infection occurring in people with present or past HBV infection; delta hepatitis is a complication of HBV infection and can increase the severity of HBV infection.

Hepatitis, non-A non-B (NANB)   Viral hepatitis caused by a virus other than Hepatitis A or B; there are probably several viruses included under this name. NANB hepatitis is a blood-borne infection; the cause of 90% of post-transfusion hepatitis cases.

Herpes   A family of similar viruses, which can cause different diseases, including chicken pox, zoster, cold sores, and genital herpes type II.

Herpes zoster   A painful skin rash caused by recurrence of a past case of chicken pox herpes, zoster is not typically spread person-to-person; however, persons who have not had chicken pox previously can contract chicken pox after exposure to a patient with zoster.

Hertz   Unit of measurement used to express frequency; numerically equal to cycles per second (cps).

HIV   See *Human immunodeficiency virus.*

HIV antibody screening test   A blood test that reveals the presence of antibodies to HIV.

HIV antibody positive   A test result indicating that HIV antibodies are present.

HIV antigen positive   A result of antigen testing where it has been found that HIV is present; antigen testing can be useful in predicting the progression of HIV infection and monitoring treatment.

HMTA   Hazardous Materials Transportation Act (U.S.).

HMTUSA   Hazardous Materials Transportation Uniform Safety Act (U.S.).

Host   A person that can harbor or nourish a disease-producing organism; the host is said to be infected; the host is a *carrier*.

Hot tap   A procedure that involves welding on a piece of equipment.

Hot work   Work involving electric or gas welding, cutting, brazing, or similar flame-, spark-, or heat-producing operations.

HRA   See *Human reliability analysis*.

HSPD-5   Homeland Security Presidential Directive 5. This directive (February 28, 2003) enhances the ability of the United States to manage domestic incidents by establishing a single, comprehensive national incident management system (NIMS).

HSPD-8   Homeland Security Presidential Directive 8. This directive (December 17, 2003) establishes policies to strengthen the preparedness of the United States to prevent and respond to threatened or actual domestic terrorist attacks, major disasters, and other emergencies by requiring a national domestic all-hazards preparedness goal, establishing mechanisms for improved delivery of federal preparedness assistance to state and local governments, and outlining actions to strengthen preparedness capabilities of federal, state, and local entities.

HTLV   Human T-cell lymphotropic virus, the former name for the AIDS virus; now called human immunodeficiency virus or HIV.

Human immunodeficiency virus (HIV)   The causative agent of AIDS; HIV type 1 (HIV-1) causes most cases of AIDS; a second virus, HIV-2, is a less common cause of the disease.

Human reliability analysis   A systematic evaluation of the factors that influence the performance of personnel; uses one of several types of task analyses and produces a systematic listing of the errors likely to be encountered during normal or emergency operations, a list of factors contributing to such errors, and proposed modifications to reduce the likelihood of such errors.

Human tissue burn tolerance   In testing thermal protective clothing, the capacity to withstand the amount of thermal energy that causes a second degree burn in human tissue.

Hurricane warning   Condition of alert that a hurricane will hit land within 24 hours; immediate precautions should be taken.

Hurricane watch   Condition of alert that a hurricane is possible within 24 to 36 hours.

Hypersensitivity   A phenomenon in which an individual demonstrates an abnormally high sensitivity to chemical exposure when compared with the median sensitivity of the population; hypersensitive individuals react to very small doses of the chemical.

Hyposensitivity A phenomenon in which an individual demonstrates and abnormally low sensitivity to chemical exposure when compared with the median sensitivity of the population; hyposensitive individuals react only to very large doses of the chemical.

I

IARC International Agency for Research on Cancer.

Iatrogenic Caused by the doctor; a complication, injury, or disease state resulting from medical treatment.

IC program See *Infection control program*.

ICP See *Incident command post*; integrated contingency plan.

IDLH See *Immediately dangerous to life and health*.

Ignitable Capable of burning.

Ignition temperature The lowest temperature at which a substance will burn.

Immediately dangerous to life and health The maximum level at which a healthy individual can be exposed to a chemical for 30 minutes and escape without suffering irreversible health effects of impairing systems.

Imminent hazard An act or condition that is judged to present a danger to persons or property that is so urgent and severe that it requires immediate corrective or preventive action.

Immune status The state of the body's immune system; factors affecting immune status include heredity, age, diet, and physical and mental health.

Immunization The process of rendering a person immune or highly resistant to a disease.

Immunosuppressed A condition or state of the body in which the immune system does not work normally.

Incendiary device Any mechanical, electrical, or chemical device used intentionally to start a fire.

Incident An occurrence (caused by either human or natural phenomena) that requires action by emergency service personnel to prevent or minimize loss of life or damage to property and/or natural resources.

Incident action plan Contains objectives reflecting the overall incident strategy and specific tactical actions and supporting information for the next operational period of emergency response. The plan may be oral or written; when written, the plan may have a number of forms as attachments (e.g., traffic plan, safety plan, communications play).

Incident area In ICS, geographical area of the incident, including effect area and traffic route to corresponding storage and disposal sites.

Incident base In ICS, location at the incident where the primary logistics functions are coordinated and administered; there is only one base per incident.

Incident command post In ICS, the location at which the primary command functions are executed; may be collocated with the incident base or other incident facilities.

Incident command system See *Incident management system*.

Incident commander In ICS, the person responsible for the overall coordination and direction of all activities at the incident scene.

Incident communications center   In ICS, the location of the communications unit and the message center at the incident site.

Incident management system   An organized system of roles, responsibilities, and standard operating procedures used to manage emergency operations; often referred to as incident command system or ICS.

Incident management team   In ICS, the incident commander and appropriate command and general staff personnel assigned to an incident.

Incident objectives   Statements of guidance and direction necessary for the selection of appropriate strategy(s) and the tactical direction of resources; incident objectives are based on realistic expectations of what can be accomplished when all allocated resources have been effectively deployed; incident objectives must be achievable and measurable, yet flexible enough to allow for strategic and tactical alternatives.

Incidents of national significance   High-impact events that require a coordinated and effective response by an appropriate combination of federal, state, local, tribal, private-sector, and nongovernmental entities in order to save lives, minimize damage, and provide the basis for long-term community recovery and mitigation activities.

Incubation period   The time from exposure to a disease vector until the first appearance of symptoms.

Indirect disease transmission   Transmission of a communicable disease from one person to another without direct contact.

Infection control officer   A person assigned specific responsibility for infection control practices, including immunizations and post-exposure follow-up protocols.

Infection control practitioner   A medical professional with a specialty interest in infection control.

Infection control program   IC Program; an oral or written policy and implementation of procedures related to the control of infection disease hazards where employees may be exposed to direct contact with body fluids.

Infectious   Capable of causing infection in a suitable host.

Infectious disease   An illness or disease resulting from invasion of a host by disease-producing organisms such as bacteria, viruses, fungi, or parasites.

Infectious waste   Blood and blood products, pathological wastes, microbiological wastes, and contaminated sharps.

Information officer   In ICS, the member of the command staff responsible for interfacing with the public and media or with other agencies requiring information directly from the incident; there is only one information officer per incident.

Infrared   In firefighting, a heat detection system used for fire detection, mapping, and identification of hot spots.

Infrared ground-link   A capability through the use of special mobile ground station to receive air-to-ground infrared imagery at an incident.

Initial actions   The action taken by personnel who are the first to arrive at an incident.

Initial attack   See *Initial actions* and *Initial response.*

Initial response   Resources (personnel, materials, and equipment) initially committed to an incident.

Integrated contingency plan   An emergency response plan designed in conformance with technical guidelines provided by the U.S. National Response Team to consolidate multiple plans developed by facilities in compliance with various U.S. regulations, including those pursuant to the Oil Pollution Act of 1990.

Interface area   An area of the body not protected by a protective garment, helmet, gloves, footwear, or SCBA face-piece; the area where the protective garment and the helmet, gloves, footwear, or SCBA face-piece meet.

Interface component   Item(s) designed to provide limited protection to an interface area.

Intravenous drugs   Drugs injected by needle directly into a vein.

Intubation   The introduction of a tube into a hollow organ (e.g., trachea).

IR   See *Infrared.*

Irritant   Any noncorrosive material that causes itching, soreness, or inflammation of exposed skin, eyes, or mucous membranes.

Irritating agent   A chemical agent (also known as control agent or tear gas) that causes respiratory distress and tearing designed to incapacitate (e.g., chloropicrin, MACE, tear gas, pepper spray, dibenzoxazepine).

ISO   International Standards Organization.

J

Jet fire   An ignited pressurized release of gases and/or liquids.

Joint advisory notice   A list of recommendations developed to assist employers in implementing Centers for Disease Control (CDC) guidelines.

Jurisdictional agency   The agency having jurisdiction and responsibility for a specific geographical area or for a mandated function.

L

L   Blister agent lewisite.

Ladder shank   Reinforcement to the shank area of protective footwear designed to provide additional support to the instep when the wearer is standing on a ladder rung.

Lassa fever   Often fatal viral disease endemic to West Africa; characterized by high fever, headache, ulcers of the mucous membranes, and disturbances of the gastrointestinal tract.

Latency   A period when a virus is in the body but is inactive.

LAV   See *Lymphadenopathy-associated virus.*

LC50   Lethal concentration of a toxic chemical (in ambient air or water) for 50% of the population (or the ambient concentration that presents an individual with a 0.5 probability of dying).

Leader   The ICS title for an individual responsible for a task force, strike team, or functional unit.

Leak-proof bag   A bag designed for disposal of potentially infectious substances, color coded and labeled in accordance with applicable laws.

LD50   Lethal dose of a toxic chemical for 50% of the population (or the dose that presents an individual with a 0.5 probability of dying).

LEPC   Local emergency planning commission established by the U.S. Emergency Planning and Community Right-to-Know Act.

LEPD   Local emergency planning district.

Liaison officer   In ICS, member of the command staff responsible for coordinating with representatives from cooperating and assisting agencies.

Linear pattern   Straight-line search pattern in area of probability when datum point is not known.

Line rescue   Water rescue using water rescue line, rescue skin diver, and line tender.

Line signal   Communication between diver and tender on search line by means of pulls.

Line tender   Person who directs search from shore via search line or pulls rescuer and victim to shore in surface rescue by means of water rescue line.

Local exhaust   A system for capturing and removing airborne contaminants at the point at which they are produced.

Logistics section   In ICS, the section responsible for providing facilities, services, and materials for the incident.

Lower explosive limit   The concentration of a gas or vapor in air below which a flame will not propagate if the mixture contacts an ignition source.

Lymphadenopathy-associated virus   An early name for the virus that causes AIDS; now called human immunodeficiency virus (HIV).

Lymphoma   Cancer of the lymph nodes or lymph tissues.

M

MAC   See *Multiagency coordination*.

MACS   See *Multiagency coordination system*.

Management-by-objective   In ICS, this is a top-down management activity that involves a three-step process to achieve the incident goal: establishing the incident objectives, selection of appropriate strategy(s) to achieve the objective, and the tactical direction associated with the selected strategy. Tactical direction includes selection of tactics, selection of resources, resource assignments, and performance monitoring.

Management of change   A program to identify and review all modifications of process equipment, procedures, raw materials, and processing conditions (other than replacement-in-kind) prior to implementing a change in order to minimize the possibility of a hazardous consequence of that change.

Manager   Individual within ICS organizational units who is assigned specific managerial responsibilities, e.g., staging area manager or camp manager.

Management control   See *Administrative control*.

MAP   See *Member assistance program*.

Material safety data sheet   A chemically specific compendium of health and safety information and data. In the United States, OSHA requires industrial users of chemicals to obtain MSDSs from chemical manufacturers or importers; MSDSs are also used to fulfill part of the hazardous chemical inventory reporting requirements under the Emergency Planning and Community Right-to-Know Act.

MBO   See *Management-by-objective*.

Measles   A vaccine-preventable viral communicable disease causing a skin rash; usually occurs in childhood.

Mechanical exhaust   A powered device for exhausting contaminants from a workplace, vessel, or enclosure (e.g., motor-driven fan, air/stream venture tube).

Medical unit   In ICS, functional unit within the service branch of the logistics section responsible for the development of the medical emergency plan, and for providing emergency medical treatment of incident personnel.

Member assistance program   A program designed to provide assistance to employees regarding personal problems (including drug and alcohol abuse) that may interfere with on-the-job performance; may also be used to provide training to employees and their families on health-related issues; sometimes know as an employee assistance program.

Meningitis   An infection of the meninges, the covering layers of the brain and spinal cord; may be caused by a bacterium or virus; considered a communicable disease.

Message center   In ICS, part of the incident communications center and is co-located or placed adjacent to it; receives, records, and routes information about resources reporting to the incident, resource status, and administrative and tactical traffic.

Mobilization   The process and procedures used by all organization s for activating, assembling, and transporting all resources that have been requested to respond to or support an incident.

Mobilization center   In ICS, an off-incident location at which emergency service personnel and equipment are located temporarily pending assignment, release, or reassignment.

MMWR   See *Morbidity and mortality weekly report.*

Morbidity and mortality weekly report   A weekly publication form the U.S. Centers for Disease Control presenting up-to-date information on communicable diseases.

Motion detector   An integral portion of the personal alarm safety system (PASS) that senses movement, or alternatively, lack of movement, and activates the alarm signal under a specified sequence of events.

MSDS   See *Material safety data sheet.*

MSHA   Mine Safety and Health Administration of the U.S. Department of Labor.

Mucous membrane   The lining of the nose, mouth, eyes, vagina, and rectum; mucous membranes are not as durable as other skin; contact of infected body fluids with intact mucous membranes may transmit disease.

Multiagency coordination   A generalized term that describes the function and activities of representatives of involved agencies and/or jurisdictions who come together to make decisions regarding the prioritizing of incidents and the sharing and use of critical resources. The MAC organization is not a part of the on-scene ICS and is not involved in developing incident strategy or tactics.

Multiagency coordination system   The combination of personnel, facilities, equipment, procedures, and communications integrated into a common system. When activated, the MACS has the responsibility for coordination of assisting agency resources and support in a multiagency or multijurisdictional environment. A MAC group functions within the MACS.

Multiagency incident   An incident where one or more agencies assist a jurisdictional agency or agencies; may be single or unified command.

Multijurisdiction incident   An incident requiring action from multiple agencies that have a statutory responsibility for incident mitigation. In ICS, these incidents will be managed under unified command.

Mutual aid agreement   Written agreement between agencies and/or jurisdiction in which they agree to assist one another upon request by furnishing personnel and/or equipment.

Mumps   A vaccine-preventable communicable disease caused by a virus; usually occurring in children; may cause serious complication in adult cases.

Musculoskeletal   Pertaining to muscles, bones, and joints.

Mutagen   Any material that causes changes in genetic information that is inherited from generation to generation.

Myositis   Inflammation of a muscle caused by heavy use or repeated use of a muscle with inadequate time for recovery.

N

NAERG   North American Emergency Response Guidebook.

NANB Hepatitis   See *Hepatitis, non-A non-B.*

National incident management system   The nation's first standardized management approach that unifies federal, state, and local lines of government for incident response; provides nationwide template enabling federal, state, local, and tribal governments and private-sector and nongovernmental organizations to work together effectively and efficiently to prevent, prepare for, respond to, and recover from domestic incidents regardless of cause, size, or complexity.

National interagency incident management system   An NWCG-developed program consisting of five major subsystems that collectively provide a total systems approach to all-risk incident management. The subsystems are the incident command system, training, qualifications and certification, supporting technologies, and publications management.

National preparedness goal   Establishes measurable readiness priorities and targets that appropriately balance the potential threat and magnitude of terrorist attacks, major disaster, and other emergencies with the resources required to prevent, respond to, and recover from it; also includes readiness metrics and elements that support the national preparedness goal, including standards for preparedness assessments and strategies, and a system for assessing the nation's overall preparedness to respond to major events, especially those involving acts of terrorism.

National response center   The central U.S. clearinghouse for information involving emergency spills and other releases of oil and hazardous substances.

National response plan   Establishes a comprehensive all-hazards approach to enhance the ability of the United States to manage domestic incidents; it provides mechanisms for expedited and proactive federal support to ensure critical life-saving assistance and incident containment capabilities are in place to respond quickly and efficiently to catastrophic incidents. These are high-impact, low probability incidents, including natural disasters and terrorist attacks that result in extraordinary levels of mass casualties, damage, or disruption severely affecting the population, infrastructure, environment, economy, national morale, and/or government functions.

National response team   A consortium of 14 U.S. federal agencies devoted to preparing for and responding to significant releases of oil and hazardous substances to the environment; maintains and operates the National Response Center for receiving reports of releases of oil and hazardous substances.

National wildfire coordinating group   A group formed under the direction of the Secretaries of the U.S. Departments of the Interior and Agriculture to improve the coordination and effectiveness of wildland fire activities, and provide a forum to discuss, recommend appropriate action, or resolve issues and problems of substantive mature. The NWCG has been a primary supporter of ICS development and training.

Needle-stick   A parenteral exposure with a needle contaminated from patient use.

Nerve agent   A substance that interferes with the central nervous system (e.g., Sarin, Soman, tabun, and VX agent); exposure is primarily through contact with the liquid and, secondarily, through inhalation of the vapor; three distinct symptoms associated with nerve agents are pinpoint pupils, extreme headache, and severe tightness in the chest.

Neutralization   A decontamination process involving the addition of acids or bases to a contaminant in order to lessen its corrosive nature.

NFIRS   National fire incident reporting system.

NFPA   National Fire Protection Association.

NIIMS   See *National interagency incident management system.*

NIMS   See *National incident management system.*

NIOSH   National Institute for Occupational Safety and Health (U.S. government).

NIOSH/MSHA   Certified tested and certified jointly by the National Institute for Occupational Safety and Health (NIOSH) of the U.S. Department of Health and Human Services and the Mine Safety and Health Administration (MSHA) of the U.S. Department of Labor.

NITROX   Oxygen-enriched gas mixture as defined by NOAA standards for increased physiological or time advantage in dive operations.

NOAA   National Oceanic and Atmospheric Administration (U.S.).

NOAA weather station   A mobile weather data collection and forecasting facility (including personnel) provided by the National Oceanic and Atmospheric Administration that can be utilized within the incident area.

N.O.S.   Not otherwise specified (used extensively in U.S. Department of Transportation regulations pursuant to Hazardous Materials Transportation Act).

Nosoacusis   Hearing loss caused by medical abnormalities (e.g., hereditary progressive deafness; diseases such as mumps, rubella, Meniere's disease; ototoxic drugs and chemicals; blows to the head).

Nosocomial   Originating in the hospital; a disease spread by contact with the healthcare system.

Notification (procedure)   Inclusive of all procedures designed to inform all responsible persons in the event of a potential or actual incident, and to provide those persons with information required for their proper performance of emergency-related functions.

NRC   See *National response center.*

NRP   See *National response plan.*

NRT   See *National response team.*

NWCG   See *National wildfire coordinating group.*

O

OCC   See *Operations coordination center.*

Occupational exposure   Reasonably anticipated skin, eye, mucous membrane, or parenteral contact with blood or other potentially infectious materials that may result from the performance of an employee's duties; this definition excludes incidental exposures that may take place on the job, that are neither reasonable nor routinely expected, and that the worker is not required to incur in the normal course of employment.

Occupational illness   An illness or disease contracted through or aggravated by the performance of job-related duties.

Odor threshold   The lowest concentration of a substance's vapor that can be smelled; high variable, depending on the individual's sensitivity.

OECD   See *Organization for Economic Cooperation and Development.*

Officer   The ICS title for the personnel responsible for the command staff positions of safety, liaison, and information.

Olfactory   Relating to the sense of smell.

Open-circuit SCBA   SCBA in which exhalation is vented to the atmosphere and not rebreathed. There are two types of open-circuit SCBA: negative pressure (or demand type) and positive pressure (or pressure demand type).

Operational period   In ICS, the period of time scheduled for execution of a given set of operational actions as specified in the incident action plan; operational periods can be of various lengths, although usually not over 24 hours.

Operations section   In ICS, the section responsible for all tactical operations at the incident; includes branches, divisions and/or groups, task forces, strike teams, single resource, and staging areas.

Operations coordination center   The primary facility of the multiagency coordination system housing the staff and equipment necessary to perform MACS functions.

Opportunistic infection   Infection that usually is warded off by a healthy immune system.

Oral   Relating to the mouth.

Organic peroxide   Any material that spontaneously explodes due to the formation of unstable peroxides.

Organization for Economic Cooperation and Development   An intergovernmental organization in which 24 industrialized countries from North America, Western Europe, and the Pacific meet to compare, coordinate, and, where appropriate, harmonize national policies, discuss issues of mutual concern, and wok together to respond to problems with international dimensions.

ORM   Otherwise regulated material; acronym used by U.S. DOT in regulations pursuant to the Hazardous Materials Transportation Act.

ORM-A U.S. DOT classification applied to a material that has an anesthetic, irritating, noxious, toxic, or other similar property and that can cause extreme annoyance or discomfort to passengers and crew in the event of leakage during transportation.

ORM-B U.S. DOT classification applied to a material (including a solid when wet with water) capable of causing significant damage to a transport vehicle or vessel by leaking during transportation.

ORM-C U.S. DOT classification applied to a material that has other inherent characteristics not described by ORM-A or ORM-B classes, but that make it unsuitable for shipment unless properly identified and prepared for transportation.

ORM-D U.S. DOT classification applied to a material such as a consumer commodity that, though otherwise subject to DOT regulations, presents a limited hazard during transportation due to its form, quantity, and packaging.

Orthophoto map Aerial photograph connected to scale such that geographic measurements may be taken directly from the print; may contain graphically emphasized geographic features and may be provided with overlays of such features as water systems and important faculty locations.

OSHA U.S. Occupational Safety and Health Administration (in Department of Labor).

Ototoxins Any chemical agent that may cause damage to or cause a decrease in hearing ability.

Out-of-service resources Resources assigned to an incident but unable to respond for mechanical, rest, or personnel reasons.

Overhead personnel In ICE, personnel who are assigned to supervisory positions, which include incident commander, command staff, general staff, directors, supervisors, and unit leaders.

Oxidation The removal of electron (or hydrogen) from a molecule, with consequent transformation of that molecule into another chemical type; a type of chemical process that can be used in certain types of chemical detoxification.

Oxidizer Any material that promotes or initiates the burning of combustible or flammable materials.

P

Parenteral Through the skin barrier, as in a parenteral exposure, which is an exposure that occurs through a break in the skin barrier (e.g., injections, needle-sticks, human bites, and cuts contaminated with blood).

PASS Personal alert safety system.

Pathogen A microorganism that can cause disease, including bacterial, fungal, parasitic, or viral pathogens.

Pathogenic Capable of causing disease.

Patrol unit In firefighting, any light, mobile unit having limited pumping and water capacity.

PCP See *Pneumocystis pneumonia*.

PDD-39 See *Presidential decision directive 39*.

PEL See *Permissible exposure limit*.

Percutaneous   Entering the body through the skin (e.g., by needle-stick or through broken skin).

Pericardial fluid   A clear fluid contained in the thin membranous sac that surrounds the heart.

Peritoneal fluid   Fluid contained in the membrane lining of the abdominal cavity.

Permanent threshold shift   A permanent hearing impairment as determined by comparing a baseline audiogram with subsequent audiograms.

Permissible exposure limit   In the United States, a legally enforceable maximum airborne exposure limit for a specific chemical.

Personal Protective Equipment   Specialized clothing or equipment worn by an employee for protection from a hazard; general work clothes not intended to function as protection against a specific hazard are not considered to be personal protective equipment.

PFD   Personal flotation device (U.S. Coast Guard approved, Class III).

Phlebotomist   Any healthcare worker who draws blood samples.

PIH   Poison inhalation hazard.

Planning meeting   In ICS, a meeting held as needed throughout the duration of an incident to select specific strategies and tactics for incident control operations, and for service and support planning. On larger incidents, the planning meeting is a major element in the development of the incident action plan.

Planning section   In ICS, responsible for the collection , evaluation, and dissemination of tactical information related to the incident, and for the preparation and documentation of incident action plans. The section also maintains information on the current and forecasted situation, and on the status of resources assigned to the incident; includes the situation, resource, documentation, and demobilization units, as well as technical specialists.

Plan of action   A written document that consolidates all the operational actions to be taken by various personnel in order to stabilize an incident.

Pleural fluid   Fluid contained in the membrane that covers the lung and lines the chest cavity.

Plume   A visible or measurable discharge of a contaminant to air, water, and/or soil from a given point of origin.

Polymerization   A chemical reaction that forms a compound (polymer) from the bonding together of the molecules of a particular chemical.

Pneumocystis pneumonia (PCP)   A type of pneumonia caused by a parasite; seen in patients with impaired immune systems.

Poison   Any material that causes life-threatening damage to tissues or internal organs in very small amounts (e.g., several teaspoons or less).

Poison, Class A   U.S. DOT classification applied to an extremely dangerous poison.

Poison, Class B   U.S. DOT classification applied to any non-Class A poison that nonetheless presents a hazard to health during transportation.

Polio   Poliomyelitis; a vaccine-preventable viral disease not commonly seen in the United States.

Polymerization   A chemical process in which many molecules of the same type (called monomers) are joined together in long chains; used in certain types of chemical detoxification.

Positive pressure SCBA   SCBA in which the pressure inside the face-piece, in relation to the pressure surrounding the outside of the face-piece, is positive during both inhalation and exhalation.

Post-traumatic stress syndrome   Persistent and profound state of psychological dysfunction attributed to previously experienced traumatic stress.

PPD   See *Purified protein derivative*.

PPE   See *Personal protective equipment*.

ppb   Parts per billion.

ppm   Parts per million.

Presbycusis   Natural hearing loss associated with aging.

Presidential decision directive 39   Issued in June 1995, this directive (United States Policy on Counterterrorism) directs a number of measures to reduce the nation's vulnerability to terrorism, to deter and respond to terrorist acts, and to strengthen capabilities to prevent and manage the consequences of the terrorist use of nuclear, biological, and chemical weapons; provisions of PDD-39 that affect the FRP are included in the FRP Terrorism Incident Annex.

Pressure demand SCBA   See *Positive pressure SCBA*.

Process safety information   A compilation of written information that describes safety-related attributes of chemical substances, the processes in which they are used, and the equipment used for those processes.

Procurement unit   In ICS, a functional unit within the finance/administration section responsible for financial matters involving vendor contracts.

Prophylaxis   Any substance, material, or steps taken to prevent something from happening (e.g., condoms, vaccine).

Proximity clothing   Reflective protective clothing that is designed to provide protection against conductive, convective, and radiant heat.

PTS   See *Permanent threshold shift*.

Purified protein derivative   A skin test for exposure to tuberculosis.

Pyrolysis   The breaking apart of a complex molecule into a simpler molecule by the use of heat.

Pyrophoric   Any material that ignites spontaneously in air at temperatures of 130°F or lower.

Pyrotechnics   Any combustible or explosive compositions or manufactured articles designed and prepared for the purpose of producing audible or visible effects (i.e., fireworks).

Q

Q Fever   Infectious disease caused by *Coxiella Burnetii* rickettsia; characterized by fever, general malaise, and muscular pains.

Quayside search   A semicircular, surface-directed search pattern conducted alongside a quay or dock.

R

Radiant heat   Heat that is emitted by one material and absorbed by another over a distance.

Radio cache   A supply of radios stored in a predetermined location for assignment to incidents; used to disperse radioactive particles over an extensive area.

Radiological dispersal device   Any conventional explosive that incorporates nuclear materials.

RAM   Radioactive materials.

Rated service time   The period of time that SCBA supplies air to the breathing machine when tested according to the requirements of 30 CFR II; recorded on the SCBA's certification label.

Recombinant vaccine   A vaccine produced by genetic manipulation (gene splicing).

Recorders   Individuals within ICS organizational units who are responsible for recording information; may be found in planning, logistics, and finance/administration units.

RDD   See *Radiological dispersal device.*

Reduction   The addition of electrons (or hydrogen) to a molecule, with consequent transformation of that molecule into another chemical type; a type of chemical process that can be used in certain types of chemical detoxification.

Reinforced attack   See *Reinforced response.*

Reinforced response   Those resources requested in addition to the initial response.

Reporting locations   Location or facilities where incoming resources can check-in at the incident.

Regional response team   Any of 13 regional teams designated by the U.S. national response team to provide planning, policy, and coordination assistance to incident response organizations.

Reportable quantity   An amount of a hazardous chemical that, if released to the environment, must be reported (in the Emergency Planning and Community Right-to-Know Act, the Hazardous Materials Transportation Act (HMTA), and/or the Comprehensive Emergency Response, Compensation and Liability Act (CERCLA)).

Rescue incident   An emergency incident that primarily involves the rescue of persons subject to physical danger, and may include the provision of emergency medical services.

Residual risk   The risk still remaining after the implementation of risk management practices.

Resources   Personnel and equipment available or potentially available for assignment to incidents. Resources are described by kind and type (e.g., ground, water, air) and may be used in tactical support or overhead capacities at an incident.

Resources unit   In ICS, functional unit within the planning section responsible for recording the status of resources committed to the incident; also evaluates resources currently committed to the incident, the impact that additional responding resources will have on the incident, and anticipated resource needs.

RESTAT   Resources status; in ICS, used in reference to a unit within the planning section responsible for tracking resources assigned to an incident.

Retro-reflective   Descriptive of a material that reflects and returns a relatively high proportion of light in a direction close to the direction from which it came.

Ricin   Mixture of poisonous proteins produced by the castor oil plant; plant gene controlling the production of ricin has been successfully transferred to the bacterium *Escherichia coli*; ricin interferes with the body's normal synthesis of proteins; symptoms include decreased blood pressure, with death occurring most often through heart failure.

Risk   The probability that an individual (or group) will actually experience a hazard; usually expressed as a decimal between 0.0 and 1.0 with values approaching 1.0 denoting increasing risk.

Risk assessment   The value judgment of the significance of the risk, identified by a risk analysis taking into account relevant criteria.

Risk management   Actions taken to achieve or improve the safety of an installation and its operation.

Root cause analysis   A technique of incident investigation that includes developing a chronology of events, listing deviation from normal conditions or circumstances, and possible causes based on these deviations.

Route-of-entry   The pathway that a chemical takes to enter or come into contact with the body (includes inhalation, ingestion, skin or eye contact, absorption, and puncture).

RPR   A blood test for syphilis.

RQ   See *Reportable quantity*.

RRT   See *Regional response team*.

RTECS   The U.S. Registry of Toxic Effects of Chemical Substances.

Rubella   A vaccine-preventable viral disease; rubella infection during pregnancy can cause birth defects.

S

Safety diver   A back-up diver, geared up, on shore, and available to assist.

Safety interlock system   An automatic system that inhibits the operation of critical equipment until certain process parameters are within acceptable ranges.

Safety officer   In ICS, a member of the command staff responsible for monitoring and assessing safety hazards or unsafe situations, and for developing measures for ensuring personnel safety.

Salmonellosis   Infection caused by intestinal bacteria of the genus *Salmonella*; characterized by nausea, abdominal pains, diarrhea, and fever; it can lead to death, especially in people with impaired immune systems.

SAFA   Superfund Amendments and Reauthorization Act (U.S.).

Sarcoma   Cancer of the skin or other connective tissue.

Saxitoxin   Toxin produced by marine blue-green alga (i.e., *cyanobacterium*), which serves as food supply for various shellfish that are immune to effects of the toxin, but pass it on to higher order consumers (e.g., humans); in humans, toxin acts on central nervous system to produce paralysis; at high doses, death can occur in less than 15 minutes.

SCBA   See *Self-contained breathing apparatus*.

Search line   Water rescue line; 100-ft three-strand twisted quarter-inch polypropylene with a monkey's fist at one end and 30-inch loop with snap at the other end.

Search pattern   Logical means of covering a search area while maintaining diver accountability.

SEB   See *Staphylococcus enterotoxin type B*.

Second-degree burn   A burn that is more severe than a first-degree burn and typically characterized by blistering, reddening of the skin, swelling, and destruction of the superficial tissue.

Section   In ICS, that organizational level with responsibility for a major functional area of the incident (e.g., operations, planning, logistics, finance/administration). The section is organizationally between branch and incident commander.

Sector   Term used in some applications to describe an organizational level similar to an ICS division or group; sector is not a part of ICS terminology.

Segment   In ICS, a geographical area in which a task force/strike team leader or supervisor of a single resource is assigned authority and responsibility for the coordination of resources and implementation of planned tactics. A segment may be a portion of a division or an area inside or outside the perimeter of an incident. Segments are identified with Arabic numbers.

Self-contained breathing apparatus   A respirator that supplies a respirable atmosphere that is either carried in or generated by the apparatus and is independent of the ambient environment.

Sensitizer   A material that causes allergic reaction after repeated exposures, with possibly severe or even life-threatening consequences.

SERC   State emergency response commission; established by the U.S. Emergency Planning and Community Right-to-Know Act.

Seroconversion   A change in the status of one's serum test.

Serologic test   Any number of tests that are performed on blood; usually refers to a test that measures antibodies to a virus.

Seropositive   A condition in which antibodies to a disease-causing agent are found in the blood; a positive reaction to a blood test; the presence of antibodies indicates that a person has been exposed to the agent.

Service branch   In ICS, a branch within the logistics section responsible for service activities at the incident, including communication, medical, and food units.

Sexually transmitted disease   A disease spread through sexual contact; HIV and HBV are both blood-borne and sexually transmitted diseases.

Sharps   Any object that can penetrate the skin including, but not limited to, needles, lancets, scalpels, and broken capillary tubes.

Shingles   Common term for *Herpes zoster* infection, resulting in painful rash.

Single resource   In ICS, an individual, a piece of equipment and its personnel complement, or a crew or team of individuals with an identified work supervisor that can be used on an incident.

Site work zone   A designated work area at a hazard waste site; may include one or more designated exclusion zones, contamination reduction zones, and support zones.

SITSTAT   Situation status; in ICS, used in reference to the unit in the planning section responsible for keeping track of incident events.

Situation analysis   The process of evaluating the severity and consequences of an incident and communicating the results.

Situation unit   In ICS, a functional unit within the planning section responsible for the collection, organization, and analysis of the situation as it progresses; reports to the planning section chief.

Smallpox   Highly contagious, often fatal, viral disease; characterized by a high fever and successive stages of severe skin eruptions.

Sociocusis   Hearing loss associated with everyday noises (e.g., lawn mowers, load music, traffic).

Sound pressure level   The measurement of loudness of sound in decibels.

Span-of-control   The supervisory ratio of from three-to-seven individuals, with five-to-one being established (in ICS) as optimum.

Staging areas   The locations set up at an incident where resources can be placed while awaiting a tactical assignment. Staging areas (in ICS) are managed by the operations section.

Standard threshold shift   A change in hearing capacity (relative to baseline audiogram) of 10dB or more at 2000, 3000, and 4000Hz in either ear.

Staphylococcus interotoxin type B   Toxin produced by bacterium *Staphylococcus aureus*; most commonly associated with food poisoning; characterized by stomach cramps, diarrhea, and vomiting.

State EOP   In the United States, state emergency operations plan; framework within which local EOPs are created and through which the federal government becomes involved.

Sterilization   The use of a physical or chemical procedure to destroy all microbial life, including highly resistant bacterial endospores.

STD   See *Sexually transmitted disease*.

STEL   See *TLV-STEL*.

Strategic goals   Broad, general statements of intent with respect to incident command.

Strategy   The general plan or direction selected to accomplish incident objectives.

Strike team   In ICS, specified combinations of the same kind and type of resources, with common communications and a leader.

Subcutaneous   Beneath the skin (e.g., subcutaneous injection).

Substance P   A protein closely related to normally produced proteins in the body; may cause pain, act as anesthetic, or affect blood pressure; rapid loss of blood pressure in victim may cause unconsciousness.

Supervisor   The ICS title for individuals responsible for command of a division or group.

Supply unit   In ICS, functional unit within the support branch of the logistics section responsible for ordering equipment and supplies required for incidents operations.

Support branch   In ICS, the branch within the logistics section responsible for providing personnel, equipment, and supplies to support incident operations; includes the supply, facilities, and ground support units.

Support zone   Uncontaminated area at a hazardous waste site where response personnel should not be exposed to hazardous contamination.

Supporting materials   In ICS, the several attachments that may be included with an incident action plan, e.g., communications plan, map, safety plan, traffic plan, and medical plan.

Support resources   In ICS, nontactical resources under the supervision of the logistics, planning, finance/administration sections, or the command staff.

Surfactant   A chemical added to reduce the force of adhesion between a contaminant and the surface to which the contaminant is attached by molecular forces; household detergents are common examples of surfactants.

Suspect carcinogen   A substance that might cause cancer in humans or animals but which has not been proven to do so.

Syndrome   A collection of signs and symptoms that occur together.

Synergy   A phenomenon in which two or more chemicals interact to produce an effect (e.g., toxicity, lethality) in a magnitude that is an unpredictable multiple of the individual magnitude of each chemical's solitary effect; synergistic interactions may involve other than chemicals; a general term denoting a multiplication of consequences as opposed to additive consequences.

Syphilis   A sexually transmitted infectious disease; commonly transmitted through blood exposure or transfusion.

T

Table-top exercise   Generally, a conference-based exercise in which members of the emergency management group discuss their responsibilities and how they would react to different emergency scenarios; may often involve the acting out of specific scenarios, with specific information about a hypothetical emergency being parceled out to the group In a scheduled manner; may also involve simulated computer-assisted data/information processing as well as voice/telephonic and radio communication.

Tactical direction   In ICS, direction given by the operations section chief, which includes the tactics appropriate for the selected strategy, the selection and assignment of resources, tactics implementation, and performance monitoring for each operational period.

Target Capabilities List (TCL)   Designed to assist jurisdictions and agencies in understanding and defining their roles in a major event, the capabilities required to perform a specific set of tasks, and where to obtain additional resources if needed; the first version of the TCL contains 36 target capabilities.

Task analysis   A method of evaluating components and task sequences of jobs to identify potential sources of error, performance inefficiencies, and safety hazards.

Task force   In ICS, a combination of single resources assembled for a particular tactical need, with common communications and a leader.

TB   See *Tuberculosis*.

TB bacteria   Bacteria that cause tuberculosis.

TCL   See *Target Capabilities List*.

Team   See *Single resource*.

Technical specialists   Personnel with special skills that can be used anywhere within the ICS organization.

Temporary flight restrictions   Temporary airspace restrictions for nonemergency aircraft in the incident area; in the United States, TFRs are established by the Federal Aviation Administration to ensure aircraft safety, and are normally limited to a five-nautical-mile radius and 2000-feet altitude.

Temporary threshold shift   A short-term hearing impairment as determined by comparing the baseline audiogram with subsequent audiograms.

Tendonitis   Inflammation of a tendon, often caused by repetitive, forceful exertions involving rotations around a joint.

Teratogen   A material that causes malformation of the developing fetus.

TFR   See *Temporary flight restrictions*.

Thermal degradation   A decontamination procedure in which heat is used to degrade or destroy the molecular structure of a contaminant.

Thermal shrinkage   The change in dimensions of a fabric specimen when exposed to heat.

Third-degree burn   The most serious type of burn, characterized by charring (blackening) of the skin and by skin necrosis (tissue death).

Threshold planning quantity   The amount of an extremely hazardous substance present at a facility above which the facility must give emergency planning notification to the state emergency response commission and local emergency planning committee as required by the U.S. Emergency Planning an Community Right-to-Know Act.

Threshold shift   A short-term or permanent hearing impairment as determined by comparing a baseline audiogram with subsequent audiograms.

TLV   Threshold limit value; concentration limit for prolonged exposure to a particular hazardous chemical.

TLV-C   Threshold limit value–ceiling; concentration that should not be exceeded by exposure, even instantaneously.

TLV-STEL   Threshold limit value–short-term exposure; concentration that should not be exceeded by an exposure of 15 minutes.

TLV-TWA   Threshold limit value-time weighted average; the time-weighted average concentration to which persons can be exposed for eight hours per day (of a 40-hour workweek) without adverse effect.

Top event   The culminating event of a fault tree from which basic causes are deduced to determine the cause and probability of the top event.

Tornado warning   Condition of alert that a tornado has been sighted in the area or is indicated by radar; should immediately take shelter.

Tornado watch   Condition of alert that tornadoes are likely; prepare to take shelter.

Torpedo buoy   Towable surface rescue float.

Torpedo buoy rescue   Water rescue using a torpedo buoy and a rescue skin diver.

Toxic   Any material that causes life-threatening damage to tissues or internal organs, but in amounts greater than a poison (e.g., greater than several teaspoons).

Toxic release inventory   A national (U.S.) inventory of annual toxic chemical releases from manufacturing facilities.

TPQ   See *Threshold planning quantity*.

TRACEM   Acronym to identify six types of harm from a terrorist incident: thermal, radioactive, asphyxiation, chemical, etiological, and mechanical.

Trans-frontier damage   Any damage to human health or the environment, including property, in the event of an accident suffered by a country other than the country where the accident originated.

Traveler's advisory   Condition of alert that conditions may make driving difficult or dangerous.

TRI   See *Toxic release inventory*.

TTS   See *Temporary threshold shift*.

Tuberculocidal   Capable of killing tuberculosis (TB) bacteria.

Tuberculosis   A communicable disease caused by the bacterium *Mycobacterium tuberculosis*; usually affecting the lungs.

Tularemia   Infectious disease caused by *Francisella tularensis* bacterium; chiefly affects rodents but can also be transmitted to human beings by the bite of various insects or contact with infected animals; characterized by intermittent fever and swelling of the lymph nodes.

TWA   See *TLV-TWA*.

Type   In ICS, refers to resource capability. A type 1 resource provides a greater overall capability due to power, size, capacity, and so on than would be found in a type 2 resource. Resource typing provides managers with additional information in selecting the best resource for the task.

U

UN   United Nations.

Unassisted rescue   Water rescue conducted by a rescue skin diver without assistance of a torpedo buoy or water rescue line.

Unconfined vapor cloud explosion   A high-velocity flame front created by the ignition of a released flammable vapor that has mixed with air to form a vapor cloud.

Unified area command   An incident management system established when incidents under an area command are multijurisdictional.

Unified command   In ICS, unified command is a unified team effort that allows all agencies with responsibility for the incident (either geographical or functional) to manage an incident by establishing a common set of incident objectives and strategies. This is accomplished without losing or abdicating agency authority, responsibility, or accountability.

Unit   In ICS, the organizational element having functional responsibility for a specific incident planning, logistics, or finance/administration activity.

Unity of command   The concept by which each person within an organization reports to one and only one designated person.

Universal precautions   A system of infectious disease control that assumes that every direct contact with body fluids is infectious and requires every employee exposed to direct contact with body fluids to be protected as though such body fluids were HBV or HIV infected; intended to prevent healthcare workers from parenteral, mucous membrane, and nonintact skin exposures to blood-borne pathogens and should be used by emergency response personnel.

Unstable/reactive   Any material that spontaneously explodes with production of pressure, gas, heat, and possibly toxic fumes.

Universal Task List (UTL)   Defines the essential tasks to be performed by federal, state, and local governments and the private sector to prevent, respond to, and recover from a range of threats from terrorists, natural disasters, and other emergencies.

Upper explosive limit   The concentration of a chemical in air above which a flame will not propagate if the mixture contracts an ignition source.

USCG U.S Coast Guard.

USFA U.S. Fire Administration.

UTL See *Universal Task List*.

V

Vaccine A substance that produces or increases immunity and protection against a particular disease.

Vapor density The weight per unit volume of a vapor; commonly given in comparison to air when the density of air is defined as 1.0.

Vapor pressure The pressure created within a closed container by the gaseous phase of contained liquid.

VDRL A blood test (Venereal Disease Research Laboratory) for syphilis.

Venereal Due to or propagated by sexual contact.

Venezuelan equine encephalitis Viral infection of the central nervous system, with potentially fatal swelling of the brain.

Vesicant A chemical agent (also called blister or mustard agent) that causes severe burns to eyes, skin, and tissues of the respiratory tract.

Virulence The disease-provoking power or potency of a microorganism in a live host.

Virus A microorganism usually visible only with the electron microscope; viruses normally reside within other living (host) cells, and cannot reproduce outside of a living cell.

Volatility A measure of how quickly a substance forms vapor at ordinary temperatures.

Vulnerability analysis Any procedure used to integrate facility-specific information on (a) types of potential emergency, (b) likely impacts and probability of occurrence of each type, and (c) resources available for use in an actual emergency response.

W

Walk-through drill A training activity in which the emergency management group and response teams actually perform their emergency response functions in a simulation; generally involves more people and is more thorough than a table-top exercise, although elements of both types of exercises are often integrated.

Water reactive Any material that reacts with water to form a flammable or toxic gas.

Water rescue line See *Search line*.

Water sector That part of a water rescue scene that involves divers and dive rescue activity; in ICS, it is the area of responsibility of the senior rescue diver.

Water tender Any ground vehicle capable of transporting specified quantities of water.

Western blot A test for HIV, used to confirm a positive ELISA test; more expensive and time consuming to perform than ELISA, but more specific. Diagnosis of HIV infection requires two positive ELISA tests, confirmed with a positive Western blot test.

What-if-analysis A hazard evaluation method that uses a brainstorming approach to develop a list of questions or concerns addressing hazards or specific accidents that could produce undesirable consequences in a system or a process.

Whitlow   A fingertip infection commonly caused by herpes virus; spread by contact with respiratory secretions.

WHO   United Nations World Health Organization.

Window phase   The time from exposure to a disease until a positive diagnostic test for its presence.

Worst-case analysis   Any analysis of a situation, circumstance, or process in which it is assumed that all or some of the relevant factors are at their most negative or adverse value.

Wristlet   The circular, close fitting part of the glove, usually made of knitted material, that extends beyond the opening of the glove body to over the wrist area; also used to describe connection of a safety line to the wrists, which can be used to extricate a person out of a narrow confined space.

Z

Zoster   See *Herpes zoster*.

INDEX

CPSIA information can be obtained at www.ICGtesting.com
Printed in the USA

236450LV00004BB/1/P